Forest Genomics and Biotechnology

Forest Genomics and Biotechnology

Richard Meilan

Department of Forestry and Natural Resources, Purdue University, West Lafayette, USA

and

Matias Kirst

School of Forest Resources and Conservation, University of Florida, Gainesville, USA

CABI is a trading name of CAB International

CABI	CABI
Nosworthy Way	745 Atlantic Avenue
Wallingford	8th Floor
Oxfordshire OX10 8DE	Boston, MA 02111
UK	USA
Tel: +44 (0)1491 832111	T: +1 (617)682-9015
Fax: +44 (0)1491 833508	E-mail: cabi-nao@cabi.org
E-mail: info@cabi.org	
Website: www.cabi.org	

A catalogue record for this book is available from the British Library, London, UK.

Library of Congress Cataloging-in-Publication Data

Names: Meilan, Richard, author. | Kirst, Matias, author.
Title: Forest genomics and biotechnology / Richard Meilan, Matias Kirst.
Other titles: CABI biotechnology series ; 9.
Description: Boston, MA : CAB International, 2019. | Series: CABI
  biotechnology series ; 9 | Includes bibliographical references and
  index. | Summary: "There have been major recent advances in genomics and
  biotechnology of trees, including the sequencing and annotation of the
  first tree genomes. Biotechnology applications have now moved from the
  lab to the field, and this book considers in detail the impact of these
  technologies in forest productivity and the environment"-- Provided by
  publisher.
Identifiers: LCCN 2019030922 (print) | LCCN 2019030923 (ebook) | ISBN
  9781780643502 (hardback) | ISBN 9781780643519 (ebook) | ISBN
  9781786394071 (epub)
Subjects: LCSH: Forest genetics. | Plant genomes. | Forestry biotechnology.
Classification: LCC SD399.5 .K55 2019 (print) | LCC SD399.5 (ebook) | DDC
  634.9/56--dc23
LC record available at https://lccn.loc.gov/2019030922
LC ebook record available at https://lccn.loc.gov/2019030923

ISBN-13: 978 1 78064 350 2 (hardback)
         978 1 78639 407 1 (ePDF)
         978 1 78639 407 1 (ePub)

Commissioning Editor: David Hemming
Editorial Assistant: Emma McCann
Production Editor: Shankari Wilford

Typeset by SPi, Pondicherry, India
Printed and bound in the UK by Severn, Gloucester

Contents

Part III: Commercialization

Dedication

This book is dedicated to the memory of James ("Jake") A. Eaton (1958–2013). Jake contributed greatly to the adoption of *Populus* spp. as a short-rotation woody crop that is being grown commercially on a large scale for a variety of end uses (e.g. fiber for papermaking, dimensional lumber, and biofuel feedstock). He worked tirelessly and fervently for causes he believed in, and his love for life and his work was infectious. Because of his involvement with projects in North and South America, Asia, and Europe, Jake's influence was felt worldwide. Jake was also a kind, gentle, caring, thoughtful, and loyal friend. He continues to be missed by the people whose lives he touched.

Jake evaluating a rust-resistance screening trial with Vic Steenackers, Jon Johnson, and Marijke Steenackers in Geraardsbergen, Belgium (2002).

Preface

Over the last few decades, there has been an exceptional transformation in the way genetics is studied and how hereditary information is used in forestry. This transformation was catalyzed by the creation of two new fields of science: forest genomics and forest biotechnology. Genomics was a logical evolution of traditional genetic studies, which focused on a single or a few genes, into the analysis of the entire ensemble of genetic elements that define a species. Biotechnology involves the exploitation of this information and the biological processes they control for the development of products that are desired by humans. Although many aspects of genomics can be considered biotechnology, for the purposes of this book we draw a distinction between the two fields.

Approaches have now been developed to identify the specific sources of genetic variation that make a tree more productive and predict the manifestation of mature traits at the seedling stage. In parallel, we now have the ability to identify regions of the genome that determine the capability of a tree to adapt to existing and new environments—critical discoveries, in view of the climate change that is now affecting our forests. This progress has been made possible by the rapid development of new technologies for genomic analyses that emerged in the years that followed the sequencing of the human genome. In the first section of this book, we review these technologies and their application to the study of the genomes of conifers and hardwoods, and the study of quantitative traits and populations.

While genomics has allowed us to discover the role of specific genetic elements in tree growth and development, tree species have not been domesticated to the same extent as annual, agronomic crops, primarily because of their long juvenile periods. Genetic engineering has allowed us to partially overcome this obstacle by inserting a small number of genes that control desired traits. What we hope to accomplish with conventional breeding is often no different from what we do using recombinant DNA technology, but we can expedite the process through genetic engineering. Moreover, genomic-assisted improvement does not preclude the future need for genetic engineering. For example, we may want to introduce a trait that is controlled by genes or alleles that do not exist in the genome of the species of interest, or sexually compatible species. In addition, genomic editing (currently) needs to be done using genetic engineering. In the second section of this book, we review the application of several biotechnologies to forest trees, but the emphasis is on genetic engineering.

Considerable effort goes into developing the techniques and final products that are eventually commercialized. It is necessary to protect this intellectual property (IP) to prevent competitors from taking unfair advantage of investments made by its developers.

These protections can also provide incentive for researchers and inventors to produce new and/or improved technologies and products. In some cases, regulatory approval is need for trees that have been genetically altered. In the final section of this book, we address both of these topics. Unlike the other chapters, Chapter 12 has additional co-authors, whose help was solicited because of their vast knowledge and experience working in the area of intellectual property protection.

Why do we need another book on genomics and biotechnology? While there are many outstanding textbooks in the field, we felt there was a need for a book that describes the different ways in which genomics and various biotechnologies have been used in the study and improvement of forest-tree species. These trees are largely undomesticated, and are the keystone species in many biomes. Thus, the issues related to genomics and biotechnology are often unique to them and have not been addressed using herbaceous, annual model plants or agricultural crops. This book is an attempt to fill part of that gap. It is intended to provide selected examples of how existing, key technologies are currently being used in trees; it is not an exhaustive review of what has been done to date.

While genomics and biotechnology have already impacted the study of trees, these areas are evolving rapidly. As a consequence, technological and methodological improvements, as well as the development of new analytical methods, are certain to arise in the near term. Still, we hope that the information conveyed in this book will serve as a useful foundation to those who are new to the field, and to those who are actively working in the area but seek a broad overview of what has been achieved in forest genomics and biotechnology.

Rick Meilan
Matias Kirst
June 2019

Acknowledgements

R.M. is grateful to Chung-Jui Tsai, Amy Brunner, Shawn Mansfield, Arie Altman, and Steve Strauss for their reviews of and helpful suggestions for Chapters 6–10, respectively. He also appreciates the comprehensive review and all of the insightful comments made by Khamkeo Vongpaseuth regarding Chapter 11. Khamkeo Vongpaseuth is a Biotechnologist at Biotechnology Regulatory Services, Animal and Plant Health Inspection Service, U.S. Department of Agriculture. The views expressed by him for that chapter are his own and do not necessarily represent those of the U.S. Government. In addition, R.M. is thankful for the support provided by the Department of Forestry and Natural Resources at Purdue University while he was on sabbatical, which allowed him to work on the initial draft of this book. Finally, R.M. is grateful to Priscilla for the many sacrifices she made while he was immersed in the writing of this book.

M.K. is grateful to Sally Aitken, Janice Cooke, Dario Grattapaglia, Nathalie Isabel, David Neale, Christophe Plomion, Ron Sederoff, Gerald Tuskan, and other scientists whose foundational work in the fields of forest genomics, and population and quantitative genomics, is the basis for the research described in Chapters 1–5. M.K. is also thankful for the support from Tim White and other administrators in the School of Forest Resources and Conservation at the University of Florida, in the preparation of this book. Finally, the time dedicated to preparing this book meant time away from assisting graduate students, post-docs, and research collaborators in their work. M.K. sincerely acknowledges their patience and contribution to discussions of many topics that are described herein.

1 Principles of Genome Sciences

Introduction

Genome science is the discipline that studies ensembles of genes and genomes, and their interactions, at scales that range from individual cells to entire populations. The creation of the discipline was an obvious progression from the characterization of structure and function of individual genes to the analysis of the entire complex of elements that are genetically inherited. Genomics encompasses a wide array of areas, including deoxyribonucleic acid (DNA) sequencing, assembly, and annotation, and all the analytical and computational approaches required to obtain, interpret, summarize, and display this information (Fig. 1.1). Genomics also includes the study of areas beyond genome characterization, such as the epistatic interactions among loci. Finally, a number of derivatives of the term genomics have emerged to represent the study of other elements of the central dogma of biology, which are now being studied at an "omics" level. These include the analysis of transcripts of all coding sequences, or "transcriptomics", as well as their translation products, termed "proteomics". The purpose of this chapter is to describe the major disciplinary areas of genomics that have been explored in and utilized for various tree species. It will serve as a foundation for the remaining chapters in Section I of this book.

Why Study the Genome of Tree Species?

Fundamentally, genetic phenomena such as recombination and segregation are similar between annual herbaceous and woody perennial plant species. However, tree species have properties that make them unique among plants.

Trees differ from most agricultural crops by their perennial growth habit, and by their ability to form secondary xylem, or wood. Some of the other distinctive characteristics of trees create unique challenges for genomic analysis. For instance, their long generation times preclude the use of traditional methods for genomic dissection of complex traits, such as fine mapping of quantitative trait loci (QTLs) in multi-generation, biparental populations. In addition, several tree species have exceptionally large and complex genomes relative to other plant species. For instance, the nuclear DNA content of conifers is several times the size of the human genome (Zonneveld, 2012), making them particularly challenging to study, compared with other plants (Fig. 1.2). Regardless of the difficulty, the beginning of the 21st century has witnessed tremendous progress in the genomic characterization and analysis of trees. This information has led to a better understanding of how the genomes of trees have evolved in natural populations (Neale and Ingvarsson, 2008; Neale and Kremer, 2011; Evans *et al.*, 2014), and in the improvement of gains in breeding (Grattapaglia *et al.*, 2009; Harfouche *et al.*, 2012). Genomic information is becoming even more critical as the threat from climate change impacts natural forests (Bonan, 2008; Allen *et al.*, 2010), and for developing genotypes that will be needed to sustainably meet society's growing demand for food and energy (IEA, 2016).

Organization of Section I: genomics of forest trees

Section I of this book is focused on describing genomic methods, their application to the characterization of woody perennials, and the results derived from these analyses.

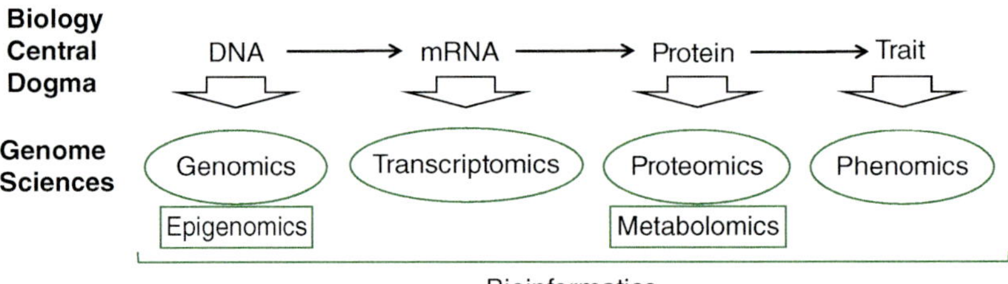

Fig. 1.1. In the biology central dogma, hereditary information flows from the DNA that is transcribed to produce mRNA, which is then translated to produce proteins. The genome sciences aim to characterize each of these levels (genomics, transcriptomics, and proteomics). As other factors that influence the biology central dogma have been recognized, other "omics" sciences have emerged to characterize them (e.g. epigenomics). Similarly, new fields that describe intermediate and later steps of the central dogma have been proposed (e.g. metabolomics and phenomics). Bioinformatics can be broadly defined as a set of tools and resources used for analysis of data derived from the genome sciences.

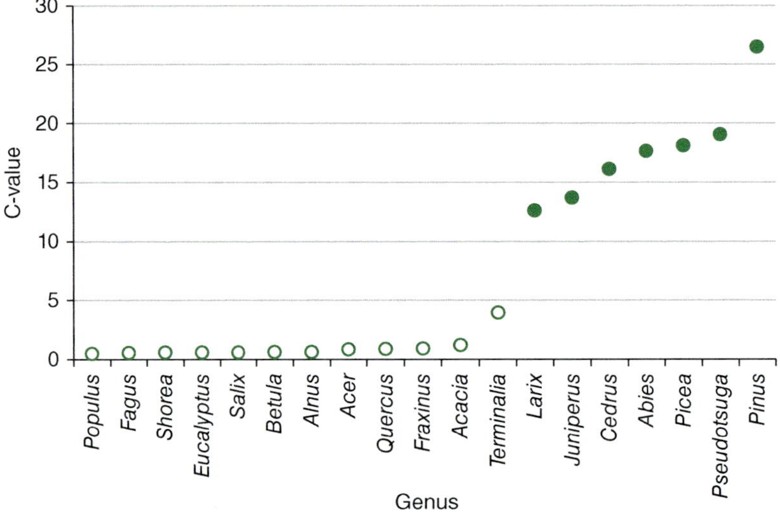

Fig. 1.2. Mean DNA amount (C-value) in the gametic nucleus of the most common genera of tree species that occur worldwide (Plant DNA C-values database; Garcia *et al.*, 2014), ranked based on their genome size. Conifers are identified by filled circles.

Chapter 1 provides an overview of the analytical approaches used to study the genomes of tree species, including the analysis of DNA, the transcriptome, and the proteome. The methods described are necessary for understanding the discoveries stemming from tree genomic analyses, as well as those from population and quantitative genetics, which are described in the remaining chapters of Section I.

Despite the challenges of studying conifers, the dramatic improvement in DNA sequencing technology has resulted in the recent characterization of three of their genomes (Birol *et al.*, 2013; Nystedt *et al.*, 2013; Neale *et al.*, 2014). Because of the significant level of synteny among conifer genomes (Krutovsky *et al.*, 2004), these draft sequences will serve as a foundation for the sequencing of other conifer species. The properties of conifer genomes are reviewed in Chapter 2.

While not significantly different in average size from the genome of other flowering plants, several woody angiosperms have been shown to possess their own unique properties. For instance, sequencing of the genome of flooded gum (*Eucalyptus grandis*) revealed the highest frequency of tandem repeats ever recorded for a plant species (Myburg *et al.*, 2014). The properties of the genomes of species within the genera *Eucalyptus*, *Populus*, and other woody angiosperms recently sequenced are described in Chapter 3.

The phenotypic diversity and remarkable adaptability of trees is a reflection of variation in their DNA composition, as well as the result of complex interactions between genetic elements and various environmental cues (Neale and Kremer, 2011). Most traits of relevance in forestry are complex and are likely to be controlled by a large number of loci of small effect. Therefore, until it is possible to characterize the entire genetic variation in a species or population, and its contribution to phenotypes, it is unlikely that the full potential of genomics will be applied to breeding and other forms of tree improvement. Similarly, studies of natural forest populations have shifted from the analysis of separate loci to a focus on the understanding of the changes in frequencies of all alleles that affect an individual's adaptation to the environment. The final two chapters of this section review the impact of genomics on the study of tree populations (Chapter 4) and breeding for genetic improvement (Chapter 5).

Genetic Linkage and Mapping of Forest Tree Species

Cytogenetic maps

Genetic maps describe the order, orientation, and distance between loci in the genome. Before genetic maps were developed, cytogenetics was used to assess the general position of certain features in chromosomes. Cytogenetic maps rely on staining so that chromosomes can be visualized under a microscope. Banding patterns generated by the staining create unique profiles that characterize individual chromosomes, or karyotypes, and allow comparisons to be made among individuals. Methods of staining have evolved significantly in the last century, allowing visualization of specific segments of chromosomes,

such as heterochromatic regions (Speicher and Carter, 2005). Further developments, such as fluorescent *in situ* hybridization, which uses fluorescently labeled probes that hybridize to chromosome preparations, identify the general location of specific sequences (Caspersson *et al.*, 1970). This and other methods provide an overview of the organization of specific elements along the chromosomes and allow them to be distinguished from one another. However, cytogenetic maps do not permit the development of precise genetic or physical maps of these genomic elements.

Genetic maps

Genetic maps characterize genomes by representing the position of various loci relative to each other. Genetic maps are developed by observing and measuring the frequency of recombination events between loci during meiosis (Box 1.1). Genetic linkage between two loci is defined by the lack of independent assortment between their alleles. Because the probability of recombination among molecular markers defines the genetic distance that separates them, it is also possible to infer their distribution and order along chromosomes (Box 1.2). Even though genetic maps don't offer the resolution of complete genome sequences, they can define the relative position of specific genetic features in genomes. This information can then be used to map QTLs and to support the assembly of genome sequences, among other applications.

Genetic mapping populations

Genetic mapping of plants has generally relied on the analysis of inbred lines, near-isogenic lines, or backcross populations, which are used in the analysis of most crop species. The development of these populations requires the ability to self or interbreed closely related individuals, typically for several generations. However, tree species usually outcross, have long generation times, and suffer from severe inbreeding depression when selfed due to high genetic load (Williams and Savolainen, 1996; Hedrick *et al.*, 2016). As a consequence, the generation of traditional genetic mapping

Box 1.1. Genetic Linkage

Mendel's law of independent assortment states that alleles at one locus segregate independently from alleles at different loci during gamete formation. In this scenario, an individual that is heterozygous at two unlinked loci (locus A = A/a, locus B = B/b) is expected to have four types of gametes (AB, Ab, aB, and ab), each with a frequency of approximately ¼.

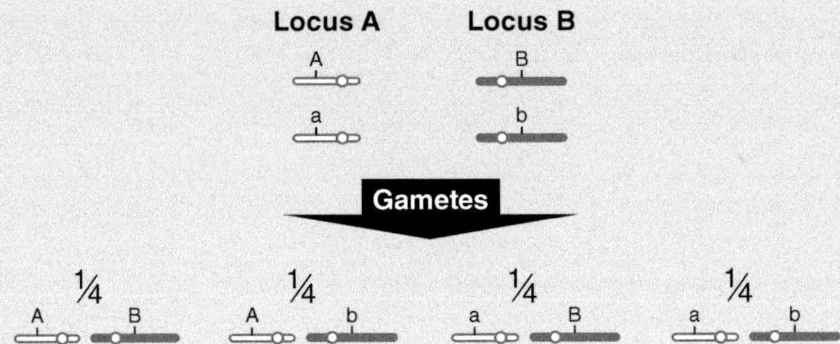

If two loci are on the same chromosome, the number of gametes in each class will deviate from ¼. A higher frequency of gametes will have the same haplotype as the parents (AB and ab). However, due to recombination during meiosis, some gametes will have the haplotype Ab and aB (recombinant gametes). The number of recombinant gametes will be proportional to the genetic distance between the A and B loci. If they are in close proximity, it is less likely that a recombination event will occur between them. Linkage mapping uses the frequency of recombination between loci to estimate their genetic distance.

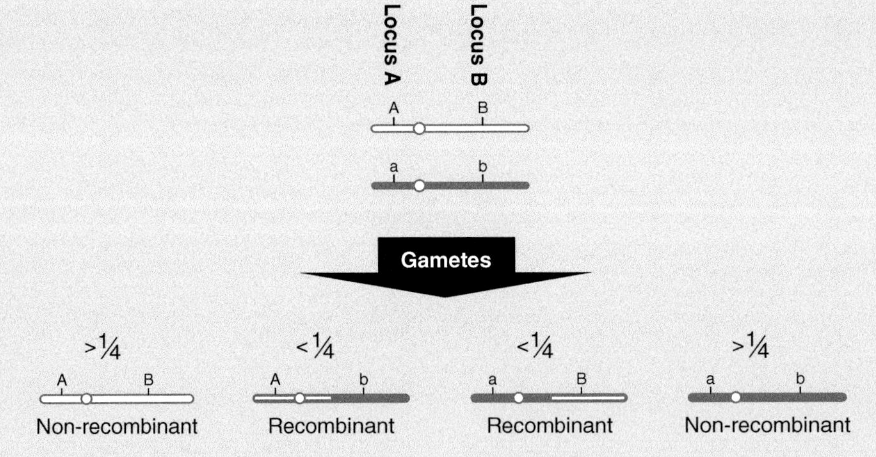

pedigrees is too time-consuming or infeasible; hence, novel approaches needed to be developed for trees. In addition to the lack of fully or near-homozygous lines or linkage-phase information, linkage analysis of forest trees is complicated by the fact that for each marker locus, up to four alleles may occur and segregate. Therefore, a mixture of segregation types resulting from the presence of heterozygous loci in one or both parents may be observed across genetic markers. On the other hand, tree populations typically carry high levels of genetic diversity among the full- and half-sib progeny that are generated. Thus, identifying variable, recombining, and segregating loci in a mapping population is more likely than in species that have gone through severe genetic bottlenecks during the multiple rounds of breeding and selection in their domestication. The constraints associated with the

Box 1.2. Genetic Mapping

A genetic map is developed by quantifying the number of recombination events between different loci. If no recombination occurs, the loci are considered to be in complete linkage. Otherwise, loci are either located on the same chromosome (partial linkage) or on different chromosomes (unlinked loci). Recombination was first hypothesized by Thomas Hunt Morgan in 1911, while studying the segregation of traits in the fruit fly (*Drosophila melanogaster*). An undergraduate student working with him, Alfred Henry Sturtevant, suggested that the frequency of recombination should be related to the distance between loci. To study this phenomenon, geneticists typically used populations derived from crosses that could easily be analyzed for segregation of distinct loci. For instance, for the loci A and B, the cross between a heterozygous individual (Aa, Bb) and an individual that is fully homozygous recessive (aa, bb; also called a tester) permits quantification of the number of recombination events between both loci during meiosis.

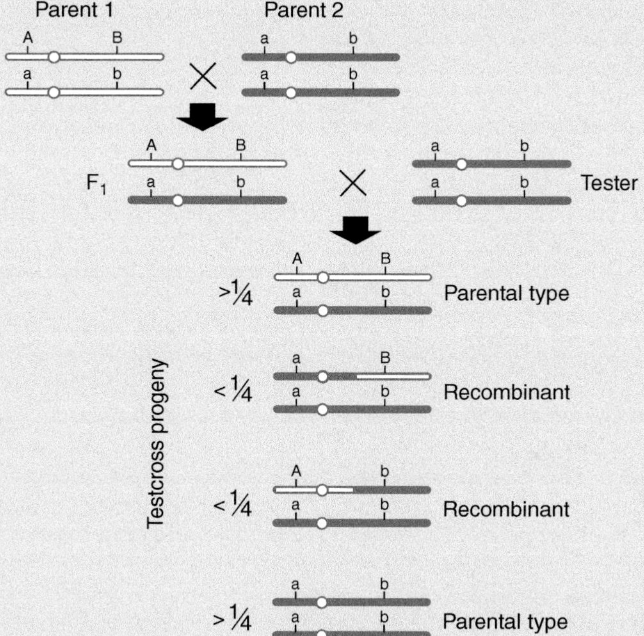

The actual genetic distance between two loci, measured in centimorgans (cM), is calculated by:

$$cM = \frac{\text{number of recombinants}}{\text{number of offspring}} \times 100$$

Thus, the maximum genetic distance that can be measured between any two loci is 50 cM, when both recombinant parental classes each represent 50% of the offspring. Loci are then considered to be located on two different chromosomes. John B.S. Haldane also proposed a correction of the map distance to reflect that the relationship between recombination and map distance is only linear within ~10 cM. Beyond a frequency of recombinants of 10%, the relationship between measured and actual map distances is described by:

$$r = 1(1 - e^{2m})$$

where *r* is equal to the proportion of recombinants and *m* is the corrected genetic distance.

development of inbred and backcross populations have resulted in the development of new types of pedigrees and segregation analyses that are more suitable for forest-tree species (Box 1.3). These are described below.

Pseudo-testcross mapping population

Traditionally, testcrosses were performed by crossing an individual displaying a dominant phenotype with one exhibiting the recessive form. Segregation in the progeny revealed whether the determinant locus was homozygous dominant or heterozygous in the dominant parent. A similar principle is applied in the "pseudo-testcross" approach, which is based on the assumption that a dominant marker will segregate in a 1:1 ratio in a testcross of heterozygous parents (Grattapaglia and Sederoff, 1994). Dominant genetic markers, such as random amplified polymorphic DNA (RAPD; Williams *et al.*, 1990) or amplified fragment length polymorphisms (AFLPs; Vos *et al.*, 1995) were widely used in the early days of genetic mapping. These markers are considered to be "dominant" because homozygous individuals for the dominant allele are not distinguishable from heterozygous individuals (for more details about genetic markers, see Chapter 4 in White *et al.*, 2007). The "pseudo-testcross" approach allows analysis of segregation of: (i) testcross markers inherited from either the male or female parent (segregating 1:1) and (ii) intercross markers inherited from both parents (segregating 3:1). Based on the parental source of the markers, the two testcross marker sets can be used to construct single-tree genetic maps of the two parental trees. This approach has been widely used in intraspecific full-sib pedigrees, or in first-generation (F_1) interspecific families, particularly when only dominant markers were available.

Pseudo-backcross mapping population

The "pseudo-backcross" follows principles similar to those underlying traditional backcross designs. However, instead of the F_1 progeny being backcrossed to the original parents, it is crossed to alternative parents of one of the two species used for the original cross in order to avoid the negative consequences of inbreeding depression (Myburg *et al.*, 2003). In this scenario, dominant alleles that are fixed in one species segregate 1:1 in the progeny. The double pseudo-backcross approach is based on the two-way pseudo-testcross design but allows comparative mapping with much higher resolution due to the higher proportion of shared marker polymorphism in the resulting pedigree (through the shared F_1 parent). This provides an excellent genetic framework for comparative mapping of genes and genetic factors involved in interspecific differentiation of the parental species.

Open-pollinated (half-sib) mapping population

This model relies on a unique feature of conifer seeds: the haploid megagametophyte. This tissue contains a mitotic derivative from one of the four cells resulting from a single meiotic event and is identical to the maternal contribution to the zygote, which develops into an embryo. Thus, each megagametophyte is genetically identical to the single recombinant gamete inherited from the maternal parent. As a consequence, each heterozygous locus in the maternal parent segregates in a 1:1 ratio. Pairs of segregating markers may, therefore, be tested for linkage and recombination. The testcross segregation pattern (1:1) of marker alleles in megagametophytes can be used to determine linkage between markers, estimate recombination distance, and assign linkage phase. This information can be used to construct single-tree genetic linkage maps of the maternal parent.

Physical maps

A physical map is another way of depicting the genome and is represented by a continuous sequence of DNA. The distance between features is described by the number of nucleotides that separate them. Physical maps are typically generated by cleaving the genome into segments using restriction enzymes, or by randomly shearing it, and cloning the resulting fragments so that they can be analyzed individually. The characterization of each

---

Box 1.3. Genetic Mapping Populations

Genetic mapping of forest trees has typically relied on the use of available populations developed by tree-breeding programs, rather than pedigrees specifically designed for that purpose.

Pseudo-testcross

A pseudo-testcross mapping population is obtained when two highly heterozygous individuals are crossed and genotyped with dominant markers. Two distinct parental marker configurations that segregate in the progeny can be observed: the testcross and the intercross. The testcross allows the generation of single-tree maps for each individual parent, based on the segregation of alleles at heterozygous loci from each parent. The intercross can be used to establish synteny between the two single-tree, parental maps.

Testcross Marker Segregation

	Parent 1	Parent 2	Progeny genotypes	
			1	2
Marker alleles	A/a	a/a	A/a	a/a
Marker phenotype	present	absent	present	absent
Frequency progeny			½	½

Intercross Marker Segregation

	Parent 1	Parent 2	Progeny genotypes		
			1	2	3
Marker alleles	A/a	A/a	A/A	A/a	a/a
Marker phenotype	present	present	present	present	absent
Frequency progeny			¼	½	¼

Pseudo-backcross

A pseudo-backcross mapping population is generated by crossing the F_1 individual derived from two heterozygous parents with a different parent. Crosses typically involved F_1 parents from different species, and primarily target loci where the dominant maker is homozygous in one species and the recessive marker is homozygous in the alternative species. As a consequence, the F_1 hybrid is heterozygous for all loci that are fixed for alternative alleles in the distinct species.

F_1 Hybrid

	Species 1	Species 2	Progeny genotype
			1
Marker alleles	A/A	a/a	A/a
Marker phenotype	present	absent	present
Frequency progeny			1

As the F_1 is crossed to an alternative parent from either of the hybrid parental species, dominant alleles that were fixed in each species are expected to segregate.

Pseudo-backcross Segregation

	F_1	Species 2'	Progeny genotypes	
			1	2
Marker alleles	A/a	a/a	A/a	a/a
Marker phenotype	present	absent	present	absent
Frequency progeny			½	½

Continued

Box 1.3. Continued.

Half-sib

This design is suitable for genetic mapping of open-pollinated populations, where haploid tissue from the known parent can be obtained. In this case, dominant alleles that are heterozygous in the known parent are expected to segregate in the progeny.

Testcross Marker Segregation

		Progeny genotypes	
	Known parent	1	2
Marker alleles	A/a	A/a	a/a
Marker phenotype	present	present	absent
Frequency progeny		½	½

segment can be done through the generation of unique restriction digestion profiles, to produce a minimum-tiling path (see below). Arranging these overlapping segments into a contiguous sequence is often the foundation of genome sequencing projects and relies on a hierarchical strategy.

Principles of DNA Sequencing

Genetic linkage maps describe the relative position of individual loci, established based on their linkage to neighboring loci. While useful, genetic maps lack the resolution of complete genome sequences – they inform the distribution of loci dispersed across the genome, but do not describe the genome sequence data between these loci. Furthermore, the genetic distance between two loci does not represent their physical distance (i.e. the number of nucleotides that separate them). Physical maps address that limitation by aiming to describe the actual physical distribution of genome features. The most detailed physical map is the complete genome sequence.

The genome is composed of DNA, which encodes the genetic information that is inherited from generation to generation, and defines the growth and development properties of all eukaryotes. DNA is formed by two complementary strands of successive nucleotides. Nucleotides are composed of the bases adenine (A), cytosine (C), guanine (G), or thymine (T), together with a deoxyribose sugar and a phosphate group. Each nucleotide is joined to the one next to it by covalent bonds between the sugar of one nucleotide and the phosphate of the next, resulting in an alternating sugar–phosphate backbone. According to base pairing rules (A with T, and C with G), hydrogen bonds bind the nitrogenous bases of the two separate polynucleotide strands to make double-stranded DNA. Further details about the DNA molecule can be found elsewhere (see Chapter 2 in White *et al.*, 2007).

DNA sequencing methods

DNA sequencing involves identification of the successive nucleotides (A, C, G, and T) that make up a DNA strand. Methods to identify the individual bases and their order have been pursued since the discovery that DNA is the inherited genetic material (Avery *et al.*, 1944). Early approaches to DNA sequencing were costly and labor-intensive, but in the past decade the throughput of sequencing technologies has matched or outpaced what is predicted by Moore's Law. This maxim refers to the doubling of capacity of computer components every 2 years. Similarly, the cost of sequencing an individual nucleotide has been reduced to less than 1/100,000 of what it was at the beginning of the 21st century (Fig. 1.3). Current methods of DNA sequencing and their commercial application differ significantly with respect to: (i) throughput, or the amount of DNA sequence that can be characterized within a certain timeframe; (ii) cost per nucleotide; (iii) error rate; and (iv) read length (for a review of the most current methods, see Goodwin *et al.*, 2016). Sequencing instruments also differ greatly with respect to their cost. An overview of the main methods of DNA sequencing, and their advantages and disadvantages, are described in the following sections. The most commonly used commercial providers that employ these methods, and the equipment and their specifications, are presented in Table 1.1.

Chain-termination (Sanger) DNA sequencing

Frederick Sanger achieved the most significant early advance in DNA sequencing technology by using primer extension coupled with a mixture of native deoxynucleotides and dideoxynucleotides (Sanger *et al.*, 1977). Incorporation of the latter terminates synthesis of a complementary strand, generating a collection of extension products whose length is dependent on the position of the dideoxynucleotide (Fig. 1.4). Numerous advances were later made in this method, including the incorporation of dideoxynucleotides with fluorescent tags, capillary electrophoresis, and general automation. While the most advanced methods of Sanger sequencing can generate reads that exceed several hundred bases and with a low error rate, the low throughput and high cost per base have led to its replacement. Since the early 2000s, chain-termination DNA sequencing has largely been substituted by

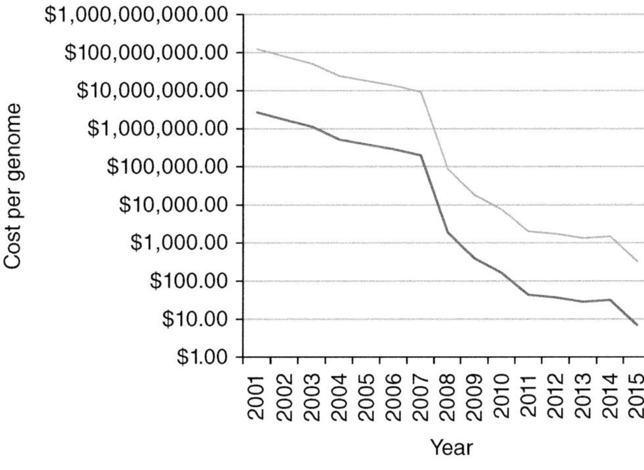

Fig. 1.3. Estimated cost (US$) of sequencing the genome of a species of *Populus* (~500 Mbp, dark green line) and *Pinus* (~23,000 Mbp, light green line). The cost estimate assumes that the genomes are sequenced with short-read, next-generation sequencing methods, with an average sequencing depth of 30×. Costs also include other expenditures beyond sequencing reagents, as described in www.genome.gov/sequencingcostsdata/ (accessed July 15, 2019).

Table 1.1. The most widely used commercial DNA sequencing platforms. With the exception of Sanger sequencing, all other platforms are referred to as next-generation DNA sequencing instruments. Cost estimates are approximate and vary depending on sequencing provider and sequencing mode.

Method	Main commercial provider (instrument)	Throughput (Gbp)	Cost (US$/Gbp)	Read length (bp)	Error (%)
Sanger sequencing	Thermo Fisher Scientific (3730XL DNA analyzer)	0.69–2.1	2,000	400–900 (depending on run module)	<1
Sequencing by synthesis	Roche (454 GS FLX Titanium XL+)[a]	0.7	10,000	~700	~1
	Illumina (HiSeq X)	800–900	7	150	<1
Single-molecule sequencing	Pacific Biosciences (Sequel)	3.5–7	250–500	Variable (8,000–12,000)	10–15

Gbp, gigabase pairs.
[a]Platform discontinued.

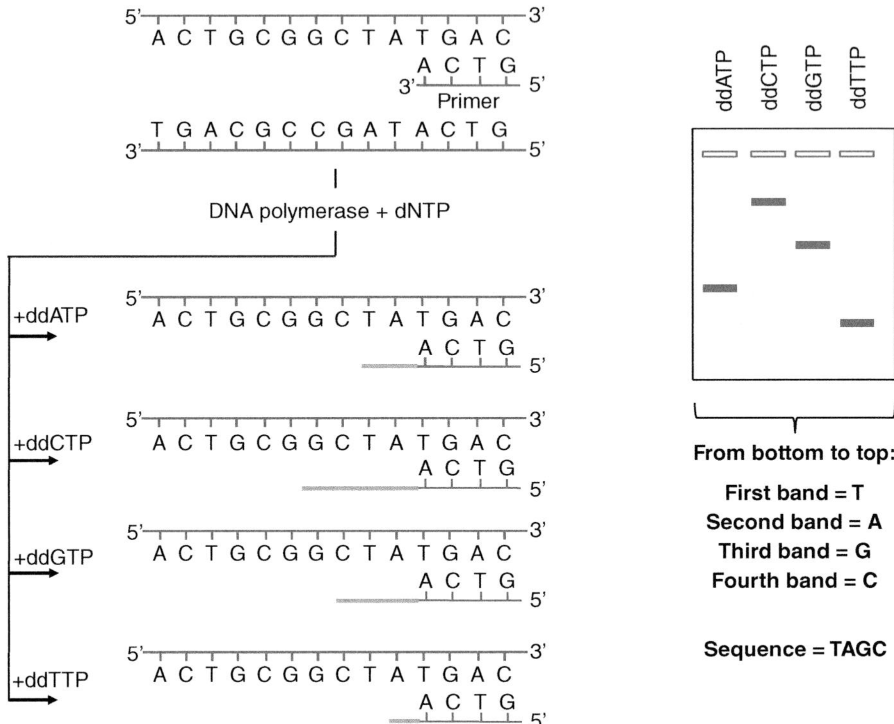

Fig. 1.4. Sanger DNA sequencing is based on the differential extension of the complementary strand by DNA polymerase using dideoxynucleotides (ddNTPs). A template is denatured and a primer anneals to one of the strands. Four separate extension reactions are carried out (left panel); each contains DNA polymerase, all four dNTPs, and one type of ddNTP (ddATP, ddCTP, ddGTP, or ddTTP). When the DNA polymerase introduces a ddNTP, the reaction stops, generating an extension product of length that corresponds to the position of the specific ddNTP in the strand. These can be determined by separating the product of each of the four separate extension reactions in a matrix, such as a polyacrylamide gel (right panel). The smallest product moves the fastest, so the sequence can be read directly from the gel.

next-generation sequencing (NGS) methods, except when low error rates and long reads for a few target sequences are desired.

Next-generation DNA sequencing

SEQUENCING BY SYNTHESIS Sanger sequencing and advances in its implementation provided the tools necessary to sequence the first human and plant genomes early in this century (Lander *et al.*, 2001; Venter *et al.*, 2001). However, the use of the Sanger method is too costly, low throughput, and labor intensive to support sequencing large numbers of individuals. To address this limitation, the US National Institutes of Health and numerous private enterprises supported efforts to develop new approaches to DNA sequencing that would overcome the limitations of Sanger sequencing. The result was a dramatic increase in DNA sequencing throughput that doubled every 7 months for the first part of the 21st century. A parallel decrease has occurred in the cost of DNA sequencing of genomes (Fig. 1.3).

Compared with Sanger sequencing, NGS uses a fundamentally different approach to identify the successive bases that define a DNA strand. Instead of inferring the position of each base from the molecular weight of a partially extended complementary strand, they are detected as the DNA polymerase incorporates them. Because the strength of the signal produced by the incorporation of a single nucleotide is limited, these approaches typically require that many copies of identical molecules be synthesized. Thus, the first critical advance of NGS was the development of methods to amplify DNA molecules in parallel, to produce multiple copies of thousands to millions of individual templates. Two variations of this approach that have been widely adopted are based on emulsion polymerase chain reaction (PCR) amplification (Tawfik and Griffiths, 1998) and bridge PCR amplification (Fedurco *et al.*, 2006) (Box 1.4).

As soon as the DNA is amplified, the next step for sequencing is the detection of the specific base incorporated into a template strand (Box 1.5). The differences between

---

Box 1.4. Methods of DNA Amplification Prior to Sequencing by Synthesis

---

Prior to sequencing, most next-generation sequencing methods require that multiple copies of the DNA template be generated. Several methods have been developed to achieve this goal, two of which have been extensively used: (a) the emulsion PCR and (b) the bridge PCR amplification.

Emulsion PCR – (a) Two different adaptor (light and mid lines) are first ligated to the ends of each DNA molecule (green line), and combined with a bead coated with oligonucleotides that are complementary to one of the two adaptors (light green lines on surface of bead). (b) After denaturation, the adaptor complementary to the oligonucleotide on the bead surface anneals to it. (c) The complementary sequence of the template that annealed to the bead is synthesized (dashed line) in the presence of DNA polymerase and dNTPs. (d) After the cycle is repeated multiple times, the bead surface contains several copies of identical single-stranded DNA templates. To be successful, emulsion PCR requires that a single DNA molecule be combined with each bead, otherwise different molecules are amplified on the same bead surface. In order to achieve this, beads and DNA are combined in an oil–aqueous emulsion containing DNA polymerase and dNTPs. The solution creates individual droplets that encapsulate the components of the reaction.

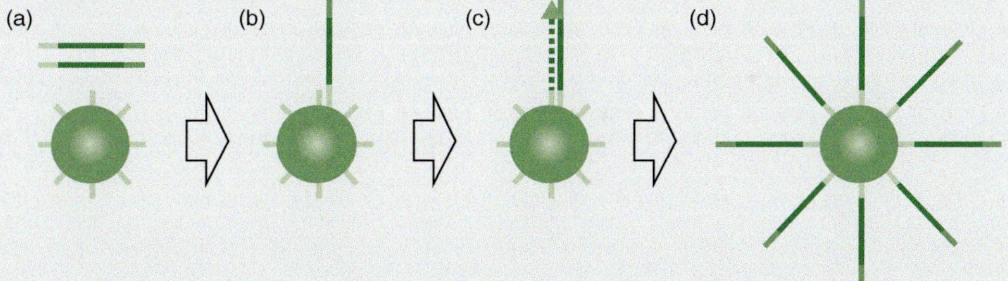

Bridge PCR amplification – Similar to emulsion PCR, bridge PCR amplification begins with (a) DNA molecules (green line) that contain two different adaptors (light and mid green lines). However, the reaction occurs on a solid surface coated with oligonucleotides that are complementary to one of the two adaptors (light green lines on a solid support). (b) After denaturation, one of the adaptors anneals to an oligonucleotide on the surface of the solid support. (c) This oligonucleotide now serves as the starting point for an extension reaction by the DNA polymerase, which generates a strand complementary to the template (dashed line). (d) This creates a complementary DNA molecule that is bound to the surface, after denaturation and release of the template strand. (e) The complementary DNA molecule can now create a "bridge" by the annealing of the adaptor on the other end, to the alternative oligonucleotide on the solid surface. (f) This oligonucleotide and the annealed DNA molecule now serve as a template for synthesis of the complementary strand (dashed line). (g) After denaturation, two complementary strands bound to the solid surface will be present. Because the reaction is repeated many times, large numbers of identical molecules are created.

Continued

Box 1.4. Continued.

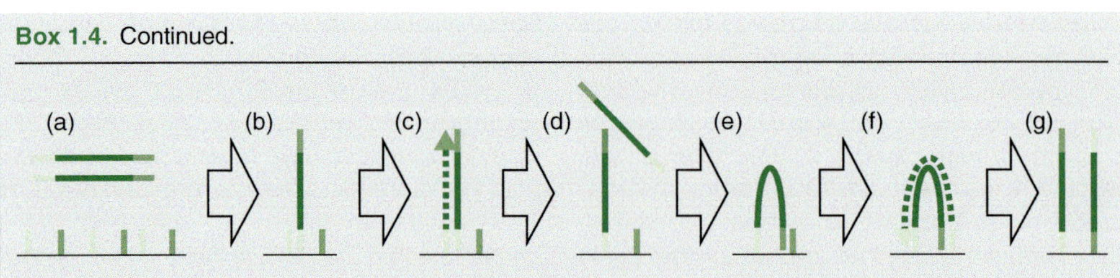

Box 1.5. Sequencing by Synthesis

After amplification of DNA templates by emulsion PCR or bridge PCR, sequencing can be initiated. Sequencing by synthesis can be carried out using different approaches: (i) cyclic reversible termination; and (ii) single-nucleotide addition. These methods also adopt different approaches to detect the incorporated nucleotides.

Cyclic reversible termination

Initially, DNA polymerase, primer (grey arrow) and modified nucleotides are added to the solid surface, where the DNA molecules that are targeted for sequencing have been immobilized (a). All four nucleotides (dATP, dCTP, dGTP, and dTTP) that are added are fluorescently labelled with base-specific and cleavable fluorophores (stars). These nucleotides also contain a reversible block on the 3′ group—as a consequence, only a single nucleotide can be added by the DNA polymerase until the block is removed. After the DNA polymerase adds the first nucleotide, the reaction stops (b). An image is taken and the fluorophore type is recorded. Next, the blocker is removed together with the fluorophore (c). The extension reaction is initiated again, with the addition of the four fluorescently labelled and cleavable nucleotides and DNA polymerase (d). As a nucleotide is incorporated in each successive cycle, an image is taken in step; thus, the DNA sequence is based on the detection of each base-specific fluorescence.

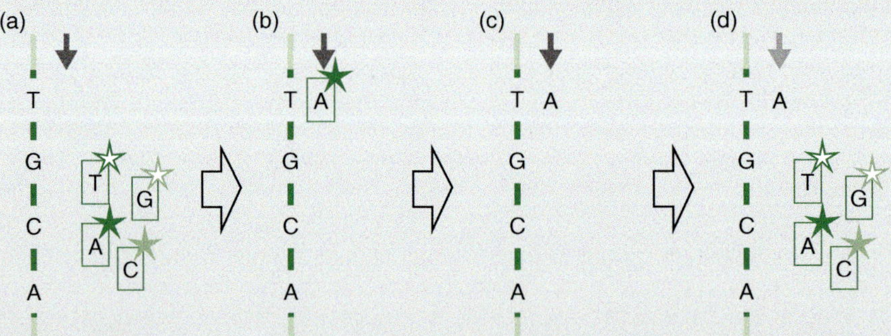

Single-nucleotide addition

This method is similar to cyclic reversible termination but differs in the way individual nucleotides are incorporated and detected. The method developed and used in the first NGS platform, commercialized by 454 Life Sciences Corp., was based on the detection of the pyrophosphate molecule (PPi) that is released when each nucleotide is incorporated by the DNA polymerase. When combined with ATP sulfurylase, PPi transforms adenosine 5′-phosphosulfate (APS) into ATP. ATP then acts as a cofactor in the conversion of luciferin to oxyluciferin—a reaction that produces light and can be detected by a charge-coupled device (CCD) camera. Thus, as each individual nucleotide is added to the elongating chain, light is detected to define the presence or absence of that nucleotide as the complementary base to the template. When many nucleotides of the same type are added, the light intensity increases to reflect their repeated incorporation. Other NGS platforms that were developed were based on similar principles. For instance, the Ion Torrent (Thermo Fisher Scientific) uses the same approach but instead of detecting the release of PPi, it detects the H^+ ion that is released in the process, and the pH change.

these approaches are related to the nature of the signal emitted as nucleotides are incorporated. The two methods most widely adopted are based on: (i) release and detection of pyrophosphates (Ronaghi *et al.*, 1996; Margulies *et al.*, 2005) or ions (Rothberg *et al.*, 2011) as a consequence of the incorporation of one or more nucleotides; or (ii) fluorescence (Turcatti *et al.*, 2008). Both rely on the direct detection of the base(s) incorporated, rather than the inference of their position in the sequence based on the molecular weight of a terminated chain.

The requirement that multiple copies of each template be generated during bridge amplification or emulsion PCR creates a limitation for sequencing by synthesis. In each cycle of nucleotide incorporation, the process will fail for some of the copies of the template being synthesized. Failure occurs because addition of nucleotides may not occur for all molecules. As the process of nucleotide incorporation is repeated, sequencing quality deteriorates. Thus, accurate sequencing is typically only achieved for a few hundred bases without significant loss in sequence quality. As a consequence, while DNA analysis using sequencing by synthesis has achieved high throughput and low cost, compared with Sanger sequencing, read lengths have remained relatively short.

SINGLE-MOLECULE DNA SEQUENCING To address the read-length limitation of sequencing by synthesis, single-molecule methods were developed. As opposed to sequencing by synthesis, where incorporation of each nucleotide or nucleotide type is detected in discrete steps, in single-molecule DNA sequencing, detection occurs as the DNA polymerase progresses, without a pause in the sequencing. As a consequence, sequencing occurs more rapidly than in existing NGS platforms. Currently, the predominant single-molecule DNA sequencing platform (Sequel; Pacific Biosciences) is based on monitoring the DNA polymerase as the individual, complementary, and fluorescently labelled nucleotides are incorporated into the copy being made of the template strand. This generates a signal pulse of the color associated with the nucleotide being incorporated. The fluorescent tag is then released, eliminating the fluorescent signal detected previously (Box 1.6).

While technically challenging, single-molecule DNA sequencing has been demonstrated in at least one commercial platform (Eid *et al.*, 2009). The primary advantage of using single-molecule sequencing is the generation of very long reads compared with the existing NGS platforms. A second advantage of this and other DNA sequencing platforms is that the incorporation of individual nucleotides is detected in real time. However, the error rate is significantly higher in the existing platforms compared with those that use sequencing by synthesis (Table 1.1).

SEQUENCING BY LIGATION This approach is fundamentally different from other NGS methods because it uses hybridization of short, labeled oligonucleotides and ligation for sequencing, instead of DNA polymerization. As with the other methods, sequencing initiates with a single-stranded DNA template that is flanked by a sequence common to all templates (typically an adaptor). In the first step, an oligonucleotide complementary to the flanking, common sequence anneals to each template. Next, a mixture of diverse oligonucleotides and DNA ligase is added. In this case, only those oligonucleotides complementary to the template and positioned immediately adjacent to the common oligonucleotide will hybridize and be ligated. These oligonucleotides are labelled, allowing the detection of those that are incorporated. However, because a limited number of fluorescent tags is available, the identification of specific nucleotides being incorporated at each position has to rely on a combination of signals. While platforms based on sequencing by ligation have been developed and commercialized, they have largely been surpassed by other methods due to its limitations with respect to throughput, accuracy, speed, and cost.

*Future prospects in
DNA sequencing advances*

Despite the significant advances in DNA sequencing, existing methods still pose certain limitations that hamper their broad

Box 1.6. Single-molecule DNA Sequencing

The single-molecule DNA sequencing method developed by Pacific Biosciences does not require amplification of DNA templates for detection of the nucleotides incorporated. Instead, it relies on a highly sensitive method of detecting the fluorescence associated with individual nucleotides as they are added by the polymerase to the copy of the template molecule. To detect fluorescent single nucleotides, the sequencing platform utilizes a microscopic well (also referred as a zero-mode waveguide), to the bottom of which a DNA polymerase is attached, as shown in (a). The DNA template is processed by the DNA polymerase, which incorporates fluorescently labeled complementary nucleotides (A, C, T, and G) on the top of the well. As the next complementary labeled nucleotide (shown in (b) as a C with a star) is added, excitation from a laser located at the bottom of the well leads to emission of a fluorescent signal from the labeled nucleotide, as shown in the graph in the lower panel. Because the well is small, the zone excited by the laser is restricted to the lower surface, where the DNA polymerase is located. After cleavage of the fluorescent tag linked to the pyrophosphate group of the complementary nucleotide, the tag diffuses away from the bottom of the well (dotted arrow), leading to a drop in signal strength.

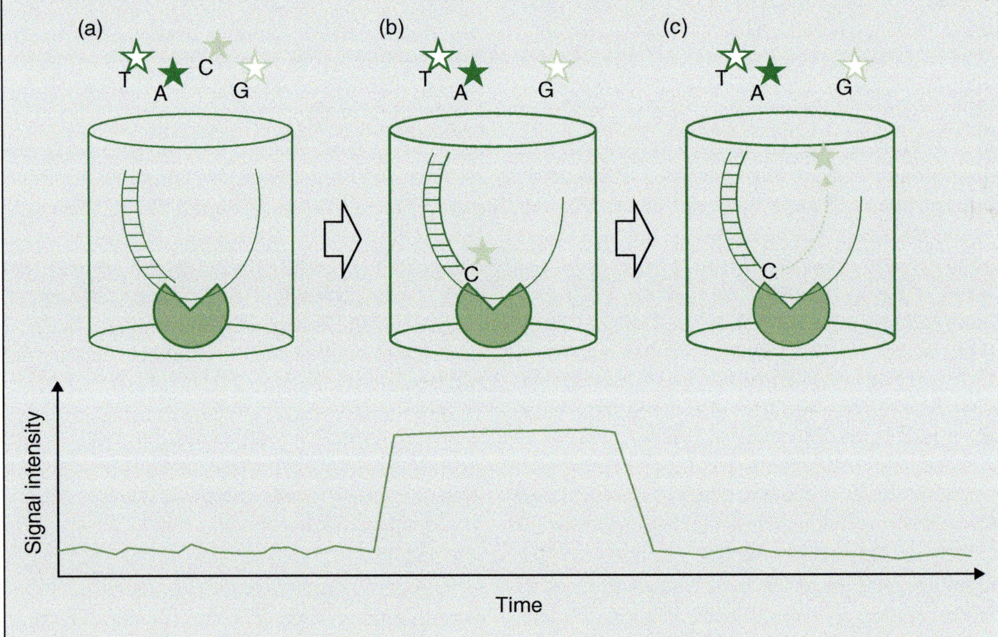

use. For instance, the read length of NGS sequencers with the highest throughput is still limited to a few hundred bases, making it computationally difficult to generate contiguous genome assemblies. Other sequencing platforms provide reads of longer length but with limited throughput and high error rates. Finally, all existing methods require relatively sophisticated techniques to prepare the DNA for analysis, along with specialized equipment and highly trained personnel. These and other limitations provide incentives for the development of even more advanced DNA sequencing platforms that depart from the current paradigms. Several potentially disruptive technologies are currently under development or in the early stages of commercialization, including electron-microscopy sequencing (Mankos *et al.*, 2014) and nanopore-based sequencing (Jain *et al.*, 2016), among others. Recent advances in these technologies have been reviewed in detail elsewhere (Goodwin *et al.*, 2016).

Principles of Genome Sequencing, Assembly, and Annotation

The genome size of plants varies over a range of at least three orders of magnitude, from less than 100 Mb in several species of Lentibulariaceae, a family of carnivorous plants (Greilhuber *et al.*, 2006), to over 100 Gbp in the monocot *Paris japonica* (Pellicer *et al.*, 2010). Forest-tree species also vary significantly in genome size, from the relatively small genome of black cottonwood (*Populus trichocarpa*) (480 Mb; Tuskan *et al.*, 2006), the first tree genome to be sequenced, to Norway spruce (*Picea abies*), the first conifer genome to be characterized, with a genome of 20 Gbp (Nystedt *et al.*, 2013). Sequencing a genome involves a series of steps that include decisions concerning the: (i) sequencing and assembly strategy to use; (ii) annotation of that sequence; and (iii) evaluation of the quality of the final product. A description of these steps is provided below.

Genome sequencing and assembly

The decision about the method to use to sequence a genome has depended primarily on its size and on the amount of repetitive DNA. Large genomes typically contain a significant fraction of repetitive, nearly identical segments of DNA. Repetitive DNA represents the most significant challenge to the assembly of a genome, because it can lead to the conclusion that a given sequencing read may be positioned in another region of the genome. Two distinct, general approaches have been used to overcome this obstacle. The first, clone-by-clone or hierarchical genome sequencing, attempts to simplify the challenge of assembling repetitive regions of the genome by dividing it into long segments (Fig. 1.5). These segments are likely to contain sequences that extend beyond the repetitive DNA. As a consequence, at least part of the segment is non-repetitive and can be positioned uniquely, relative to other sequences in the genome. The method was used to sequence the highly complex and large genome of maize (*Zea mays*) (Schnable *et al.*, 2009). The second approach,

whole-genome shotgun (WGS) sequencing, relies on generating DNA fragments of different sizes and sequencing their ends (Staden, 1979; Anderson, 1981). Because larger fragments are likely to span most repetitive regions, the end sequences are used to assemble local regions of the genome (Fig. 1.5). Since the advent of NGS methods in the early years of the 21st century, most genome sequencing projects have adopted a WGS approach. Detailed descriptions of both approaches are described in the following sections.

Clone-by-clone or hierarchical genome sequencing

The clone-by-clone sequencing approach typically involves digesting the genome with "rare-cutter" restriction enzymes to generate large DNA segments. These fragments are then cloned into large-insert vectors, such as bacterial artificial chromosomes (BACs) or fosmids. Clonal segments then undergo "fingerprinting" where they are individually digested by frequent-cutter restriction enzymes to generate an individual restriction map. As other clones are fingerprinted, their overlap is determined by the similarity in their restriction pattern. Based on this overlap, a minimum-tiling path is constructed—that path defines the minimum set of clones required for as much coverage of the genome as possible. These selected clones are then sequenced separately, and assembled. Finally, sequences from individual clones are combined with those of adjacent clones to form scaffolds (Fig. 1.5).

An advantage of this method is that the position of one clone relative to its neighbor is known. Therefore, as the assembly is completed, the amount of missing DNA sequence information can often be estimated. By defining the minimum-tiling path, the clone-by-clone approach also attempts to minimize the amount of sequencing required, a significant concern before low-cost and high-throughput NGS became available. A significant disadvantage of this approach to sequencing genomes is the extensive amount of work required for "pre-sequencing" in the fingerprinting of clones and creation of the minimum-tiling path. Furthermore, because some genomic

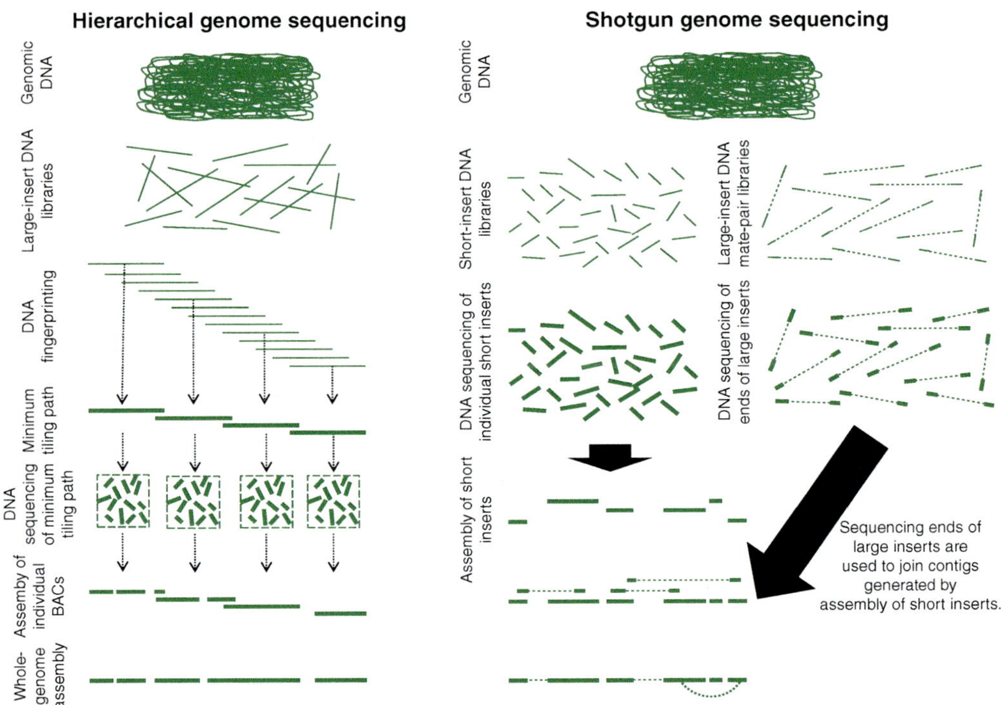

Fig. 1.5. Genomic sequencing and assembly strategies. In hierarchical genome sequencing, genomic DNA is cloned into large-insert DNA libraries. Individual large-insert clones are then fingerprinted to select a minimum number of partially overlapping inserts (minimum-tiling path) to cover the genome. Selected individual large-insert clones are sequenced and assembled individually. Based on the end sequences of the large-insert clones, they are assembled with neighboring large-insert clones. Further positioning and orientation of clones or assemblies can be done by approaches such as genetic mapping. In shotgun genome sequencing, genomic DNA is cloned into short-insert and large-insert mate-pair DNA libraries. Sequences from the short-insert libraries are assembled to generate contigs. Based on the sequenced ends of the large-insert mate-pair libraries, contigs are joined into scaffolds.

regions lack the restriction sites needed to create large-insert clones, they may not be present in the final genome assembly.

WGS sequencing

This is an alternative to the hierarchical genome sequencing approach and is based on randomly shearing the genomic DNA, cloning the sheared fragments into vectors, and then sequencing and assembling them. In contrast to the hierarchical approach, WGS sequencing does not rely on cloning the genomic DNA into large-insert libraries and fingerprinting them to select those to be sequenced. Instead, it is based on the assumption that if a sufficient number of sequencing reads is generated, most of the genome will be represented at least once among those reads. A potential drawback of the WGS method is sequencing repetitive regions of genomes, because individual reads may not span these regions. This limitation has been addressed, in part, by combining data from short-insert libraries with sequencing of mate-pair libraries, where DNA is cloned into larger-insert libraries and sequenced from both non-overlapping ends (Box 1.7). Combining data generated from mate-pair library sequencing with that from short-insert paired-end reads provides a powerful combination of read lengths for maximal coverage of the genome. In fact, studies have shown that sampling multiple libraries and fragment lengths can reduce bias in genome sequencing and result in a

Box 1.7. Paired-end sequencing and Mate-pair DNA Libraries

Paired-end sequencing is the characterization of the nucleotide sequence from both ends of a contiguous DNA segment. Because most NGS platforms generate relatively short reads (<500 bp), the contiguous segments are typically only a few hundred bases long. The genomic DNA is first fragmented into short segments of a few hundred bases (a). These segments are then ligated to adaptors (b). The DNA segments with adapters are then ready for sequencing from both ends using an NGS platform such as Illumina (c). Assuming 100 bp are sequenced from both ends, the first 100 and the last 100 bp will be known. However, the internal bases are not sequenced.

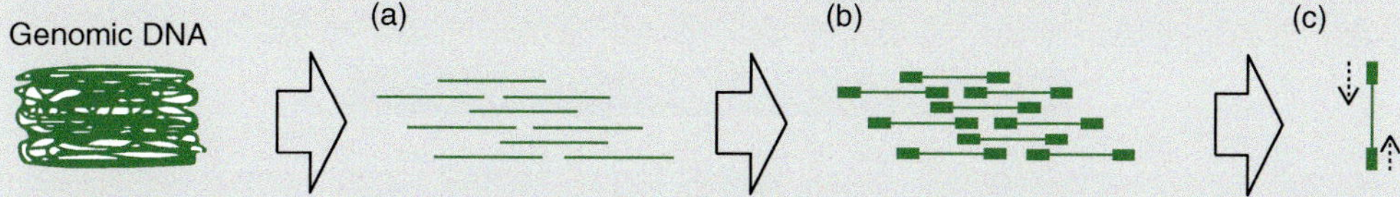

Mate-pair DNA libraries generate segments for sequencing that, in contrast to paired-end sequencing, are separated by several thousand bases. Many methods to create mate-pair libraries exist, but one of the most common involves the following steps. The genomic DNA is first fragmented into large segments of a specific, pre-determined size (a). End repair of the DNA segments using biotinylated nucleotides is then carried out (b). Next, the DNA segments are circularized, such that the junction of each end now contains biotinylated nucleotides (c). Finally, the circular DNA molecules are fragmented and the segments that contain biotinylated nucleotides are recovered (d). These segments now contain the ends of a larger DNA segment that can be sequenced after adaptor ligation, as described in the figure above (step c).

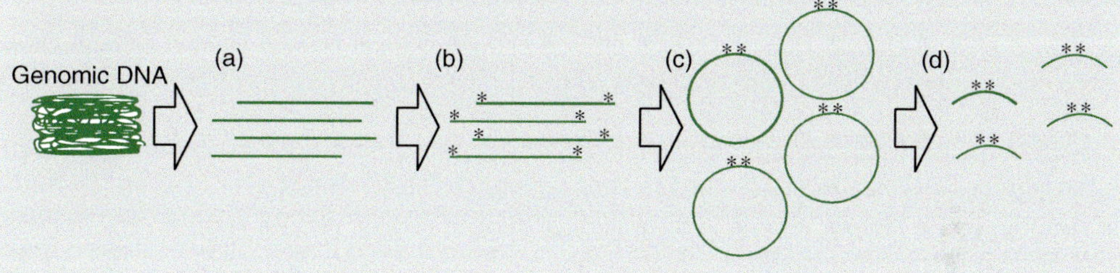

more consistent representation. With the rapid increase in DNA sequencing throughput and reduction in costs, random-shotgun sequencing has become the method of choice for sequencing most plant genomes, including the large conifer genomes of Norway spruce and loblolly pine (*Pinus taeda*) (Chapter 3).

Measuring the quality of a genome assembly

The assembly of a genome sequence usually results in two types of sequence assemblies: contigs and scaffolds. A contig is a continuous sequence of DNA, where bases are known and there are no gaps. A scaffold is a set of sequences or contigs that have a defined order and orientation (Fig. 1.6); however, there are gaps of known or unknown size among them. Several summary statistics have been used to define the completeness of genome assembly (Yandell and Ence, 2012). The most common is the contig or scaffold N50 and L50 (Fig. 1.7). N50 is generated by ordering the assembled segments from the longest to the shortest. The number of segments with a cumulative size that is greater than 50% of the assembly size, corresponds to N50. Thus, a longer N50 indicates that a larger proportion of the genome sequence has been assembled and, consequently, that a higher-quality outcome has been obtained. Because N50 is calculated relative to the total assembly size, comparing N50 values between different

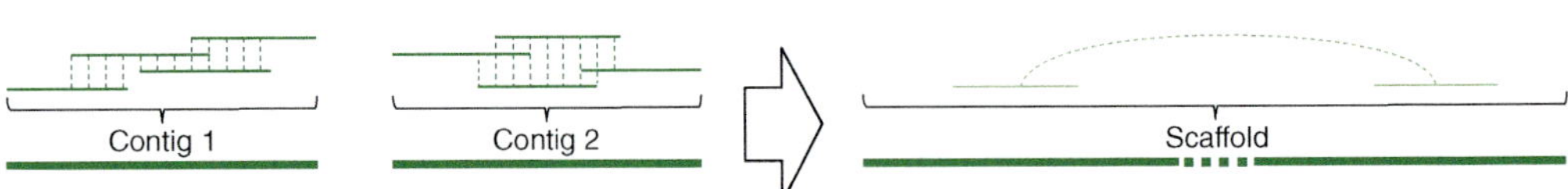

Fig. 1.6. DNA sequencing contigs and scaffolds. Left panel: contig sequence assembly. The overlap between multiple DNA sequencing reads (narrow horizontal lines joined by dashed vertical lines) is used to determine the sequence of contigs (thick lines, contig 1 and contig 2). Right panel: scaffold sequence assembly. Sequencing reads (narrow horizontal lines) derived from mate-pair libraries can have their ends overlapping two contigs, allowing them to be joined (thick dashed line) into a scaffold.

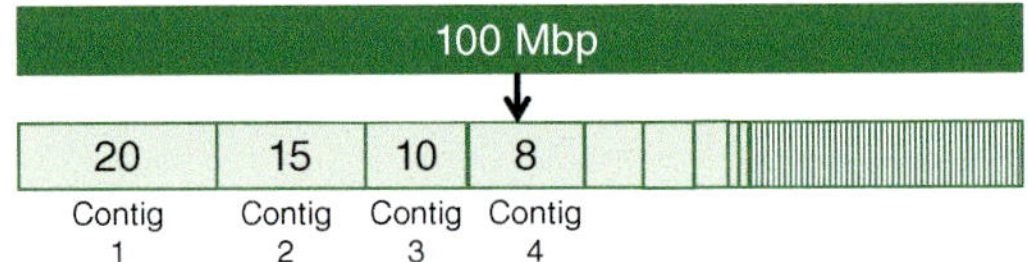

N50 = 8 Mbp (20 Mbp + 15 Mbp + 10 Mbp + **8 Mbp** ≥ 50 Mbp)
L50 = 4 contigs

Fig. 1.7. Genome assembly quality measures (N50 and L50). N50 is the length of the shortest contig or scaffold that represents more than 50% of the genome sequence, when ordered from largest to smallest. In the figure, a 100 Mbp genome (green box) is assembled into multiple contigs (pale green boxes), with sizes 20, 15, 10, and 8 Mbp, and below. When summing the length of the contigs from larger to smaller, as the fourth contig is added, the length reaches 53 Mbp, which is larger than 50% of the genome sequence length. Thus, N50 is 8 Mbp. L50 is the number of contigs or scaffolds that were required to reach N50 (four contigs in this example).

assemblies is often not meaningful because of variability in their sizes. To address this limitation, NG50 was proposed as an alternative. NG50 adheres to the same principles as N50 but is based on the estimated genome size, not the assembly size. Another measure related to N50 is L50, which refers to the number of contigs or scaffolds required to generate an assembly that contains more than 50% of the genome (Fig. 1.7). Finally, the quality of a genome assembly can also be estimated by the number of gaps in the contigs and scaffolds, as well as the percentage of coverage of the gene and genome. The latter measures are defined by counting the number of genes represented in the assembly or the total length of the assembly represented in contigs or scaffolds, relative to the theoretical expectation. The quality of tree genome assemblies released to date varies greatly depending mainly on the size of the genome and the strategy and platform used for sequencing. The N50 of contigs and scaffolds of the large loblolly pine genome (22 Gbp), which was assembled largely from short (< 200 bp)

reads, is 8.2 and 66.9 kilobase pairs (kbp), respectively (Zimin *et al.*, 2014). In the much smaller *E. grandis* genome, which was shotgun sequenced, using longer, higher-quality Sanger sequencing, the N50 of the contigs and scaffolds extends over significantly longer distances (Myburg *et al.*, 2014).

Genome annotation

The sequence derived from a genome assembly has limited value unless its "landmarks," such as genes and repetitive sequences, are properly identified and annotated. Genome annotation follows a series of logical steps. This process normally begin with the identification of regions of repetitive DNA, which are detected computationally using bioinformatics tools that compare the sequence of the genome with libraries or databases of repetitive DNA sequences. Because repetitive DNA is largely the product of known transposable elements, such as retrotransposons, many of them can be identified and annotated.

However, because these repetitive sequences are not under selective pressure, ancient events of transposition may diverge in sequence to the point where they are not computationally recognized. In the process of detecting repetitive DNA, other low-complexity sequences, such as homopolymers, are also identified. After categorizing these sequences into their distinct classes, they are commonly "masked" (i.e. they are made unavailable to further analysis) in order to simplify the identification of genes and other features, and to avoid their spurious annotation as other genomic features.

The second major step in the description of a genome involves the identification of genes and other associated features, such as promoters. This step takes advantage of known expressed sequence tags and protein sequences from the same or phylogenetically related species that are aligned with the genome to identify homologous coding sequences—these are commonly referred to as homology-based methods. Gene predictions can also be made using *ab initio* methods, which utilize statistical approaches to define the probability of a gene occurring in a certain DNA segment, including its untranslated regions, exons, and introns. Most bioinformatics tools used for identification of genes combine evidence-based and *ab initio*-based methods to support the conclusion that a gene is present in a given sequence (Yandell and Ence, 2012).

Identification of repetitive sequences, genes, and their associated features is only the beginning of the full description of a genome. Next, genes and repetitive sequences that were identified have to be placed into specific categories. For genes, this may include classification into a specific molecular function, based on homology with previously characterized genes, the detection of certain motifs, or predicted protein structure. In order to organize the way genes are categorized, specific categories have been developed by the Gene Ontology (GO) Consortium (http://geneontology.org/, accessed July 15, 2019; Ashburner *et al.*, 2000; Gene Ontology Consortium, 2015). This approach categorizes each gene based on: (i) its cellular component (i.e. the cellular environment within which the gene product acts, such as the nucleus or cell wall); (ii) its biological process (i.e. the general process the gene contributes to, such as development or response to stimuli); and (iii) its molecular function (i.e. the specific molecular activity that the gene product carries out, such as binding or catalyzing a reaction). A database of GOs can be searched using a browser (AmiGO), which relates genes and their products to specific ontologies. Furthermore, tools that automatically assign genes to GO categories have been developed (BLAST2GO) and are widely utilized for the functional annotation of genomes (Conesa *et al.*, 2005). Much more information can be incorporated into the annotation of a gene, and is available in dispersed databases. This includes data about the interaction of a gene with other genes (e.g. STRING, http://string-db.org, accessed July 15, 2019); its domains (e.g. Prosite, http://prosite.expasy.org, accessed July 15, 2019); and secondary or tertiary structures (e.g. PredictProtein, www.predictprotein.org/, accessed July 15, 2019).

Genome annotations typically require the combined use of various bioinformatics tools to identify features of interest and assign them a role, whether structural or functional, or both. In order to integrate these tools, annotation pipelines that combine their distinct capabilities have been developed and are now commonly used in genome annotation. Expectedly, the quality of the annotation of a genome depends primarily on the quality of the assembly of its sequence. Short contigs or scaffolds limit the ability to detect complete genome features. However, genome properties can also limit the quality of the assembly. For instance, there is direct relationship between genome size and the length of its genes—larger genomes have longer genes while smaller genomes contain shorter genes. As a consequence, for an assembly of similar quality, it will be more difficult to annotate the genome of a 20 Gbp conifer than that of the 480 Mb black cottonwood. Also, certain genomes may have an exceedingly large number of tandem copies of genes, as is the case with flooded gum. These repeated sequences can limit the computational assembly of genomes, which may fail to recognize the occurrence of two or more copies of the same gene and instead collapse them into a single gene. Thus, because multiple genes are not recognized, the annotation will not represent a complete description of the genome.

Discovery and Genotyping of DNA Sequence Variation

DNA sequence variation contributes to the phenotypic diversity observed among individuals and populations. Thus, identifying and genotyping that variation is critical for understanding the genetic basis of trait differences. When DNA sequence variation is genotyped in a segregating population, it can be used to generate genetic maps and to map simple or complex traits. Genotyping also provides a tool for uniquely identifying individuals (fingerprinting) and the dispersal of their genetic material in the environment (gene flow). Other uses include: (i) identifying genes under natural and artificial selection; (ii) uncovering genetic structure in natural and breeding populations; and (iii) screening for resistance to plant diseases.

Types of DNA sequence variation

Variation in DNA sequence can occur at the chromosome level. These mutations can in-volve deletions, translocations (i.e. a segment of one chromosome is transferred to another), inversions (the orientation of a chromosome segment is reversed), and duplications. Such changes can impact large parts of the chromosome.

While there may be differences in large chromosome segments among samples, these are typically very rare. Instead, most DNA sequence variation observed between any two samples affects one or a few nucleotides. These most common DNA variants can be categorized into two non-exclusive classes (Fig. 1.8). First, they can be classified based on their functional consequences. A synonymous substitution occurs when the mutation does not result in a change in the amino acid during the translation of a codon (i.e. there is degeneracy in the genetic code). A non-synonymous substitution leads to a change in amino acid, resulting in a missense mutation that may lead to loss of function of the protein. Non-synonymous substitutions can also result in the formation of a premature stop codon, which is commonly referred to as a nonsense mutation.

Second, DNA variation is classified based on the structural consequence of the variation, which is generally one of three kinds: (i) single-nucleotide variants (SNVs), which arise by the substitution of one nucleotide by another; (ii) insertion and deletion variants (indels), which are characterized by the addition or removal of one or more nucleotides at a given position; and (iii) sequence-repeat

Original sequence	AGT methionine	TAT tyrosine	AGT serine	Original sequence	AGTTATAGT	
Synonymous substitution	AGT methionine	TA**C** tyrosine	AGT serine	Single-nucleotide variant/polymorphism	AGTT**C**TAGT	
Non-synonymous substitution (missense)	AGT methionine	**CAC** histidine	AGT serine	Insertion Deletion	AGTT**AA**TAGT AGTT-TAGT	
Non-synonymous substitution (nonsense)	AGT methionine	TA**A** STOP		Repeat variant	AGTT**ATTA**TAGT	

Fig. 1.8. Functional and structural DNA sequence variation. DNA sequence variation that may have functional consequences can be classified as synonymous and non-synonymous substitutions (left panel). Synonymous substitutions occur when a nucleotide change results in a codon for the same amino acid, while non-synonymous substitutions lead to a different amino acid (missense mutation) or a stop codon (nonsense mutation). Structural variation in the DNA sequence (right panel) can be classified as single-nucleotide variant or polymorphism, insertion and deletion, and repeat variants. Functional and structural DNA sequence variations are non-exclusive; for example, an insertion or deletion may lead to a non-synonymous substitution.

or copy-number variants, which are defined by variation in the number of consecutive copies of the same sequence. The most common type of genetic variation in eukaryotic genomes is SNVs. When their frequency in a population exceeds 1%, SNVs are often referred to as single-nucleotide polymorphisms (SNPs). Because of their abundance, SNVs/SNPs are the preferred type of genetic variant used in genome-wide analyses. However, variants in the number of repetitive sequences can be highly valuable when the unique identification (fingerprint) of an individual is desired, or when an individual's parental origin is sought. This occurs because the multi-allelic nature of these variants results in a higher probability that any two unrelated individuals will not share the same profile at a given locus. As a consequence, at the level of an individual locus, a sequence-repeat variant may be more informative than an SNV. However, given the abundance of SNVs in genomes, a larger number of loci may need to be sampled to obtain the same information content.

Factors that contribute to DNA sequence variation

The abundance of DNA sequence variation in an individual or population is dependent on the species' mutation rate and its evolutionary history. The average mutation rate of eukaryotes has been shown to be directly related to the size of its genome; species with larger genomes have a higher frequency of mutation, per nucleotide, than species with smaller genomes (Lynch, 2010). This may be due to a higher frequency of errors during DNA replication or lower efficiency of DNA repair. However, the mutation rate is not consistent across the genome, and is dependent on the base composition and genome context (Baer *et al.*, 2007). DNA sequence variation is constantly being removed because of the contribution of evolutionary forces, such as natural selection and random genetic drift. Natural selection imposes a limitation on the number of mutations that may accumulate, particularly in genomic regions

that are critical for growth and development, by removing alleles that may be detrimental to the fitness of an individual. Random genetic drift—the effect of random sampling of gametes that will contribute to the next generation—also results in the removal of DNA sequence variation. The elimination of DNA sequence variation by random genetic drift is particularly significant when the frequency of the new allele is low in the population, or when the size of the population is small or undergoing a reduction in size.

Identification and genotyping DNA sequence variation

Numerous methods of characterizing genetic sequence variation have been developed since the discovery of DNA. Detection of DNA variation began with the analysis of allozymes. Allozymes are enzymes that perform the same function but differ in amino acid sequence, and can be separated by their molecular weight because of their variable migration rates through a solid matrix (e.g. starch gel) in the presence of an electrical current (i.e. gel electrophoresis). While allozymes offered the opportunity for analysis of segregation and detection of genetic linkage, and established the foundation for the first tree genetic maps (Guried and Ledig, 1978; Guries *et al.*, 1987), there are limits to the number of loci that can be assessed using this type of marker. Genotyping allozymes is also labor-intensive and throughput is low; thus, it is not useful for analyzing a large number of samples. Hence, the analysis of genetic variation moved towards characterizing variation in the DNA sequence itself, rather than the reflection of that variation in a protein. DNA sequence variation can be identified and genotyped in two ways: (i) indirectly, by detecting a difference in mobility of a DNA fragment or (ii) directly, by observing the variant in a sequencing read or by single-base extension. In the indirect approach, the actual variant that causes the change in DNA mobility is usually not known, but its presence results in changes in DNA fragment length, charge, or conformation, allowing alternative alleles to be distinguished.

Indirect methods of genotyping

Until the development of NGS methods, most analyses of DNA sequence variation were based on approaches that detected variants by gel electrophoresis. In the following sections, the genetic markers commonly used in the analysis of forest-tree species are briefly reviewed. A more detailed description of indirect genotyping methods widely used in forestry can be found elsewhere (e.g. Chapter 4 in White *et al.*, 2007).

Restriction fragment length polymorphism (RFLP) markers

Prior to the development of genetic markers that utilized PCR (see below), RFLP markers were the only alternative to allozymes for genotyping. RFLP analysis is done by digesting the genomic DNA with a restriction enzyme, followed by separation of the resulting fragments by gel electrophoresis, transfer of the resulting DNA bands to a membrane and detection of specific loci using a labelled DNA probe that is complementary to the target region. DNA sequence variation at the restriction recognition site, or the occurrence of insertions or deletions within the digested fragment, results in a shift in the migration pattern, which can be visualized by the probe. While RFLP markers can be designed to target unique regions of the genome, the procedure is labor-intensive and costly, explaining why most tree genotyping studies using them assayed a very limited number of loci (<100) (Devey *et al.*, 1991; Bradshaw *et al.*, 1994).

Random amplified polymorphic DNA (RAPD) markers

The most significant improvement in early methods of genotyping occurred with the development of PCR. This is a method for amplifying target DNA, based on the use of one or more pairs of primers that are complementary to sequences flanking the region of interest. The technique is performed by first denaturing the DNA using heat, followed by annealing the primers to their targets and extending them with a thermal-stable DNA polymerase, using the genomic DNA as a template. By repeating this process many times, the region between the primers is amplified exponentially, resulting in the generation of large numbers of copies of one or more target sequences. These amplicons can then be separated and visualized by gel electrophoresis for the detection of variation among individuals. While PCR was rapidly adopted for the analysis of specific target sequences, the lack of genome sequence information, which was necessary for primer design, initially limited its application.

The dearth of genome sequences and the difficulty in developing primers targeting large numbers of specific loci led to the development of RAPD markers. This method utilizes a pair of short (~10 nt) primers to amplify one or more positions in the genome. If a sequence variant is present in one or both primer annealing sites, or if a large insertion occurs between them, the amplification will fail or will result in a shift in the migration rate of the amplified product. Because the short RAPD primers are not selected based on genome sequence information, primers with a large number of different nucleotide combinations were typically screened in a subsample to identify those that led to amplification of variable loci.

Amplified fragment length polymorphism (AFLP) markers

After the development of PCR and RAPD markers, numerous other approaches to genotyping soon emerged. AFLP markers became one of the most widely used methods. This technique is based on a combination of restriction digestion and PCR, and uses primers that would anneal to a subset of the restriction fragments. Basically, after genomic DNA is digested with one or more restriction enzymes, an adaptor is ligated to the "sticky ends" left by the digestion. In the next step, PCR primers that anneal to the adaptor sequence, and one or more specific bases beyond the restriction site, are used to amplify a subset of fragments. This "selective" amplification ensures that only a fraction of the fragments produced by restriction are detectable, resulting in higher reproducibility and an increase in the number of loci that can be analyzed, compared with other methods, such as RAPD.

Microsatellites or simple sequence repeats (SSRs)

Most of the methods described above share the characteristic that the loci being genotyped are "unknown" until the amplified region is sequenced or positioned on a genetic map. In other words, restriction enzymes or RAPD/AFLP primers were not selected based on prior information but rather were tested randomly, and the ones that yielded the most variable number of markers were used for analysis. Microsatellite, or simple sequence repeat (SSR), markers differ in that assays are developed to target each locus individually. Analysis of microsatellites are based on PCR amplification of sequences with differences in short, repetitive sequences (e.g. CAA versus CAACAACAA) by using DNA markers specifically designed to bind to each occurrence of the chosen repetitive DNA sequence. Due to high variability in the number of repeats among individuals, this class of markers is highly informative, and is used primarily for forensic and identification purposes. However, the labor intensity and high cost of microsatellite development meant that, typically, only a few hundred to a few thousand loci were developed for any species. While valuable for identification and mapping purposes, these markers are of limited value for applications requiring high-density markers that are distributed genome-wide.

Direct methods of genotyping: highly parallel assays of individual loci

One limitation of the early methods of genotyping was the difficulty in carrying out parallel assays to characterize thousands of loci simultaneously. Another shortcoming was that early genotyping methods, such as RFLPs or RAPDs, only sampled a fraction of the existing genetic variation. This is insufficient for some applications, such as genome-wide association studies, which are used to dissect complex traits (see Chapter 4, this volume).

The most abundant form of genetic variation in eukaryotic genomes is SNVs. These variants may be detected indirectly, as with other types of genetic markers (e.g. a SNP at a restriction enzyme recognition site may result in an RFLP marker), but until recently the tools to analyze, in parallel, a large number of such variants were unavailable. This limitation has been overcome since the development and reduction in cost of methods of DNA synthesis on two-dimensional surfaces. This led to the creation of the first methods of high-throughput genotyping of DNA polymorphisms using DNA "chips" and other approaches described in the following sections. Most high-throughput methods of genotyping rely on the same few principles of polymorphism detection, including differential hybridization and ligation, and primer extension (Fig. 1.9). A comparison of approaches and how they have been applied in commercial platforms developed for high-throughput analysis of DNA variants in forest trees is described below.

Genotyping by differential hybridization and ligation

Methods of polymorphism detection based on differential hybridization rely on the presence or absence of annealing of an oligonucleotide, depending on whether or not its sequence is perfectly complementary to a region of the genome (Fig. 1.9a). For a genotyping assay based on differential hybridization, oligonucleotides that are complementary to alternative allelic forms of a locus have to be designed. Differential hybridization was the principle behind the first high-throughput method of DNA genotyping developed for analyzing over 10,000 human SNV loci (Matsuzaki *et al.*, 2004). While the complexity of the procedure used in this first approach to SNV detection was not applicable to species without a high-quality genome sequence, it showed the feasibility of using DNA chips and the sensitivity of differential hybridization as a method of SNV detection, and resulted in the creation of improved approaches.

The first method of highly parallel genotyping used in a forest-tree species was the GoldenGate platform from Illumina, which

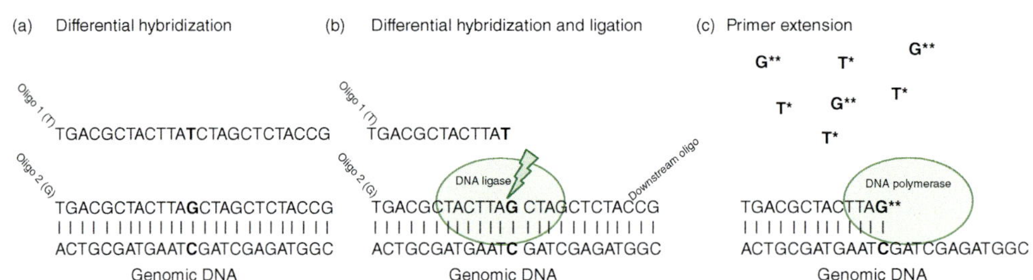

Fig. 1.9. Principles of high-throughput DNA polymorphism detection assays. Methods of detecting DNA polymorphisms in high-throughput assays are generally based on: (a) differential hybridization, (b) differential ligation, (c) primer extension, or a combination of them. In this example, the allele to be detected is a cytosine (C) in the genomic DNA. As shown in (a), using differential hybridization, two oligonucleotides (oligo 1 and oligo 2) compete for hybridization with the same segment of genomic DNA. Only oligo 2 anneals because it is perfectly complementary to the genomic sequence. In (b), using differential hybridization and ligation, only the complementary oligonucleotide anneals to the genomic DNA, allowing ligation of a downstream oligonucleotide. In (c), using primer extension, one oligonucleotide complementary to the upstream position of the SNV anneals to the genomic sequence. Distinct nucleotides labelled with alternative signal molecules are added to the reaction (G**, T*), but only the nucleotide complementary to the SNV (G**) is added by DNA polymerase, producing a signal that reflects the genotype.

utilized differential hybridization followed by ligation (Fig. 1.9b) (Pavy *et al.*, 2008). The method is based on oligonucleotides synthesized on beads, which are then affixed to a two-dimensional surface. Two oligonucleotides are synthesized to contain a sequence complementary to the upstream regions of DNA variants. However, these oligonucleotides differ in their last position, so that each is complementary to one of the two possible alleles. A third oligonucleotide anneals downstream of the variable site. Extension of these oligonucleotides and ligation of the product allows the identification of the genotype of each individual (i.e. the two alleles are distinguished by the differential hybridization of alternative oligonucleotides).

Genotyping by single-base primer extension

This method of genotyping relies on primer extension based on an oligonucleotide that anneals adjacent to a polymorphic site. After annealing the primer to genomic DNA, alternative nucleotides labelled with distinct signaling molecules (i.e. various fluorescent tags) are added and DNA polymerase incorporates the nucleotide that is complementary to the variable site. Because the nucleotides are modified to avoid incorporation

of additional bases, the extension reaction is stopped, creating a molecule that fluoresces at a distinctive wavelength that defines the allele at the variable position (Fig. 1.9c).

The genotyping principle of single-base primer extension is currently used in the most widely commercialized genotyping platforms, such as the Infinium II from Illumina. Its main advantage is the simplicity of the assay compared with previous high-throughput genotyping methods, which were based largely on differential hybridization. Single-base primer extension platforms are also able to test a much larger number of loci in each genotyping assay.

Direct methods of genotyping: genome resequencing and reduced genome representation

The rapid decline in the cost of sequencing and the dramatic increase in throughput have created the possibility of genotyping by simply "resequencing" genomes from samples of interest, and comparing them to a reference to call variable DNA loci. While resequencing multiple genomes is now feasible for several species, including tree species (Evans *et al.*, 2014), there are significant drawbacks to using this approach due to at least two major limitations. First, several forest species, including all conifers, have very large genomes. Thus, the effort to sequence, assemble, and detect sequence variants for large numbers of individuals can be computationally demanding and cost-prohibitive. Second, several tree species harbor significant amounts of genetic diversity, which can limit the value of a reference for the purpose of aligning reads and identifying DNA variants.

An alternative approach to whole-genome resequencing and genotyping is to focus the analysis on a reduced representation of the genome. The effort can be targeted to regions where the most relevant DNA sequence variation may occur, avoiding the expenditure of a significant sequencing effort on regions of limited interest. Genomic regions typically avoided include the highly repetitive, non-coding fraction of the genome. Most plant genomes, and particularly conifers, have large amounts of highly repetitive sequences (de la Torre *et al.*, 2014). Several approaches to generate a reduced genomic representation have been proposed, and can be separated into four main categories: (i) C_0t-based cloning; (ii) methyl filtration; (iii) sequence capture; and (iv) restriction digestion-based methods. The most commonly used strategies are described in the following sections.

After sequencing, the reads are aligned to a reference genome, and DNA variants are identified using standard bioinformatics tools. The most widely used bioinformatics tools for DNA variant detection from sequencing include the Genome Analysis Toolkit (GATK; McKenna *et al.*, 2010) and SAMTools (Li *et al.*, 2009). However, numerous other methods have since been developed and are regularly being published and compared for their performance (e.g. Sandmann *et al.*, 2017).

C_0t-based cloning

The product of C_0 (the initial concentration of DNA) and time (t) is represented by C_0t. C_0t-based cloning and sequencing is based on the principle that, after denaturation, the highly repetitive fraction of a genome is expected to renature more rapidly than the single-copy fraction (Waring and Britten, 1966).

This occurs because complementary strands of the highly repetitive fraction are more likely to encounter their complement than single-copy or rare sequences. Renaturation is also affected by the concentration of DNA and time. This method has been used to quantify the fraction of the genome that is highly and moderately repetitive, or a single copy, in several plant species (Flavell *et al.*, 1974). In a genome-sequencing context, this approach has been used to reduce genome representation, to sequence the low-copy, gene-rich fraction of complex plant genomes (Peterson *et al.*, 2002; Yuan *et al.*, 2003).

Methylation filtration

DNA methylation involves the addition of a methyl group to cytosine or adenine nucleotides, and is a mechanism used by several eukaryotes to regulate gene expression in the genome (Law and Jacobsen, 2010). Genome complexity reduction based on methylation filtration relies on the observation that expressed genes are generally hypomethylated, whereas the non-expressed and the repetitive fraction of genomic DNA from plants is often methylated. To reduce genome complexity, genomic DNA is cloned into strains of *Escherichia coli* that restrict methylated DNA. As a consequence, only the hypomethylated fraction of the genome is obtained. As with Cot-based cloning, methods to filter the genomic DNA and capture the fraction that is hypomethylated have been applied to characterize large plant genomes (Martienssen *et al.*, 1999; Palmer *et al.*, 2003; Whitelaw *et al.*, 2003).

Sequence capture

A significant limitation of Cot-based cloning and methylation filtration is that the fraction of the genomic DNA retrieved is dependent exclusively on its abundance or methylation state, rather than the sequence per se. However, when genotyping based on sequencing a reduced genome representation, there is often interest in retrieving not only the general fraction that is low-copy or hypomethylated but also a specific set of genes or genomic regions. Other than through direct PCR of individual regions of the genome, methods for obtaining a large pool of defined regions were not available until the development of sequence capture. This approach relies on the use of a set of probes (generally 50–120 nt) that are complementary to the regions of interest. The probes are hybridized to sheared genomic DNA and the fraction that anneals to it is then retrieved for sequencing. Since its introduction, sequence capture has undergone a series of refinements to simplify the process and increase its efficiency. Sequence capture was done initially by hybridizing genomic DNA to microarrays containing probe sequences (Albert *et al.*, 2007). The procedure was later simplified by being carried out in solution, where the genomic DNA is combined with probes that are biotinylated (Gnirke *et al.*, 2009). After hybridization to the biotinylated probes, the genomic DNA is retrieved using streptavidin beads (Fig. 1.10).

Other approaches to genotyping by sequencing a reduced genome representation

With the recent reduction in sequencing costs, a suite of alternative methods of genotyping based on sequencing has been proposed. The most popular methods rely on a general approach that resembles the RFLP and AFLP methods of genotyping, using DNA digested by endonucleases. The difference is that DNA variants are identified based on sequencing instead of differences in the migration of DNA fragments in a gel (Fig. 1.11). These methods include: reduced-representation libraries (RRLs; Altshuler *et al.*, 2000); complexity reduction of polymorphic sequences (CRoPS, van Orsouw *et al.*, 2007); restriction-site-associated DNA sequencing (RAD-seq; Davey *et al.*, 2011); and genotyping-by-sequencing (GBS, Elshire *et al.*, 2011). All of these methods follow a similar general strategy, where genomic DNA is digested with one or more restriction enzymes, adaptors are ligated to the digested fragments, which are then sequenced. However, they differ in the stage at which restriction digestion occurs and the adaptors are ligated, as well as the types of restriction digestion enzymes that are frequently used. An overview of the most common approaches

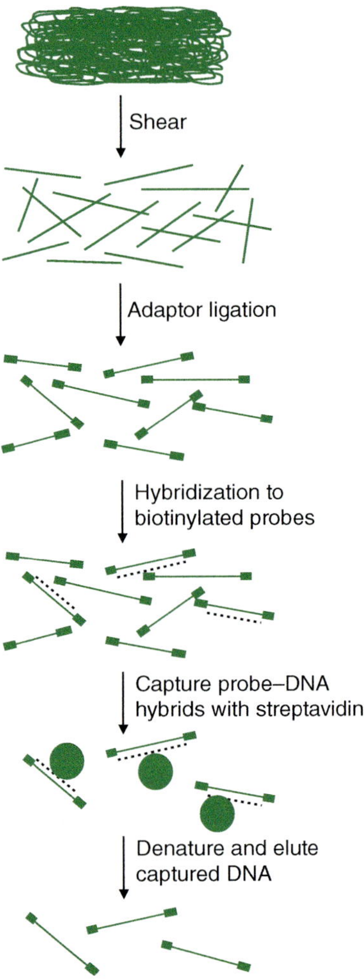

Fig. 1.10. Sequence-capture approach to reduce genome representation. Sequence capture begins with shearing of the genomic DNA and addition of adaptors (thick lines) to the ends of the DNA fragments (thin lines). These are then hybridized to probes (dashed line) that have a sequence complementary to the segment of the genome that is being targeted for analysis. Probes can be biotinylated to facilitate their later separation and removal. The hybrid molecules composed of the probe and the complementary genome fragment are then captured using streptavidin magnetic beads (circles). The magnetic beads are retrieved using a magnet, allowing the separation of probes and complementary DNA fragments from the rest of the genome. After probe–DNA denaturation, genomic fragments that correspond to the regions of interest are obtained. The presence of common adaptors added at the beginning of the process allow universal PCR amplification and sequencing of these fragments.

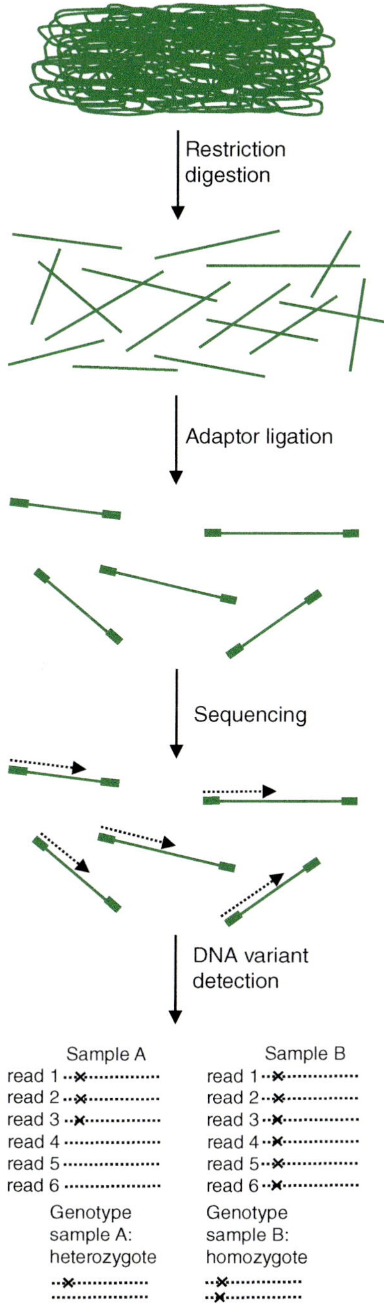

Fig. 1.11. General principles of genotyping based on restriction digestion and NGS. This approach to genotyping involves the restriction digestion of genomic DNA with endonucleases and the addition of adapters (thick lines); variation in the order that these reactions are carried out depends on the specific method used (reviewed by Davey *et al.*, 2011). The adaptor sequence provides a site for a sequencing primer, necessary for NGS of each sample (dashed arrow). Following sequencing, reads originating from each sample are aligned, and variants are detected based on differences among them or relative to a reference genome sequence. In the figure, sample A has three sequencing reads that contain a DNA variant relative to a reference (× on dashed line), while three reads do not contain the variant. These data suggest that the genotype of sample A is heterozygous. In contrast, because all reads of sample B have the DNA variant, it is expected that it is a homozygote that carries the alternative allele, relative to the reference.

that use restriction digestion-based genotyping from sequencing reads and their differences has been provided elsewhere (e.g. Davey *et al.*, 2011). A comparison between bioinformatics pipelines used for DNA variant detection in the datasets resulting from these approaches revealed that there is extensive variation in the number of polymorphisms detected and in the accuracy of the genotyping call (Torkamaneh *et al.*, 2016).

Transcriptome Analysis

The DNA sequence represents the most fundamental genomic information. However, DNA needs to be transcribed into mRNA and translated into protein so that the information contained in the genome sequence can result in a phenotype. Until the mid-1990s, the tools available to monitor gene-expression variation were northern blotting and (after the discovery of PCR) quantitative real-time PCR (qRT-PCR). Northern blotting requires the separation of mRNA by gel electrophoresis, followed by transfer to a nylon membrane and hybridization of radioactively or chemiluminescently labeled probes that are complementary to the target mRNA. After stringent washes, the signal, which is captured on X-ray film, serves as an indication of the presence and abundance of the target mRNA molecule. Quantitative RT-PCR was introduced later and is based on the principle that the most abundant mRNA molecules in a sample will reach exponential amplification in fewer cycles than rare mRNAs. The procedure requires synthesis of complementary DNA (cDNA) from mRNA, followed by PCR using specific fluorescent dyes that bind to double-stranded DNA. Alternatively, a sequence-specific oligonucleotide probe labeled with a fluorescent reporter is detected after hybridization of the probe to a complementary sequence. While northern blotting and quantitative PCR have become established methods to quantify gene expression, both methods have significant limitations in that they are restricted to the analysis of one or a few genes in parallel. Thus, they are unsuitable for analysis of gene expression on a transcriptomic scale.

A few alternative approaches to quantifying mRNA have been developed to allow highly parallel analysis of gene expression, including serial analysis of gene expression (SAGE; Velculescu *et al.*, 1995). In this case, mRNA is transcribed into cDNA, which is then cleaved by a restriction endonuclease. Adapters that contain restriction sites are ligated to the cDNA fragments, creating two tag sequences, or "ditags." After further enzymatic digestion, these ditags can be concatenated and amplified prior to sequencing. While SAGE effectively permits the analysis of thousands of tags, representing a snapshot of the genes transcribed in a sample at a given time point, the laborious nature of this method, as well as the high cost of DNA sequencing that prevailed in the 1990s, limited the extent of its use.

Microarrays

Complimentary DNA microarrays provided the first platform to analyze gene expression on a transcriptomic scale, dramatically expanding the number of genes that could be analyzed, beyond the few that could be done with qRT-PCR or northern blotting. Microarrays were first developed based on the relatively simple principle that cDNAs representing a wide variety of transcribed genes could be cloned, amplified, and attached to a solid surface (Schena *et al.*, 1995). This provides a platform for transcriptome analysis of any sample derived from individuals of the same species, by labeling their mRNA or cDNA and hybridizing

it to the microarray (Fig. 1.12). As with northern blots, the signal detected, after stringent washes, provided an estimate of transcript abundance. While microarrays were, in principle, not much different from northern blots, they revolutionized gene expression analysis by miniaturizing the platform to detect and quantify a massive number of transcripts in parallel. As a consequence, analysis of the expression of hundreds to thousands of genes and, eventually, the full suite of known genes in a genome, became possible.

Although cDNAs provided an easily accessible substrate for the production of microarrays, they have several drawbacks. Inconsistent PCR amplification of individual cDNAs, and the potential for cross-hybridization, were some of the limitations that led to the development of microarrays based on synthetic oligonucleotides. In this case, each oligonucleotide is designed to complement only a specific transcript and is synthesized in defined amounts. Therefore, regions of a transcript that have a sequence similar to that found in other genes in the genome could be avoided, and concentration of oligonucleotides deposited in the microarray could be standardized (Fig. 1.12).

The quality of early microarrays was also inconsistent because of the approach used to "print" them on a slide. Essentially, the method required that a hollow pin be immersed in the cDNA or oligonucleotide solution. The pin would withdraw a small fraction of that solution, which was then deposited on and fixed to the slide. Numerous pitfalls were possible in this process, including clogged or damaged pin tips, which would result in inconsistent printing within or between slides. These limitations were addressed by another significant advance in microarray technology: *in situ* synthesis of the oligonucleotides directly on the surface of the slide (Gao *et al.*, 2004). This approach permitted the production of a consistent platform for analysis of any set of genes or any genome.

RNA sequencing

Microarrays were the tool of choice for analysis of the transcriptome until the rapid

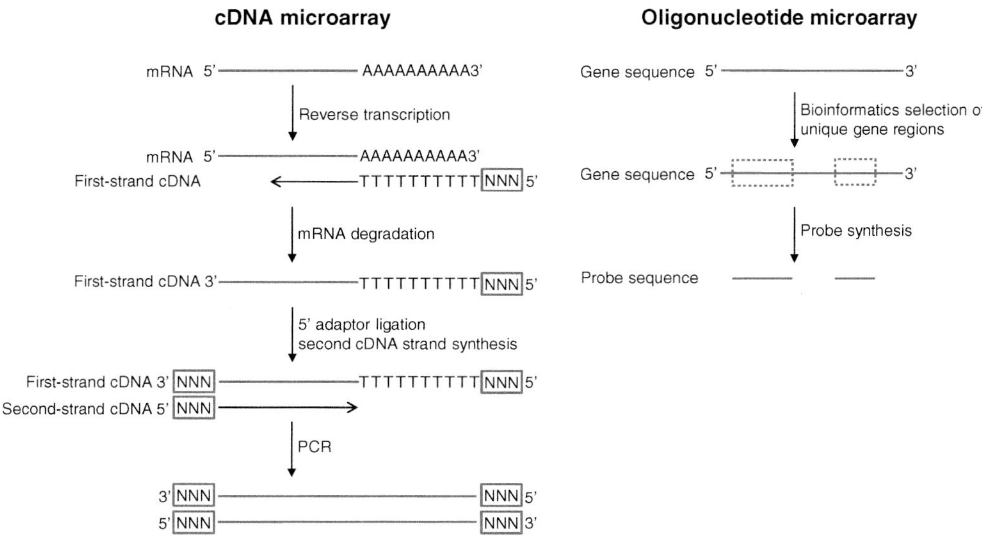

Fig. 1.12. Principles of cDNA and oligonucleotide microarray design. Early microarray technology to detect transcriptome changes were based on the use of cDNA (left panel). In this technology, mRNA extracted from the species of interest is reverse-transcribed using an oligo(dT) primer to synthesize the first strand of the cDNA. The oligo(dT) primer may also contain an adaptor sequence (NNN), which becomes incorporated into the first strand of cDNA. Thus, the product of the reaction is the complement of the mRNA, with an adaptor at the 5′ end. Next the mRNA is degraded, leaving only the cDNA molecule. Another adaptor (NNN) can be added to the 3′ end, prior to second-strand cDNA synthesis using a primer that anneals to the adaptor sequence. Finally, the double-stranded cDNA can be amplified by PCR before immobilization on the microarray. More advanced technologies for microarray design were established based on synthesized oligonucleotides (right panel). In oligonucleotide microarrays, regions that are unique to genes to be analyzed are first identified bioinformatically. Based on these sequences, oligonucleotides are synthesized and printed on the microarray surface.

development of DNA sequencing methods led to RNA sequencing (RNA-seq). RNA-seq relies on the principle that, by producing cDNA from mRNA extracted from a sample of interest and sequencing it, one can infer the abundance of gene expression. RNA-seq has a number of advantages relative to microarrays for transcriptome analysis (Wang *et al.*, 2009). First, RNA-seq does not require prior knowledge about the content of the transcriptome. In other words, rather than interrogating the specific elements present in the microarray, the investigator can measure the abundance of any transcript present in the transcriptome. Another significant improvement lies in the fact that RNA-seq allows inferences about the occurrence of alternative splicing, based on the presence or absence of exons among the sequencing reads for a transcript. RNA-seq also provides estimates of

expression that span a wider dynamic range than microarrays, including rare transcripts whose signal may be confounded by background signals from microarrays. Considering all these advantages, RNA-seq has rapidly become the predominant technology for transcriptome analysis. However, several challenges remain, particularly with respect to the most suitable approach to the analysis of RNA-seq data. More specifically, numerous methods to normalize the data across samples and declare differences in gene expression among treatments have been proposed, but no consensus approach that addresses the method's limitations has yet emerged. Other questions related to the method of sample preparation and amount of sequencing required for a suitable estimate of gene expression also remain largely unknown. To address these limitations, best-practice workflows have been developed to standardize the analytical methods (e.g. Conesa *et al.*, 2016).

Proteome Analysis

Proteomics is the survey and quantification of the ensemble of proteins present in a sample. Proteomics complements genomics and transcriptomics by providing an additional level of information concerning gene expression. Transcript abundance reflects, to some extent, the quantity of each protein in a sample. However, numerous studies have shown that this correlation is inexact, demonstrating the need to analyze the proteome for a complete understanding of the flow of information from genotype to phenotype. Furthermore, proteins may undergo numerous modifications that can change their properties and are not accounted for by sequencing DNA or detecting transcripts.

Proteome analysis by mass spectrometry

The first and most immediate application of proteomic analysis was to catalogue the proteins present in a given sample. To achieve this goal, the proteins were extracted, solubilized, and denatured, to make them accessible to site-specific digestion. Digestion was typically achieved by treating the protein extract with trypsin, although other methods were applied, including the use of multiple proteolytic enzymes, which can increase the range of the proteins characterized. Digestion was followed by fractionation and analysis of the resulting peptides via liquid chromatography/mass spectrometry. In this process, the masses of individual peptides are recorded as they are eluted from the column used to separate them. In the final stage, the mass spectra are analyzed to determine the primary structure of the protein.

Recently, proteomics has evolved from the simple detection of protein species in a sample, to the actual quantification of each polypeptide. For quantitative proteomics studies, samples are labeled with distinct isotopes

to discriminate sample origin. Methods of proteomic analysis have evolved dramatically since the first development of mass spectrometry. Improvements in the accurate estimation of mass of peptides and the speed at which the mass spectra are collected mean that a much larger fraction of the proteome can now be sampled and quantified.

While the number of protein-coding genes is limited in most genomes, genetic variation that leads to amino acid substitutions, alternative splicing, protein modifications, and other phenomena that affect protein structure and function results in numerous alternative forms. These changes are dictated by modifications in the mass and composition of proteins.

Summary

Genomics is the science that studies the structure and function of genomes, including the elements that contribute to the way in which the genetic information encoded in DNA is manifested as a phenotype. Genomics encompasses many areas, from genome sequencing to transcriptome and proteome analysis, and beyond. Because forest-tree species have properties that are of critical societal value and that make them unique among plants (e.g. perennial growth habit and wood formation), they have been a focus of genomic studies since the development of the first methods to characterize genetic variation. However, studying tree species has been challenging because many have very large and complex genomes, and their long generation times and physical size can hinder many approaches commonly used in genomic analysis of annual plants.

The genomic study of most plant species started with the development of genetic maps. However, because genetic mapping theory was largely developed based on analysis of populations that are impractical to create for forest trees (e.g. backcross populations), new types of mating designs, such as pseudo-backcrosses and mapping theory, had to be developed.

Progress in tree genomics accelerated with the development of NGS methods and, in particular, the sequencing-by-synthesis approach. Later, sequencing by synthesis was complemented by the creation of other advanced methods of sequencing based on the analysis of single, long DNA molecules, which now allows the characterization of sequences of 1 Mbp or more (although with a high error rate).

The rapid decrease in cost and the increase in throughput of DNA sequencing have popularized the sequencing of plant genomes, including those of tree species. For genome sequencing, most scientists use the shotgun sequencing approach. Shotgun sequencing is based on shearing the genome into fragments of different sizes, sequencing the ends of these fragments, and assembling them using extensive computational resources that take into consideration the similarity between overlapping reads and their length.

Despite the advances in DNA sequencing, for many tree species it remains difficult or impractical to sequence their genome. In particular, the genome of one of the most important groups of forest-tree species, the conifers, is among the largest of land plants. Sequencing of conifers required the development of new assembly strategies and computational tools, but the quality of conifer genome sequence assemblies remains significantly lower than those of angiosperm tree species. In the future, the expectation is that sequencing methods will continue to increase their throughput and read length. In combination with better computational resources, the quality of the genome sequences of conifers is expected to continue improving.

Many studies in forestry aim to characterize genetic variation in large populations, requiring the development of approaches to identify DNA polymorphisms in many samples. While numerous methods of genotyping have been created since the discovery of DNA (e.g. RAPD and SSR genetic markers), they did not achieve genome-wide coverage until NGS and DNA chips were developed. NGS is now being used for genotyping by aligning DNA from multiple individuals to reveal their differences. However, DNA chips rely on the detection of individual bases or probes, as they are incorporated into an oligonucleotide on a two-dimensional surface. Both approaches have their advantages and disadvantages in terms of cost and accuracy, which have to be carefully considered before selecting the genotyping tool.

While genome sequencing and analysis is a fundamental component of the genomic sciences, other levels of heritable information are also relevant and have been extensively studied, such as mRNA and proteins. Tools for genome-wide analysis of all transcripts, the transcriptome, first emerged with the development of DNA microarrays, where fragments of genes were printed on a surface and their expression abundance quantified by hybridizing labelled sequences derived from mRNA. The rapid evolution in DNA sequencing methods is now resulting in the replacement of microarrays by direct sequencing of mRNA derivatives (e.g. cDNA). Characterization and quantification of all proteins, or proteomics, is also of interest, and has been achieved largely by mass spectrometry. Improvements in methods to identify and estimate the abundance of protein in a sample have evolved rapidly, although obtaining a survey of the full spectrum of peptides in a sample remains challenging. Research on the characterization of other "omics" components (e.g. metabolomics) has also advanced and is becoming common in forest-tree species.

References

Albert, T.J., Molla, M.N., Muzny, D.M., Nazareth, L., Wheeler, D., *et al.* (2007) Direct selection of human genomic loci by microarray hybridization. *Nature Methods* 4, 903–905.

Allen, C.D., Macalady, A.K., Chenchouni, H., Bachelet, D., McDowell, N., *et al.* (2010) A global overview of drought and heat-induced tree mortality reveals emerging climate change risks for forests. *Forest Ecology and Management* 259, 660–684.

Altshuler, D., Pollara, V.J., Cowles, C.R., van Etten, W.J., Baldwin, J., *et al.* (2000) An SNP map of the human genome generated by reduced representation shotgun sequencing. *Nature* 407, 513–516.

Anderson, S. (1981) Shotgun DNA sequencing using cloned DNase I-generated fragments. *Nucleic Acids Research* 9, 3015–3027.

Ashburner, M., Ball, C.A., Blake, J.A., Botstein, D., Butler, H., *et al.* (2000) Gene Ontology: tool for the unification of biology. *Nature Genetics* 25, 25–29.

Avery, O.T., MacLeod, C.M., and McCarty, M. (1944) Studies on the chemical nature of the substance inducing transformation of pneumococcal types. *Journal of Experimental Medicine* 79, 137–158.

Baer, C.F., Miyamoto, M.M., and Denver, D.R. (2007) Mutation rate variation in multicellular eukaryotes: causes and consequences. *Nature Reviews Genetics* 8, 619–631.

Birol, I., Raymond, A., Jackman, S.D., Pleasance, S., Coope, R., *et al.* (2013) Assembling the 20 Gb white spruce (*Picea glauca*) genome from whole-genome shotgun sequencing data. *Bioinformatics* 29, 1492–1497.

Bonan, G.B. (2008) Forests and climate change: forcings, feedbacks, and the climate benefits of forests. *Science* 320, 1444–1449.

Bradshaw, H.D. Jr., Villar, M., Watson, B.D., Otto, K.G., Stewart, S., and Stettler, R.F. (1994) Molecular genetics of growth and development in *Populus*. III. A genetic link-age map of a hybrid poplar composed of RFLP, STS, and RAPD markers. *Theoretical and Applied Genetics* 89, 167–178.

Caspersson, T., Zech, L., and Johansson, C. (1970) Analysis of human metaphase chromosome set by aid of DNA-binding fluorescent agents. *Experimental Cell Research* 62, 490–492.

Conesa, A., Gotz, S., Garcia-Gomez, J.M., Terol, J., Talon, M., and Robles, M. (2005) Blast2GO: a universal tool for annotation, visualization and analysis in functional genomics research. *Bioinformatics* 21, 3674–3676.

Conesa, A., Madrigal, P., Tarazona, S., Gomez-Cabrero, D., Cervera, A., *et al.* (2016) A survey of best practices for RNA-seq data analysis. *Genome Biology* 17, 13.

Davey, J.W., Hohenlohe, P.A., Etter, P.D., Boone, J.Q., Catchen, J.M., and Blaxter, M.L. (2011) Genome-wide genetic marker discovery and genotyping using next-generation sequencing. *Nature Reviews Genetics* 12, 499–510.

de la Torre, A.R., Birol, I., Bousquet, J., Ingvarsson, P.K., Jansson, S., *et al.* (2014) Insights into conifer giga-genomes. *Plant Physiology* 166, 1724–1732.

Devey, M.E., Jermstad, K.D., Tauer, C.G., and Neale, D.B. (1991) Inheritance of RFLP loci in a loblolly pine three-generation pedigree. *Theoretical and Applied Genetics* 83, 238–242.

Eid, J., Fehr, A., Gray, J., Luong, K., Lyle, J., *et al.* (2009) Real-time DNA sequencing from single polymerase molecules. *Science* 323, 133–138.

Elshire, R.J., Glaubitz, J.C., Sun, Q., Poland, J.A., Kawamoto, K., *et al.* (2011) A robust, simple genotyping-by-sequencing (GBS) approach for high diversity species. *PLoS One* 6, e19379.

Evans, L.M., Slavov, G.T., Rodgers-Melnick, E., Martin, J., Ranjan, P., *et al.* (2014) Population genomics of *Populus trichocarpa* identifies signatures of selection and adaptive trait associations. *Nature Genetics* 46, 1089–1096.

Fedurco, M., Romieu, A., Williams, S., Lawrence, I., and Turcatti, G. (2006) BTA, a novel reagent for DNA attachment on glass and efficient generation of solid-phase amplified DNA colonies. *Nucleic Acids Research* 34, e22.

Flavell, R.B., Bennett, M.D., Smith, J.B., and Smith, D.B. (1974) Genome size and the proportion of repeated nucleotide sequence DNA in plants. *Biochemical Genetics* 12, 257–269.

Gao, X., Gulari, E., and Zhou, X. (2004) In situ synthesis of oligonucleotide microarrays. *Biopolymers* 73, 579–596.

Garcia, S., Leitch, I.J., Anadon-Rosell, A., Canela, M.Á., Gálvez, F., *et al.* (2014) Recent updates and developments to plant genome size databases. *Nucleic Acids Research* 42, D1159–D1166.

Gene Ontology Consortium (2015) Gene Ontology Consortium: going forward. *Nucleic Acids Research* 43, D1049–D1056.

Gnirke, A., Melnikov, A., Maguire, J., Rogov, P., LeProust, E.M., *et al.* (2009) Solution hybrid selection with ultra-long oligonucleotides for massively parallel targeted sequencing. *Nature Biotechnology* 27, 182–189.

Goodwin, S., McPherson, J.D., and McCombie, W.R. (2016) Coming of age: ten years of next-generation sequencing technologies. *Nature Reviews Genetics* 17, 333–351.

Grattapaglia, D. and Sederoff, R. (1994) Genetic linkage maps of *Eucalyptus grandis* and *Eucalyptus urophylla* using a pseudo-testcross: mapping strategy and RAPD markers. *Genetics* 137, 1121–1137.

Grattapaglia, D., Plomion, C., Kirst, M., and Sederoff, R.R. (2009) Genomics of growth traits in forest trees. *Current Opinion in Plant Biology* 12, 148–156.

Greilhuber, J., Borsch, T., Müller, K., Worberg, A., Porembski, S., and Barthlott, W. (2006) Smallest angiosperm genomes found in Lentibulariaceae, with chromosomes of bacterial size. *Plant Biology* 8, 770–777.

Guries, R.P. and Ledig, F.T. (1978) Inheritance of some polymorphic isoenzymes in pitch pine (*Pinus rigida* Mill.). *Heredity* 40, 27–32.

Guries, R.P., Friedman, S.T., and Ledig, F.T. (1978) A megagametophyte analysis of genetic linkage in pitch pine (*Pinus rigida* Mill.). *Heredity* 40, 309–314.

Harfouche, A., Meilan, R., Kirst, M., Morgante, M., Boerjan, W., *et al.* (2012) Accelerating the domestication of forest trees in a changing world. *Trends in Plant Science* 17, 64–72.

Hedrick, P.W., Hellsten, U., and Grattapaglia, D. (2016) Examining the cause of high inbreeding depression: analysis of whole-genome sequence data in 28 selfed progeny of *Eucalyptus grandis*. *New Phytologist* 209, 600–611.

IEA (2016) World Energy Outlook 2016. Available at: www.iea.org/Textbase/npsum/WEO2016SUM.pdf (accessed June 25, 2019).

Jain, M., Olsen, H.E., Paten, B., and Akeson, M. (2016) The Oxford Nanopore MinION: delivery of nanopore sequencing to the genomics community. *Genome Biology* 17, 239.

Krutovsky, K.V., Troggio, M., Brown, G.R., Jermstad, K.D., and Neale, D.B. (2004) Comparative mapping in the Pinaceae. *Genetics* 168, 447–461.

Lander, E.S., Linton, L.M., Birren, B., Nusbaum, C., Zody, M.C., *et al.* (2001) Initial sequencing and analysis of the human genome. *Nature* 409, 860–921.

Law, J.A. and Jacobsen, S.E. (2010) Establishing, maintaining and modifying DNA methylation patterns in plants and animals. *Nature Reviews Genetics* 11, 204–220.

Li, H., Handsaker, B., Wysoker, A., Fennell, T., Ruan, J., *et al.* (2009) The Sequence Alignment/ Map format and SAMtools. *Bioinformatics* 25, 2078–2079.

Lynch, M. (2010) Evolution of the mutation rate. *Trends in Genetics* 26, 345–352.

Mankos, M., Shadman, K., Persson, H.H.J., N'Diaye, A.T., Schmid, A.K., and Davis, R.W. (2014) A novel low energy electron microscope for DNA sequencing and surface analysis. *Ultramicroscopy* 145, 36–49.

Margulies, M., Egholm, M., Altman, W.E., Attiya, S., Bader, J.S., *et al.* (2005) Genome sequencing in microfabricated high-density picolitre reactors. *Nature* 437, 376–380.

Martienssen, R.A., Rabinowicz, P.D., Schutz, K., Dedhia, N., Yordan, C., *et al.* (1999) Differential methylation of genes and retrotransposons facilitates shotgun sequencing of the maize genome. *Nature Genetics* 23, 305–308.

Matsuzaki, H., Loi, H., Dong, S., Tsai, Y.-Y., Fang, J., *et al.* (2004) Parallel genotyping of over 10,000 SNPs using a one-primer assay on a high-density oligonucleotide array. *Genome Research* 14, 414–425.

McKenna, A., Hanna, M., Banks, E., Sivachenko, A., Cibulskis, K., *et al.* (2010) The Genome Analysis Toolkit: a MapReduce framework for analyzing next-generation DNA sequencing data. *Genome Research* 20, 1297–1303.

Myburg, A.A., Griffin, A.R., Sederoff, R.R. and Whetten, R.W. (2003) Comparative genetic linkage maps of *Eucalyptus grandis, Eucalyptus globulus* and their F1 hybrid based on a double pseudo-backcross mapping approach. *Theoretical and Applied Genetics* 107(6), 1028–1042.

Myburg, A.A., Grattapaglia, D., Tuskan, G.A., Hellsten, U., Hayes, R.D., *et al.* (2014) The genome of *Eucalyptus grandis*. *Nature* 510, 356–362.

Neale, D.B. and Ingvarsson, P.K. (2008) Population, quantitative and comparative genomics of adaptation in forest trees. *Current Opinion in Plant Biology* 11, 149–155.

Neale, D.B. and Kremer, A. (2011) Forest tree genomics: growing resources and applications. *Nature Reviews Genetics* 12, 111–122.

Neale, D.B., Wegrzyn, J.L., Stevens, K.A., Zimin, A.V., Puiu, D., *et al.* (2014) Decoding the massive genome of loblolly pine using haploid DNA and novel assembly strategies. *Genome Biology* 15, R59.

Nystedt, B., Street, N.R., Wetterbom, A., Zuccolo, A., Lin, Y.-C., *et al.* (2013) The Norway spruce genome sequence and conifer genome evolution. *Nature* 497, 579–584.

Palmer, L.E., Rabinowicz, P.D., O'Shaughnessy, A.L., Balija, V.S., Nascimento, L.U., *et al.* (2003) Maize genome sequencing by methylation filtration. *Science* 302, 2115–2117.

Pavy, N., Pelgas, B., Beauseigle, S., Blais, S., Gagnon, F., *et al.* (2008) Enhancing genetic mapping of complex genomes through the design of highly-multiplexed SNP arrays: application to the large and unsequenced genomes of white spruce and black spruce. *BMC Genomics* 9, 21.

Pellicer, J., Fay, M.F., and Leitch, I.J. (2010) The largest eukaryotic genome of them all? *Botanical Journal of the Linnean Society* 164, 10–15.

Peterson, D.G., Schulze, S.R., Sciara, E.B., Lee, S.A., Bowers, J.E., *et al.* (2002) Integration of Cot analysis, DNA cloning, and high-throughput sequencing facilitates genome characterization and gene discovery. *Genome Research* 12, 795–807.

Ronaghi, M., Karamohamed, S., Pettersson, B., Uhlén, M., and Nyrén, P. (1996) Real-time DNA sequencing using detection of pyrophosphate release. *Analytical Biochemistry* 242, 84–89.

Rothberg, J.M., Hinz, W., Rearick, T.M., Schultz, J., Mileski, W., *et al.* (2011) An integrated semiconductor device enabling non-optical genome sequencing. *Nature* 475, 348–352.

Sandmann, S., de Graaf, A.O., Karimi, M., van der Reijden, B.A., Hellström-Lindberg, E., *et al.* (2017) Evaluating variant calling tools for non-matched next-generation sequencing data. *Scientific Reports* 7, 43169.

Sanger, F., Nicklen, S., and Coulson, A.R. (1977) DNA sequencing with chain-terminating inhibitors. *Proceedings of the National Academy of Sciences USA* 74, 5463–5467.

Schena, M., Shalon, D., Davis, R.W., and Brown, P.O. (1995) Quantitative monitoring of gene expression patterns with a complementary DNA microarray. *Science* 270, 467–470.

Schnable, P.S., Ware, D., Fulton, R.S., Stein, J.C., Wei, F., *et al.* (2009) The B73 maize genome: complexity, diversity, and dynamics. *Science* 326, 1112–1115.

Speicher, M.R. and Carter, N.P. (2005) The new cytogenetics: blurring the boundaries with molecular biology. *Nature Reviews Genetics* 6, 782–792.

Staden, R. (1979) A strategy of DNA sequencing employing computer programs. *Nucleic Acids Research* 6, 2601–2610.

Tawfik, D.S. and Griffiths, A.D. (1998) Man-made cell-like compartments for molecular evolution. *Nature Biotechnology* 16, 652–656.

Torkamaneh, D., Laroche, J., and Belzile, F. (2016) Genome-wide SNP calling from genotyping by sequencing (GBS) data: a comparison of seven pipelines and two sequencing technologies. *PLoS One* 11, e0161333.

Turcatti, G., Romieu, A., Fedurco, M., and Tairi, A.-P. (2008) A new class of cleavable fluorescent nucleotides: synthesis and optimization as reversible terminators for DNA sequencing by synthesis. *Nucleic Acids Research* 36, e25.

Tuskan, G.A., Difazio, S., Jansson, S., Bohlmann, J., Grigoriev, I., *et al.* (2006) The genome of black cottonwood, *Populus trichocarpa* (Torr. & Gray). *Science* 313, 1596–1604.

van Orsouw, N.J., Hogers, R.C.J., Janssen, A., Yalcin, F., Snoeijers, S., *et al.* (2007) Complexity reduction of polymorphic sequences (CRoPS™): a novel approach for large-scale polymorphism discovery in complex genomes. *PLoS One* 2, e1172.

Velculescu, V.E., Zhang, L., Vogelstein, B., and Kinzler, K.W. (1995) Serial analysis of gene expression. *Science* 270, 484–487.

Venter, J.C., Adams, M.D., Myers, E.W., Li, P.W., Mural, R.J., *et al.* (2001) The sequence of the human genome. *Science* 291, 1304–1351.

Vos, P., Hogers, R., Bleeker, M., Reijans, M., van de Lee, T., *et al.* (1995) AFLP: a new technique for DNA fingerprinting. *Nucleic Acids Research* 23, 4407–4414.

Wang, Z., Gerstein, M., and Snyder, M. (2009) RNA-Seq: a revolutionary tool for transcriptomics. *Nature Reviews Genetics* 10, 57–63.

Waring, M. and Britten, R.J. (1966) Nucleotide sequence repetition: a rapidly reassociating fraction of mouse DNA. *Science* 154, 791–794.

White, T.L., Adams, W.T., and Neale, D.B. (2007) *Forest Genetics*. CAB International, Wallingford, UK.

Whitelaw, C.A., Barbazuk, W.B., Pertea, G., Chan, A.P., Cheung, F., *et al.* (2003) Enrichment of gene-coding sequences in maize by genome filtration. *Science* 302, 2118–2120.

Williams, C.G. and Savolainen, O. (1996) Inbreeding depression in conifers: implications for breeding strategy. *Forest Science* 42, 102–117.

Williams, J.G., Kubelik, A.R., Livak, K.J., Rafalski, J.A., and Tingey, S.V. (1990) DNA polymorphisms amplified by arbitrary primers are useful as genetic markers. *Nucleic Acids Research* 18, 6531–1635.

Yandell, M. and Ence, D. (2012) A beginner's guide to eukaryotic genome annotation. *Nature Reviews Genetics* 13, 329–342.

Yuan, Y., SanMiguel, P.J., and Bennetzen, J.L. (2003) High-Cot sequence analysis of the maize genome. *Plant Journal* 34, 249–255.

Zimin, A., Stevens, K.A., Crepeau, M.W., Holtz-Morris, A., Koriabine, M., *et al.* (2014) Sequencing and assembly of the 22-gb loblolly pine genome. *Genetics* 196, 875–890.

Zonneveld, B.J.M. (2012) Conifer genome sizes of 172 species, covering 64 of 67 genera, range from 8 to 72 picogram. *Nordic Journal of Botany* 30, 490–502.

2 Genomics of Conifers

Introduction

The gymnosperms are a group of seed-producing plants that separated from the angiosperms approximately 300 million years ago, in the late Carboniferous period (Clarke *et al.*, 2011). The current classification of gymnosperms divides them into four groups (Fig. 2.1), comprising approximately 1,000 species, of which the Pinophyta, or conifers, is the largest. A recent conservative classification recognized six conifer families and 546 species (Eckenwalder, 2009). However, numerous alternative classifications have been proposed (Farjon, 2010; Wang and Ran, 2014). Nonetheless, in general, the conifers can be separated into two major clades, the Pinaceae and the remaining non-Pinaceae. This chapter addresses primarily the progress made in the genome sciences of the Pinaceae (with a few exceptions), which is the largest and most economically and ecologically important taxonomic group among the conifers.

All conifers are seed plants with vascular tissues, known as woody perennials and, for the most part, are evergreen. Conifers dominated most terrestrial communities in the Mesozoic, when drier conditions prevailed. They remained the dominant group until the rapid diversification and radiation of the angiosperms occurred in the late Cretaceous (Bell *et al.*, 2010), when warmer and wetter conditions became predominant. Although there are far fewer species of conifers and other gymnosperms, compared to angiosperms, they still represent the most abundant plant taxa in several terrestrial habitats. For instance, conifers dominate the boreal forests of the Northern Hemisphere, where they contribute to the largest terrestrial carbon sink (Dixon *et al.*, 1994).

Conifers have significant morphological and developmental differences that distinguish them from more primitive land plants, such as mosses and ferns. As opposed to spore-producing plants, conifers produce seeds, which provide a protective layer and nutrition for the developing embryo. However, in contrast to the angiosperms, conifers and other gymnosperms do not produce flowers. Instead, conifers produce male and female cones (strobili). Eggs of gymnosperms are not enclosed in an ovary, and seeds that develop from female cones are not enclosed in a fruit, as occurs in angiosperms. In addition to a distinct reproductive system, conifers and other gymnosperms differ significantly from angiosperms with respect to their system of water transport, which relies on tracheids. Tracheids are specialized, four- to six-sided tapered cells that range in diameter from 10 to 50 μm, and have bordered pits (pores) in their walls to permit the lateral movement of water from cell to cell. In contrast, the movement of water in flowering plants relies primarily on vessels that are comprised of stacked cells named vessel elements. Vessel elements typically are larger in diameter than tracheids (40–300 μm) and are linked to one another through open or partially open ends, forming long, tube-like structures. As a consequence, flowering plants can transport water more efficiently than conifers and other gymnosperms. Despite this disadvantage, conifers and other tracheid-bearing plants may be able to better tolerate water stress due to the higher tension required to cause cavitation leading to the formation of an embolism.

Conifers also have tremendous economic value and are an important source of softwood used for pulp and paper production. Several conifer species provide a source of numerous chemicals derived from products other than wood, such as terpenes. Finally, conifers and other woody species are considered candidate

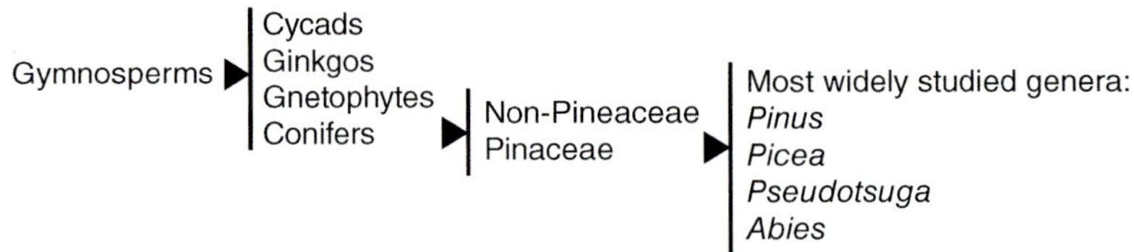

Fig. 2.1. The conifers and Pinaceae in the gymnosperm phylogeny. The gymnosperms are commonly classified into four groups, of which the conifers are the most extensively studied because of their economic and ecologic relevance. The conifers can be broadly divided into the Pinaceae and non-Pinaceae families such as Cupressaceae and Araucariaceae. The Pinaceae include most genera for which extensive genomic studies have been carried out.

species for the large-scale, global production of biomass-derived solid fuel.

Why Study the Genome of Conifers?

Studying plant genomes offers an opportunity to uncover the properties that make species and individuals unique. Genome sequences are the foundation for dissecting the genetic regulation of complex traits to assist in breeding efforts, or identifying the genes that are involved in adaptation to the environment. These aspects of genome sequence analysis will be discussed in Chapters 4 and 5 (this volume). In this chapter, the emphasis will be on characterizing conifer genomes, which offers opportunities for understanding aspects of plant biology that could not be studied otherwise.

Understanding plant evolution

Conifers and other gymnosperms represent one of two major taxonomic groups within the spermatophytes (the other being the angiosperms). Conifers are more ancient than angiosperms, and dominated most terrestrial habitats before the rise of the angiosperms in the late Cretaceous. By discovering the common features of conifer genomes and contrasting them to the genomic properties of angiosperms, it is possible to uncover properties that are unique to seed plants and those that are restricted to gymnosperms.

Understanding genome evolution

Conifers and other gymnosperms are unique in that their genome sizes are, on average, much larger than those of most other plants. The nuclear DNA content of conifers ranges from 8.3 to 71.6 pg (Zonneveld, 2012). In comparison, the modal DNA content of angiosperms is only 0.6 pg (Soltis *et al.*, 2003). Surprisingly, the gigantic genomes of conifers do not seem to have their origin in recent whole-genome duplication (WGD) events, which have been frequent among flowering plants and largely explain their genomic expansion. Instead, the large genomes of conifers and other gymnosperms appear to be mainly the consequence of ancient retrotransposon activity, suggesting that mechanisms for controlling genome expansion may not be present or are repressed (Nystedt *et al.*, 2013).

Understanding conifer unique growth and developmental properties

Conifers have unique morphological and developmental properties that can only be understood by studying their genomes. For instance, conifers develop reproductive structures that are distinct from those in angiosperms. In fact, analysis of the genome of the first conifer to be sequenced, Norway spruce (*Picea abies*), showed that it lacked several genes that are critical for the activation of flowering in angiosperms (Nystedt *et al.*, 2013).

Uncovering mechanisms of genetic adaptation to the environment

Coniferous boreal forests are the most important terrestrial carbon sink (Dixon *et al.*, 1994). As such, they play a critical role in carbon cycling and mitigation of climate change. These and other coniferous forests are under increasing pressure from new sources of biotic and abiotic stress that threaten their existence (Kurz *et al.*, 2008). Analysis of the genome sequence in individuals grown along a cline across the native range of the species can allow the identification of the genetic factors that contribute to their adaptation (Yeaman *et al.*, 2014). The ability to survey genomic variation in diverse populations is essential to understanding adaptation.

Genotyping and Genetic Mapping of Conifers

Genetic maps provide a foundation for population and quantitative genetics studies, and genome analysis. In quantitative genetics, genetic maps provide the framework by which the genetic association between polymorphisms and phenotypic traits is identified. For population genetics, the ordering of polymorphic elements can be useful to identify regions that are under natural or artificial selection when tested for signatures of selection. Finally, in genome analysis, knowledge of the relative position of individual genetic loci can be used to order and orient sequenced and assembled sections of the genome, while also serving as the basis for comparative genomic studies.

Genetic mapping of forest trees is complicated by the fact that mating designs commonly used in agricultural crops and models systems are not feasible, notably due to the fact that selfing is difficult because of high genetic loads and long generation times (Chapter 1, this volume). These challenges are particularly pronounced in conifers, where many species require several decades to reach sexual maturity. However, conifers offer the advantage of producing seeds with a large haploid (1N) megagametophyte, which serves as a nutrient source for the developing embryo. The megagametophyte is produced by mitotic cell division of spores, generated by the meiotic process that gives rise to the egg nucleus. Each haploid megagametophyte represents different recombination events and segregation of alternative alleles from the maternal parent (Box 2.1). Not surprisingly, many of the genetic mapping efforts have focused on genotyping megagametophytes from individual trees.

Isozyme marker-based conifer genetic maps

Allozymes provided the first tool to evaluate genetic inheritance and segregation in conifers. Allozymes are alternative forms of the same enzyme produced by different alleles at a given locus, which differ in amino acid sequence but catalyze the same chemical reaction (Chapter 1, this volume). The alternative forms of the same enzyme can be identified using separation by electrophoresis and detection by staining reactions to visualize the proteins. Numerous studies have resulted in over 40 allozyme genes being mapped in conifers, and linkage among their loci was reported for several conifer species between the 1970s and 1980s. Maps for Scots pine (*Pinus sylvestris*; Rudin and Ekberg, 1978); *Picea abies* (Lundkvist, 2009); balsam fir (*Abies balsamea*; Neale and Adams, 1981); Douglas-fir (*Pseudotsuga menziesii*; El-Kassaby *et al.*, 1982); and other conifers have been reported. These studies were useful for the purpose of creating unique, individual genetic fingerprints, and demonstrated linkage among genetic loci and that the segregation of alleles followed the expected Mendelian patterns.

Despite their very limited information content, early isozyme studies in mapping populations of diverse conifer species provided the first evidence that, despite millions of years of separation, the genome structure of conifers had not diverged extensively. Instead, the order and locations of homologous genes in the linkage groups in the different species appeared to be largely

---

Box 2.1. The Conifer Megagametophyte and Chromosome Recombination

The figure depicts three megaspore mother cells undergoing meiosis and mitosis to generate the megagametophyte of seeds. Megagametophyte development begins with the megaspore mother cell undergoing meiosis, resulting in four cells, one of which becomes a haploid (1N) megaspore. Because recombination occurs during meiosis, the resulting megaspore haplotypes are the result of crossing over between the parental homologous chromosomes. The haploid megaspore cells further replicate through mitosis to generate the megagametophyte (1N). By genotyping the megagametophytes from multiple seeds, the number of recombination events between loci can be counted and their genetic distance estimated.

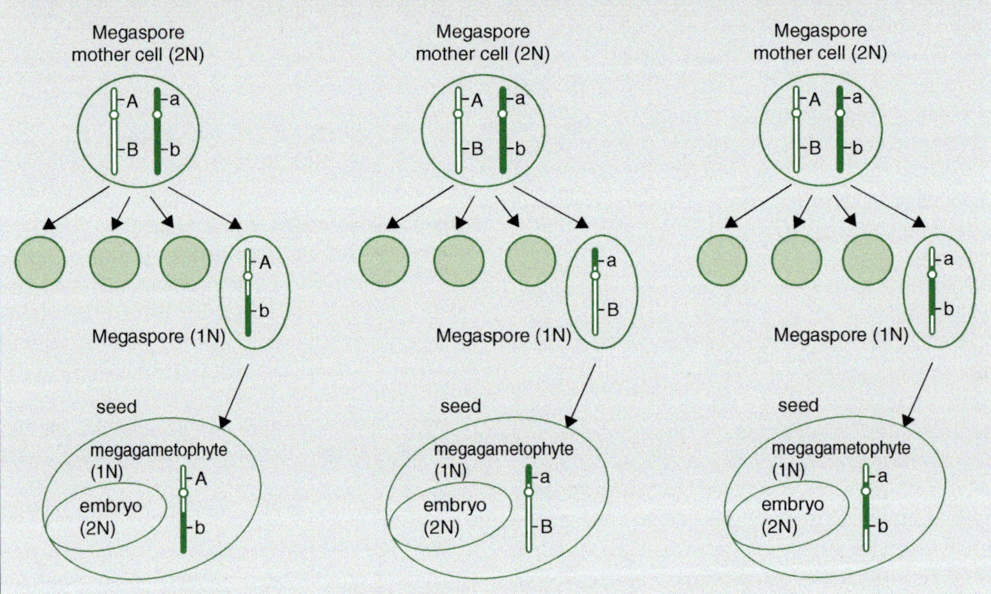

similar. For instance, mapping of 45 allozyme loci in three pine (*Pinus sylvestris*, *Pinus pallasiana*, and *Pinus pumila*) and one spruce species (*Picea abies*) showed extensive genome synteny (Goncharenko *et al.*, 1994). These observations supported earlier karyological research that indicated extensive conservation in the structure of conifer genomes. However, the restricted number of loci limited the strength of this conclusion and the use of these markers for many other applications. The development of higher-density maps and, ultimately, the sequencing of the first conifer genomes were necessary to confirm these observations.

RFLP marker-based conifer genetic maps

Genetic analysis using Restriction fragment length polymorphism (RFLP) markers in humans represented the first viable alternative to genotyping based on allozymes (Donis-Keller *et al.*, 1987). Genotyping based on RFLP required the development of radioactively labelled probes that hybridize to specific regions of the genome, which is first digested using restriction enzymes. Polymorphisms in restriction recognition sites create fragments of various sizes that are separated in a gel matrix, detected, and genotyped to create co-dominant markers. Development of RFLP markers is labor-intensive and low throughput because of the complexity of developing probes and the limitation of assaying only one or a few loci in a single analysis. However, the large number of possible probes and the diversity of restriction enzymes offered a significant improvement in the number of loci that could be analyzed in conifers or the genome of any other species. The first use of RFLP in conifer genotyping was in the analysis of segregation in loblolly pine (*Pinus taeda*) (Devey *et al.*, 1991) and construction of a genetic map with a few dozen markers (Devey *et al.*, 1994). Most important, however, was the realization that probes developed for loblolly pine were transferable to other species of *Pinus* and some other genera of the Pinaceae (Ahuja *et al.*, 1994). However, the limited genetic variation at single restriction sites, the limited number of endonucleases, and the low throughput and labor involved in assessing restriction digestion products greatly limited expansion and application of this methodology.

PCR marker-based conifer genetic maps

The development of polymerase chain reaction (PCR) as a method to amplify regions of the genome established an approach that was needed to further the characterization of genetic variants in segregating and genetically unstructured populations. The first genetic map with expanded coverage using PCR-based markers used amplified polymorphic DNA (RAPD) fragments genotyped in a set of megagametophytes from a single tree, and contained 73 RAPD markers (Nelson *et al.*, 1993). Genetic maps based on RAPD markers soon followed for slash pine (*Pinus elliottii*) (Doudrick, 1996) and loblolly pine (O'Malley *et al.*, 1996). RAPD genetic markers detect variation at PCR primer sites, or in the length of the amplified product, which is manifested by the presence or absence of a band in a gel. The diversity of RAPD PCR primer sequences that can be generated led to the rapid adoption of this approach for identifying markers linked to Mendelian disease traits, such as the white pine blister rust (*Cronartium ribicola*). An early study on the disease genotyped haploid DNA from megagametophytes, followed by inoculation with *C. ribicola* (Devey *et al.*, 1995). Because of the simple Mendelian segregation of the trait, the study allowed the detection of 10 loci linked to the gene for resistance to white pine blister rust, including one marker at only 0.9 centimorgans (cM) from the resistance locus. This study demonstrated for the first time that genetic markers could be positioned relatively close (in genetic distance) to loci of economic relevance, evidence that they could possibly be used for marker-assisted breeding in the future.

RAPD markers were later applied to multi-family populations, leading to the construction of the first integrated maps derived from crosses among various individuals. By integrating data from multiple crosses, it is possible to explore the segregation of alleles that may be variable in only one of the parents. Consequently, additional markers can be combined in one map, increasing marker density and saturation. Using this strategy, a consensus map for loblolly pine was constructed by integrating linkage data from two unrelated pedigrees derived from RAPD, RFLP, and isozyme genetic markers (Sewell *et al.*, 1999). The integration of individual maps allowed the synthesis of genetic information from independent sources on to a single consensus map with 357 genetic markers that covered approximately 1,300 cM. Interestingly, because genetic maps were also constructed to represent the male and female parents, it was possible to detect that the rate of meiotic recombination appeared to differ between them, based on their estimated genome size.

While RAPDs provided an avenue for saturating genomes with genetic markers, the detection of this type of genetic polymorphism is relatively unreliable. RAPDs are dominant markers, meaning that the absence of a PCR amplification product may be associated with a true polymorphism or with failure in the enzymatic reaction (amplification). Thus, new approaches to using PCR as a tool to detect genetic polymorphisms were necessary, and resulted in the development of new classes of genetic markers. Amplified fragment length polymorphism (AFLP) markers (Chapter 1, this volume) began replacing RAPD-derived maps rapidly after their development (Vos *et al.*, 1995). AFLPs provided the large number of markers required to construct a complete genetic map for an individual loblolly pine (Remington *et al.*, 1999). This first conifer map based on AFLPs contained 508 markers mapped into 12 linkage groups, in agreement with the number of chromosomes in the loblolly pine genome. A smaller set of 184 high-confidence markers was used to generate a framework map, for which markers could be ordered with high confidence.

The framework map provided nearly complete coverage of the genome, estimated at approximately 1,700 cM. The ease of generating AFLP makers and their robustness in genotyping conifers quickly led to the generation of maps for a suite of other conifer species, including Scots pine (Yin *et al.*, 2003); Japanese red pine (*Pinus densiflora*; Kim *et al.*, 2005); and black spruce (*Picea mariana*) (Kang *et al.*, 2010).

As the development of new methods of detecting polymorphisms evolved after the discovery of PCR, mapping efforts began integrating multiple categories of genetic markers, including microsatellites, as well as a few genes. Microsatellites are a type of short sequence, of two or more nucleotides, that occur repeatedly. Because of the high rate of error due to replication slippage in these repetitive regions (Ellegren, 2004), they tend to be highly variable and hence are useful for segregation analysis. In the meantime, the first efforts to genetically map large numbers of genes became possible with the sequencing of transcribed regions of the genome from expressed sequence tags (ESTs; see below and Chapter 1, this volume). One of the first genetic maps that integrated multiple marker types was constructed for the two parents of a Norway spruce cross using the double pseudo-test cross strategy and genotyping 73 individual F_1 progeny (Acheré *et al.*, 2004). The maps were developed primarily from AFLP markers, but also included 74 microsatellites and 18 EST polymorphisms. Other maps that included a few genes and microsatellites in conifers followed, including one for black spruce, with 816 markers in the maternal and 743 markers in the paternal map. A consensus map created from the paternal data included 1,111 markers, with 809 AFLPs, 255 selectively amplified microsatellite polymorphic loci (SAMPLs), 42 microsatellites, and five EST polymorphisms (Kang *et al.*, 2010).

SNP-based conifer genetic maps

Single-nucleotide polymorphisms (SNPs) are genetic variants characterized by an individual nucleotide substitution. These variants can occur in coding or non-coding regions of the genome, and may or may not have any effect on functionality. Their attractiveness as genetic markers comes from the fact that they are the most abundant form of DNA variation in a genome. As a consequence, if very high resolution is required, SNPs are the most suitable genetic marker because of their wide distribution and high frequency throughout the genome. Individual SNPs can be characterized using parallel assays such as DNA chips, which genotype from a few thousand to several million loci (Chapter 1, this volume). This form of genetic variation can also be identified and genotyped by direct sequencing of the regions of interest in the genome by methods such as sequence capture or by whole-genome shotgun (WGS) sequencing (Chapter 1, this volume).

The first example of the application of high-density and high-throughput genotyping approaches in conifers involved a GoldenGate SNP assay (Illumina), which characterized 768 SNPs in each of white spruce (*Picea glauca*) and black spruce (Pavy *et al.*, 2008). This genotyping effort targeted SNPs previously identified by resequencing genomic DNA from parents of segregation populations of both species. Despite the limited genome sequence information available to design the genotyping assays, 69.2% and 77.1% of genotyped SNPs had suitable quality and segregated in mapping populations of white spruce and black spruce, respectively. Failure to genotype approximately one-third of the targeted loci illustrated the challenges of designing genotyping assays in species with large and uncharacterized genomes. Following the positive outcome of the first genotyping effort using this platform, several studies followed that surveyed increasingly larger numbers of target loci. For instance, a GoldenGate assay was designed for maritime pine (*Pinus pinaster*), which evaluated 1,536 SNPs observed previously from the resequencing of a set of genes, and from EST sequencing projects (Chancerel *et al.*, 2011). Genotyping focused on segregating population and the success rate ranged from 64% to 75%, depending on the origin of the SNP locus. Fewer loci (384) were genotyped in Douglas-fir using the same genotyping strategy (Eckert *et al.*, 2009). More recently, higher-density genotyping assays that utilized an advanced genotyping method, Infinium (Illumina), and included several thousand loci, have been developed for loblolly pine (Eckert *et al.*, 2010); maritime pine (Plomion *et al.*, 2015); white spruce (Pavy *et al.*, 2013); and Japanese cedar (*Cryptomeria japonica*; Moriguchi *et al.*, 2012). To date, the highest density genetic map for a conifer was achieved using a mapping pedigree derived by crossing two unrelated white spruce parents and genotyping 1976 full-sib progeny (Pavy *et al.*, 2017). In this study 8,793 SNPs derived from transcribed genes were mapped, establishing the most complete gene mapping resource available for a conifer.

Single-nucleotide polymorphism-based genetic maps can be genotyped using a variety of strategies, including Next-generation sequencing (NGS) of targeted genomic regions (Chapter 1, this volume). In this scenario, specific stretches of the genome are captured using biotinylated oligonucleotides that are complementary to the sequences of interest, and after elution from the probes, these regions can be sequenced for identification of SNPs and other DNA polymorphisms. This approach was first used in a conifer to characterize segments of 14,729 genes in a mapping population of 72 haploid megagametophytes (Neves *et al.*, 2014). In total, 7,434 SNPs were detected in 3,787 genes, of which 2,841 could be mapped, establishing a high-density loblolly pine genetic map. The study identified large inversions in four linkage groups of previously published genetic maps; these are likely not cytological but rather the consequence of using a small mapping population.

High-density consensus genetic maps

The discovery of PCR and the development of advanced methods of genotyping led to several genetic mapping studies in conifers, most of which were carried out in segregating populations with different genetic backgrounds. However, the genotypic and mapping information generated for different families can be combined into a single consensus genetic map (Hauge *et al.*, 1993). This allows the generation of a map with higher marker density compared with individual maps, making it more suitable for other purposes, such as anchoring genome sequences. The first highly dense consensus genetic map for a conifer combined data from two three-generation outbred loblolly pine families (Martínez-García *et al.*, 2013). Individuals from these families had been genotyped with several thousand RFLPs, microsatellites, and SNP markers, and their combined analysis resulted in 2,466 of them being placed on the linkage map. Efforts to combine this consensus map with other high-density maps of loblolly pine (Echt *et al.*, 2011; Neves *et al.*, 2014) and of *Pinus taeda* × *P. elliottii* hybrids resulted in the positioning of 4,981 markers via genotyping of 1,251 individuals (Westbrook *et al.*, 2015).

Consensus genetic maps can also integrate data from species of different genera when there is extensive conservation in genome organization among them. To achieve the goal of creating a family-wide consensus map for members of the Pinaceae, data from 17 different maps that included parents from the genera *Pinus* and *Picea* were combined, resulting in a map that contained 5,927 different genes (de Miguel *et al.*, 2015).

Extensive Genome Synteny and Collinearity in the Pinaceae

Genome sequencing and assembly of conifers is particularly difficult, due to their large size and abundance of repetitive sequences. Therefore, the availability of a suitable genome model could prove useful to allow inferences to be made across conifers, as long as there is a high degree of conservation in the genome organization of these species. Synteny and collinearity provide two measures of genome conservation among species. Homologous loci are considered syntenic if they are located on homologous chromosomes in two different species, while collinearity refers to them sharing a similar order in these chromosomes (Box 2.2). The observation of

---

Box 2.2. Genome synteny and collinearity

The terms synteny and gene collinearity are used in the context of genomic studies to describe the relationship between the positions of homologous loci on chromosomes of related species. Synteny typically refers to the conservation of the localization of homologous genes on chromosomes derived from a common ancestor. Syntenic chromosomes contain genes that are collinear when their order on the chromosomes is conserved between species. Synteny and collinearity can be disrupted by a range of factors that impact chromosome organization during species evolution, such as chromosome rearrangements (deletions, duplications, inversions, and translocations) and gene movement caused by transposable elements.

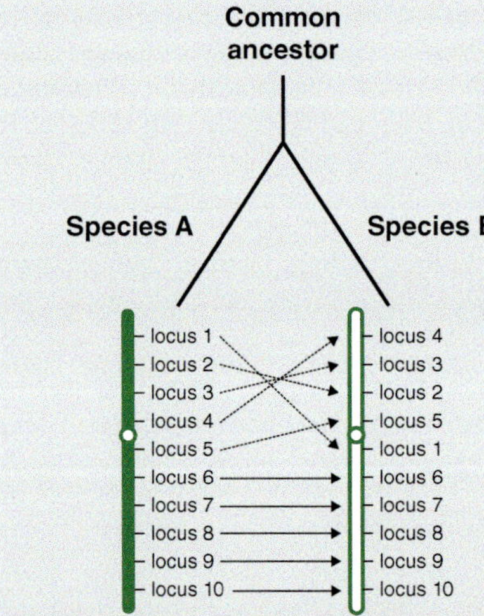

In the figure, a chromosome from a shared common ancestor is represented in species A and B. Loci 1–10 are syntenic, as both species share these loci on the same homologous chromosome. However, only loci 6–10 are collinear, because they share the same order on the homologous chromosome in species A and B.

---

synteny and collinearity between species can help infer orthology between the shared regions and inform gene function. Most importantly, a high degree of synteny and collinearity implies that genome characterization of a model species can provide useful information about related taxa (Chapter 1, this volume). The degree of synteny and collinearity varies tremendously among different organisms. For instance, mammalian chromosomes underwent only moderate shuffling as they shared a common ancestor (Ferguson-Smith and Trifonov, 2007); thus, synteny and collinearity is common. In contrast, angiosperm genomes have undergone extensive rearrangements and changes in genome size since their separation from the gymnosperms, creating a far less clear relationship about gene position among different tree species (Tang *et al.*, 2008).

In the absence of a reference genome sequence that locates contigs and scaffolds in their respective chromosome positions, inferences about synteny and collinearity among conifers has been based on comparing the genetic position of genetic elements in linkage maps. The first comparative map of conifers that comprised more than a few loci was developed for loblolly pine and Monterey pine (*Pinus radiata*) using a common set of RFLPs and microsatellite markers (Devey *et al.*, 1999). The map for loblolly pine combined data from two full-sib families and consisted of 20 linkage groups covering 1281 cM, while for Monterey pine, 14 linkage groups that covered 1,223 cM were detected. Sixty homologous RFLP loci were mapped in both species, in addition to a few microsatellites. The authors identified 12 groups of loci that were syntenic between the two species, composed of three to nine homologous loci. Interestingly, the order of these loci was largely in agreement between the species, suggesting extensive collinearity. With the subsequent sequencing of ESTs and the discovery of gene sequences from multiple conifer species, studies of synteny and collinearity migrated to the detection of genetic polymorphisms in homologous coding regions. These efforts began with the development of PCR methods to amplify homologous gene sequences in several *Pinus* spp. (Brown *et al.*, 2001). The analysis resulted in the identification of nine homologous linkage groups in loblolly pine and slash pine, and extensive syntenic regions with collinear loci. Additional small-scale comparative mapping studies between related pine species soon followed (Komulainen *et al.*, 2003; Krutovsky *et al.*, 2004) and largely supported the previous observations that synteny and collinearity are widespread in the genus *Pinus*.

The development of genetic maps for different genera allows an expansion of the hypothesis that synteny and collinearity are widespread in conifers. A confirmatory study resulted from the most complete gene map of a conifer, white spruce, which was compared with a previously developed gene-based map created for *Pinus*—these genera separated over 87 million years ago (Morse *et al.*, 2009). The study demonstrated an extremely high level of synteny among members of the Pinaceae, with over 90% of the 2,052 genes mapped found in homologous chromosomes (Pavy *et al.*, 2017). The wide dispersion of the non-syntenic genes across different linkage groups also suggested that there has not been any major chromosomal rearrangements within the genera *Pinus* and *Picea* (Fig. 2.2). In contrast, another study that compared the position of homologous genes in two distinct conifer families, Pinaceae and Cupressaceae, showed that extensive synteny cannot be assumed across these more distant lineages, despite them having a similar number of chromosomes (12 in Pinaceae, 11 in Cupressaceae). A composite map generated for the Pinaceae was compared with that of Japanese cedar (Cupressaceae; Moriguchi *et al.*, 2012), based on over 200 homologous loci identified in

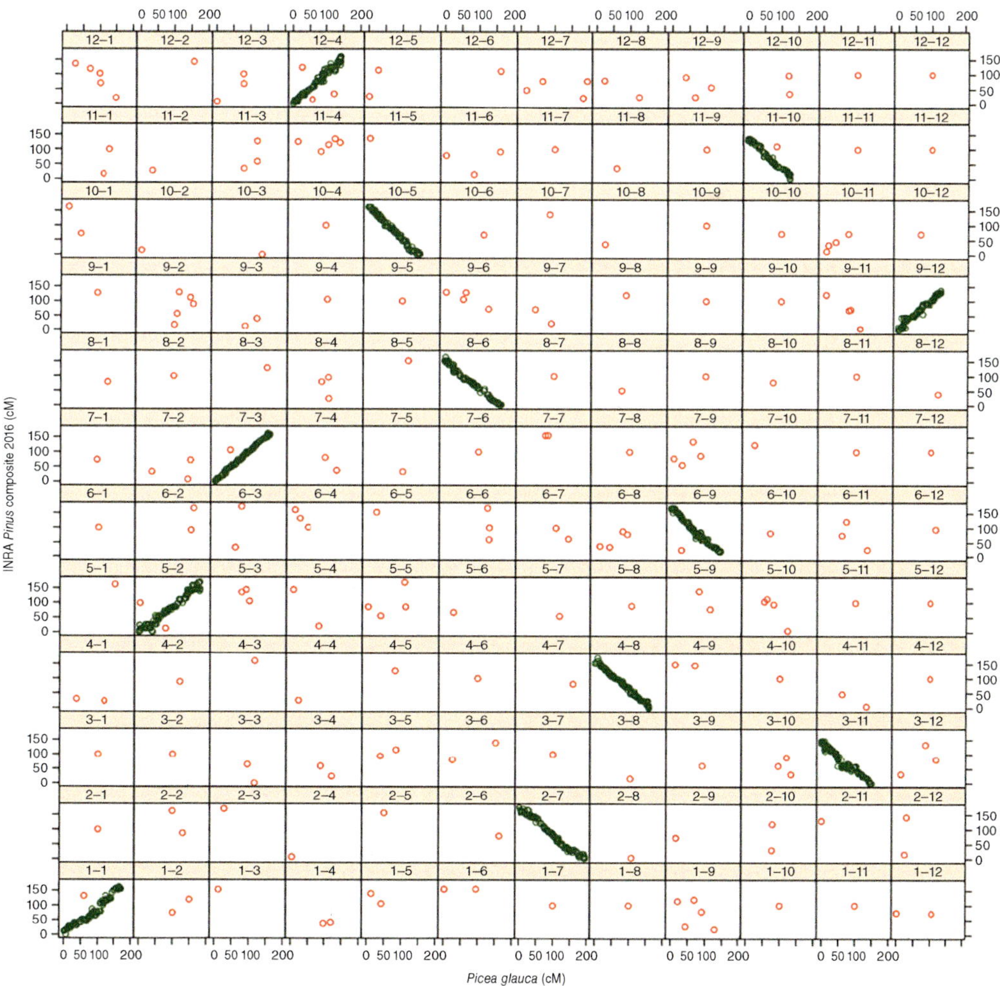

Fig. 2.2. Synteny between a composite map of *Picea glauca* and a genus composite map from *Pinus*.

these two families. Linkage maps from both families were aligned at these loci to identify common genomic regions. Surprisingly, each linkage group detected in the Pinaceae corresponded to one or two distinct segments of linkage group blocks in the Cupressaceae, while individual Cupressaceae linkage groups were related to one to three linkage groups in the Pinaceae (de Miguel *et al.*, 2015). Thus, there has been significant shuffling of chromosomes during the evolution of these families, which have been separated by almost 300 million years of evolution (Burleigh *et al.*, 2012).

Conifer Genome Size, Organization, and Evolution

All gymnosperms are recognized for having very large genomes (Fig. 2.3), with a haploid DNA content (mode 15,480 Mbp) that exceeds by over 26 times that of angiosperms (Leitch, 2001). Gymnosperm genomes dwarf the genome size of other sequenced woody perennial species. For instance, the genome of black cottonwood (*Populus trichocarpa*) has a haploid DNA content of approximately 485 Mbp (Tuskan *et al.*, 2006). The slightly larger genome of flooded gum (*Eucalyptus grandis*) has 640 Mbp (Myburg *et al.*, 2014). Among the gymnosperms, the genomes of conifers vary widely in size. A survey of 172 species from 64 genera showed that nuclear DNA content ranged from 8.3 to 71.6 pg, or 8.1 billion–70 billion bp (Zonneveld, 2012). Variation in conifer genome size, and in pines in particular, appears to depend partially on their life-history traits (Grotkopp *et al.*, 2004). An analysis of the relationship between the genome sizes of pines relative to their phylogeny suggests that the subgenus *Strobus* underwent a significant expansion, while the subgenus *Pinus* has decreased in size. Highly significant and positive correlations were also found between seed mass and DNA content ($R^2 = 0.58$), and to a lesser extent with leaf area and leaf area ratio (negative correlation). Other factors highly related to seed mass, such as seed number and dispersal mode, were also highly correlated, indirectly, with genome size.

Wide variation in genome size is not uncommon among plant species within the major plant taxonomic groups, and WGD and polyploidization are commonly considered a major cause of genome expansions.

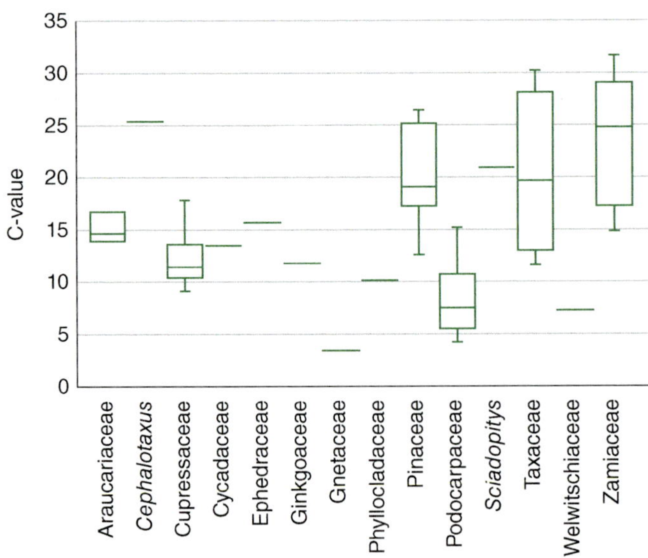

Fig. 2.3. Boxplot of DNA amount (C-value) quantified in the gymnosperm families. (Information from the Plant DNA C-values Database (http://data.kew.org/cvalues, accessed July 15, 2019; Wang and Ran, 2014.)

Polyploidization is also a critically important mechanism of genome evolution and speciation, particularly in angiosperms (Soltis *et al.*, 2015). Whole-genome duplications are triggered by the absence of separation of homologous chromosomes or sister chromatids (non-disjunction) during cell division, which may occur during meiosis or mitosis. Non-disjunction results in cells that contain additional copies of the entire genome, and can lead to polyploidization. Despite the role of WGD and polyploidization in expanding genome size, conifers do not appear to have undergone many such events since their divergence from other gymnosperms. The haploid number of chromosomes in conifers ranges from nine to 19 (Wang and Ran, 2014), but karyotypes are highly conserved across species and genera, with most having a basic number of 12 chromosomes (Sax and Sax, 1933). Deviations from that basic number are exceptions, and include the hexaploid California redwood (*Sequoia sempervirens*) and several species in the genus *Juniperus* (Khoshoo, 1961). Rare and localized examples of polyploidy have also been identified in plantations and nurseries, and are reported for Japanese cedar, European larch (*Larix decidua*), and various *Pinus* spp. (Zonneveld, 2012). Despite no evidence of WGD in conifers, a recent study suggests that three ancient events may have occurred in gymnosperms, including two in the ancestry of major conifer clades (Li *et al.*, 2015).

The extreme conservation in chromosome number, despite the long evolutionary history of the conifers, indicates that other mechanisms must be at the source of such genome "obesity." The most likely culprit is the proliferation of long-terminal-repeat retrotransposons (LTR-RTs), which are one of the two types or subclasses of retrotransposons. Retrotransposons are genetic elements that can amplify themselves through an RNA intermediate, resulting in large numbers of copies in the genome. These elements represent over half of the genome in some plant species, such as maize (*Zea mays*; SanMiguel *et al.*, 1996) and wheat (*Triticum aestivum*; Ling *et al.*, 2013). Similarly, early surveys of conifer genome sequences uncovered evidence that the *Ty3/Gypsy* and *Ty1/Copia*

LTR-RTs are highly abundant (Morse *et al.*, 2009; Kovach *et al.*, 2010; Sarri *et al.*, 2011). A detailed analysis of transposable element abundance in six conifer species, including the draft of the first conifer genome sequences to be characterized, further confirmed that LTR-RTs are the most abundant class and comprise a significant portion of the genome (Table 2.1), and that the *Ty3/Gypsy* superfamily occurs more frequently than the *Ty1/Copia* superfamily (Nystedt *et al.*, 2013; Wegrzyn *et al.*, 2014).

Despite the recent generation of the first draft genome sequence for several conifer species, the lack of a detailed assembly has left many critical questions unanswered regarding gene distribution and evolution. However, a few preliminary studies provide some tentative answers. Genetic maps of members of the Piceae that positioned thousands of genes on linkage groups and analyzed their distribution revealed distinctive gene-rich regions, some of which appeared to be comprised of functionally related genes (Pavy *et al.*, 2012, 2017). It was also shown that there are more duplicated genes shared between angiosperms and gymnosperms than conifer-specific duplicates. In combination with the extensive synteny and collinearity that has been reported between members of the genera *Picea* and *Pinus*, this suggests that conifer genomes have remained very stable, at least in the recent past. Thus, the evolution of conifer genomes appears to have been much slower when compared with that of angiosperms. An analysis of orthologous protein-coding genes from Sitka spruce

Table 2.1. Proportion of the main repetitive DNA elements in the genomes of *Picea abies* and *Pinus taeda*.

	Picea abies	*Pinus taeda*
LTRs		
Gypsy	35	11
Copia	16	9
Unclassified LTRs	7	22
LINEs	1	2
DNA transposable elements	1	1

LINE, long interspersed nuclear element.

(*Picea sitchensis*) and loblolly pine also showed that the average rate of nucleotide substitution (0.68×10^{-9} synonymous substitutions per site per year) is 15-fold lower than that observed in angiosperms (Buschiazzo *et al.*, 2012).

EST Sequencing

Over 30 years of linkage mapping in conifers showed that their genomes share a high level of synteny and collinearity, establishing a basic understanding of their genomic organization. Linkage mapping also created the foundation for genetic mapping of simple and complex traits. Despite this progress, genetic maps provide limited information regarding the genome sequence per se. Instead, they only revealed the occurrence of genetic variation, generally within broad genomic regions. As such, genetic markers developed for linkage mapping have limited relevance that can be explored in functional genomics studies. In the era immediately previous to sequencing of the first conifer genomes, uncovering polymorphisms of functional relevance, such as regulating phenotypes of commercial value, became a priority. In order for functional variants to be uncovered and analyzed, the genome—or at least specific regions more likely to contain functional variants—needed to be sequenced. The increase in DNA sequencing throughput and the reduction in cost lowered the most critical barriers to accomplishing this goal.

Sequencing of the gene-coding fraction of genomes was launched with large-scale sequencing of ESTs. These are fragments of genomic DNA generated by reverse-transcribing the mRNA into cDNA, followed by sequencing (Chapter 1, this volume). In 1991, the first large-scale EST sequencing project identified large suites of genes that were previously unknown, derived from mRNA extracted from human brain tissue (Adams *et al.*, 1991). Because the mRNA pool represents the ensemble of genes that are expressed in a given tissue or stage of development, ESTs provide not only a platform for discovery of genes but also an overview about which ones are actively being transcribed in the sample. Sequencing ESTs has limitations because the fragments sequenced correspond to only a portion of the coding sequence, and do not include intron or promoter information. Despite these limitations, EST sequencing was a logical next step toward characterizing the genome of conifers until WGS sequencing became possible.

The first attempts to sequence part of expressed genes from a conifer were carried out in loblolly pine, with the sequencing of a few hundred cDNAs obtained from whole seedlings, and cambium and phloem tissues (Kinlaw *et al.*, 1996). This first analysis detected putative homologs for a reasonable number of gene sequences (19–43%), despite the limited sequence resources in public databases in the early 1990s. An order of magnitude more ESTs (1097) were sequenced a few years later, in the first attempt to use EST data to uncover genes that regulate wood formation in conifers (Allona *et al.*, 1998). The study focused on sequencing ESTs from compression wood, which is formed on the side opposite an applied force. Compression wood has unique properties, such as lower cellulose and higher lignin content, and the purpose of one study was to uncover genes that might be involved in the developmental switch that controls carbon partitioning. Notably, comparative analysis with other databases showed that 55% of loblolly pine sequences showed similarity to previously characterized sequences, including 10% that showed similarity to proteins involved in the synthesis of cell-wall components. A few years later, a greatly expanded pine gene catalogue, based on 59,797 ESTs, showed that the gene content and composition were not dramatically distinct from that of angiosperms (Kirst *et al.*, 2003).

The initial focus of EST sequencing projects on developing wood tissues in loblolly pine soon expanded to the analysis of other plant organs and species. Conifer embryogenesis has unique properties relative to angiosperms and was, consequently, the subject of EST sequencing, which identified putative orthologs for angiosperm embryogenesis-related genes (Cairney *et al.*,

2006). Other studies focused on root-gene expression under conditions of stress, with the intent of identifying transcripts related to drought tolerance (Lorenz *et al.*, 2006). Finally, EST resources for *Picea* spp. soon rivaled those available for *Pinus,* with the execution of large sequencing projects (Pavy *et al.*, 2005; Ralph *et al.*, 2008). This and previous EST sequencing studies provided the first glimpse at the genome of conifers and their gene sequences, and led to the general conclusion that, despite over 300 million years since the separation of the angiosperms and gymnosperms, the coding fraction of their genomes has remained relatively similar. Conifers do not seem to have a large suite of novel genes that explain their developmental differences relative to angiosperms. This suggests that much of the phenotypic differences between angiosperms and gymnosperms may be driven primarily by gene-expression regulation rather than gene presence or absence, or gene structural variation. This conclusion paralleled the discovery made in the comparison of gene sequences between humans and primates three decades earlier, when it was shown that both species had a high gene sequence similarity, supporting the role of transcription regulation as a major driver of evolution (King and Wilson, 1975).

The initial EST sequencing projects in conifers provided the foundation for uncovering the coding sequences of their genomes, but were still limited in the number of sequences and tissues or stages of development that were sampled. The advances in sequencing in the early 2000s created the necessary incentive for the tree-research community to begin exploring strategies for sequencing large genome segments, as well as conifer genomes. In fact, in the years that followed the development of NGS methods, unigene sets were developed for numerous conifers, including maritime pine (Canales *et al.*, 2014); Douglas-fir (Howe *et al.*, 2013); Chinese fir (*Cunninghamia lanceolata*; Huang *et al.*, 2012); Dahurian larch (*Larix gmelinii*; Men *et al.*, 2013); Chinese pine (*Pinus tabuliformis*; Niu *et al.*, 2013); Aleppo pine (*Pinus halepensis*; Pinosio *et al.*, 2014); and dawn redwood (*Metasequoia glyptostroboides*; Zhao *et al.*, 2013).

BAC Sequencing

EST sequencing created detailed surveys of the genes expressed in several conifer species. However, EST sequences have limited value for understanding gene structure, as they do not capture introns and promoter sequences. Another limitation lies in that ESTs do not provide information regarding the organization and abundance of genetic elements in the genome, particularly in the intergenic space. For that goal to be achieved, the entire genome (or at least large segments of it) needed to be sequenced. The first detailed description of a large region of a conifer genome was made possible by the generation of a BAC library for loblolly pine (Magbanua *et al.*, 2011). Bacterial artificial chromosomes (BACs) are DNA constructs that permit fragments of tens of thousands of base pairs to be cloned and replicated. Where BACs contain a cloned genomic fragment it can be isolated and analyzed individually by sequencing (Chapter 1, this volume). Considering the enormous size of the conifer genomes, the construction of BAC libraries containing a sufficient number of clones to represent most of the nuclear DNA was considered particularly challenging. Despite the difficulties, a BAC library was constructed for the 21.7 Gbp genome of loblolly pine, consisting of 1,824,768 clones with a mean insert size of 96 kbp and a depth of 7.6×. Despite its enormous size, it was shown that the BAC library could be searched effectively for isolation of a specific clone that contained a gene of interest. More specifically, the BAC library was screened for the sequence of an EST that had similarity to the sequence for a putative late embryogenesis-abundant (LEA) protein from white spruce. The targeted *LEA* gene was successfully identified, in addition to gene-like sequences that were classified as pseudogenes. The analysis of this BAC also confirmed that a large proportion of the sequence was composed of retroelements, with LTR-RTs accounting for the majority of such sequences, and were composed of the *Gypsy* subfamily LTR element *IFG-7*. Similar conclusions were derived from sequencing 10 loblolly pine BAC clones representing almost

1 Mbp of genome sequence (Kovach *et al.*, 2010). This effort identified three putative protein-coding genes but at least another 15 possible pseudogenes. Thus, the initial sequencing of randomly selected regions of pine genomes suggested that, while conifers did not undergo the WGDs that were common in angiosperms, other mechanisms have contributed to increase the number of copies of genes. The sequencing of pine BAC clones also demonstrated that a diverse set of repetitive elements has diverged significantly in the genome. This analysis was later expanded to include 103 sequenced BACs, as well as a set of 90,954 scaffolds produced by sequencing 11 pools of fosmids derived from the loblolly pine genome (Wegrzyn *et al.*, 2013). Fosmids are another type of DNA cloning vector that are suitable for smaller fragments, of approximately 40–50 kbp. Although the 280 Mbp of genomic DNA that were assembled represented only approximately 1% of the haploid genome of loblolly pine, the sequence was the most detailed characterization of a conifer genome until the first draft became available. As expected, the analysis of sequence composition showed a wide abundance of repetitive elements, comprised primarily of full or partial LTR-RTs. However, a surprisingly extensive number of novel transposable element families were also uncovered, most of which were a single copy. Such complexity was viewed as possibly aiding in the assembly of the very complex genome.

Since the development of the first BAC library for loblolly pine, a few similar efforts have been pursued for other gymnosperms. For instance, a large BAC library was created for bald cypress (*Taxodium distichum*), a member of the Cupressaceae with a relatively small genome (1C = 9.95 pg; Hizume *et al.*, 2001) and a potential model for future studies. The bald cypress library is comprised of 606,336 BAC clones, which represents approximately 6.7× coverage of the genome (Liu *et al.*, 2011). A BAC library was also constructed for maritime pine, comprising 72,192 clones with an average size of 107 kbp (Bautista *et al.*, 2007). This library was used extensively for characterization of specific target genes in later studies (Seoane-Zonjic *et al.*, 2016).

Conifer Genome Sequencing

Strategies for sequencing the large genome of a conifer

Early efforts to sequence large plant genomes, such as that of maize, used a hierarchical approach (Schnable *et al.*, 2009). However, with the significant decrease in sequencing costs and increase in throughput of NGS platforms, the WGS sequencing approach became the preferred strategy for genome characterization. In fact, despite the enormous size of conifer genomes, sequencing them at a level of coverage and depth similar to that done to characterize the smaller angiosperm genomes seemed feasible. The main obstacle was the ability to computationally assemble the massive number of reads necessary to achieve the minimum depth of coverage required in previous sequencing projects that used a WGS strategy. The high throughput necessary to achieve the depth of coverage required implied the use of sequencing platforms that generate short (one or a few hundred bases) sequencing reads. This further complicates the assembly of a conifer genome, due to the abundance of repetitive DNA sequences that extend beyond the size of sequencing reads. To address these challenges, conifer genome sequencing projects had to follow strategies that combined WGS sequencing with sequencing of some form of genome representation that compartmentalized the sequences into large segments, and/or sequencing mate-pair, or "jumping" libraries.

Sequencing mate-pair or "jumping" libraries

Mate-pair or jumping DNA libraries are created with fragments whose length exceeds the sum of the sequencing read lengths from both ends of the fragment (Chapter 1, this volume). For instance, a jumping library with DNA fragments of 10 kbp may have 100 bp sequenced from both ends. These reads define not only the sequences of the ends of the fragments but also their length. This knowledge can be particularly valuable when assembling short sequencing reads from a large genome that has a high frequency of repetitive sequences. This occurs because the large fragments in the mate-pair library are likely to span repetitive regions. The value of sequencing large-fragment, mate-pair libraries, of various sizes, became evident when the first mammalian genome was sequenced exclusively using short sequencing reads (Li *et al.*, 2010). The value of this strategy was soon also realized for sequencing conifer genomes.

Pooling and sequencing of large DNA fragments

Another strategy to simplify the assembly of large conifer genomes is based on the sequencing of pools of large DNA fragments (Fig. 2.4). The advantage of this approach is that individual cloned DNA segments are haploid (Burgtorf *et al.*, 2003), thus avoiding allelic variants that hinder the assembly of highly heterozygous genomes like those of most conifers. The cloned fragments are also generally longer than the repetitive regions, facilitating their assembly. Finally, sequencing groups of large fragments should simplify the assembly process, because the genome is now compartmentalized into smaller segments that are computationally easier to analyze.

The pooling strategy has been evaluated in a previous study when 1,008 BACs randomly selected from a large loblolly pine library (Magbanua *et al.*, 2011) were combined and sequenced at an average depth of 220 (i.e. on average, each nucleotide was sequenced 220 times). This pool contained 103 BACs that had previously been completely sequenced to serve as a control to evaluate the completeness of the assembly. As expected, there was a steep decline in the average length of assembled contigs and coverage when the pool size was expanded (Fig. 2.5). A pool size of 700–800 maximized throughput and cost, while maintaining adequate assembly quality and coverage, relative to pools with fewer BACs.

While BAC libraries have been constructed for several conifer species, for all conifer genome sequencing projects that used a pooling strategy, the use of fosmids was preferred. Fosmids use the origin of replication of the F-plasmid for cloning in *Escherichia coli*, but packaging into the bacteriophage λ restricts the size of the fragment to a narrow interval of

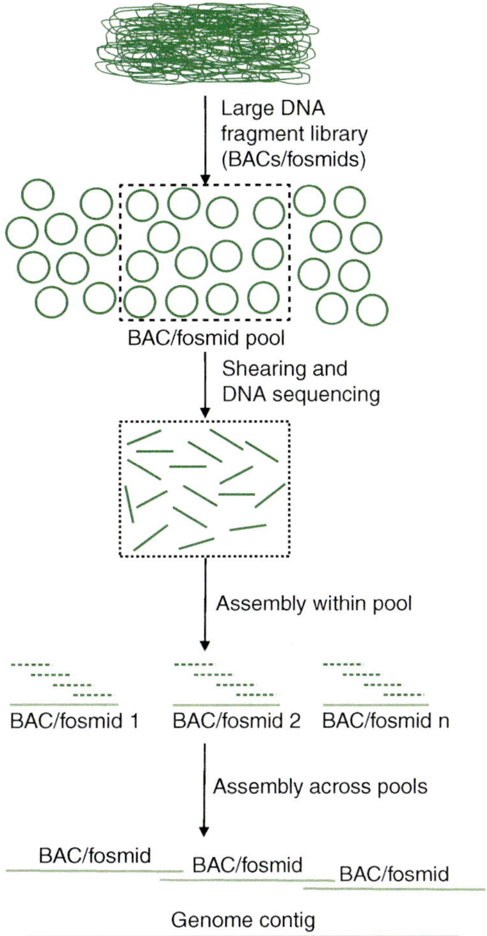

Fig. 2.4. DNA pool sequencing approach to characterize large genomes. In this approach, DNA is initially cloned into vectors that are suitable for large fragments, such as BACs or fosmids. A collection of cloned fragments is then assembled into a pool (dashed box), and the DNA is sheared and then sequenced. Following sequencing, reads from each pool are assembled with the expectation that those from individual BACs/fosmids will form contigs (pale green line). Next, the BAC or fosmid contigs from different pools are assembled into genomic contigs.

approximately 40 kbp. In addition, the process of cloning each fragment is far more efficient (over 100×), when compared with BAC library construction. Finally, each fosmid contains a single copy of a DNA fragment, and has been shown to be stably replicated over many generations. This fosmid pool sequencing approach

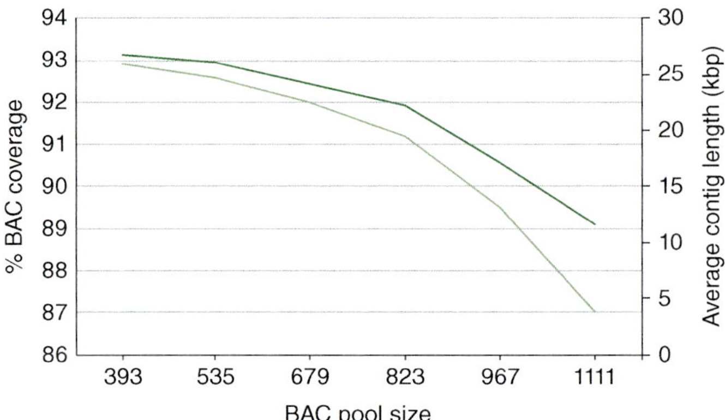

Fig. 2.5. Assembly of *Pinus taeda* BACs from pooled sequencing. The percentage coverage (pale green line) and average length of contigs (dark green line) assembled from 103 control BACs contained within pools of size 393 to 1,111 is shown. Coverage and contig length decrease with BAC pool size.

was adopted by two of the early conifer genome sequencing projects by combining hundreds to thousands of fosmids into pools, which were then sequenced (Alexeyenko *et al.*, 2014).

Conifer Genomes

The Norway spruce genome

Norway spruce (*P. abies*) is a conifer species native to Europe and is one of the most economically important forest species on that continent. The Norway spruce nuclear DNA content (2C) was estimated, through flow cytometry, to be approximately 40.7 pg (Zonneveld, 2012), the equivalent of approximately 20 Gbp. In 2013, an individual Norway spruce became the first gymnosperm to be sequenced (Nystedt *et al.*, 2013).

The Norway spruce genome was characterized using a strategy that combined the sequencing of pools of fosmids with WGS and RNA sequencing (RNA-seq). In total, 450 pools, each containing approximately 1,000 individual fosmid clones, were characterized by shotgun paired-end sequencing. In addition, random shotgun paired-end sequences representing 38× coverage of the genome were produced from DNA extracted from an individual haploid megagametophyte. For this step, library insert sizes were 180, 300, and 625 bp. Additional sequences were produced from jumping or mate-pair genomic libraries

of various sizes (2.4, 4.4, and 10.4 kbp), in order to facilitate the joining, or scaffolding, of contigs.

To integrate the sequencing reads derived from the Norway spruce genome, the sequences derived from individual fosmids pools were initially assembled and scaffolded, resulting in 6.7 Gbp of fragments larger than 1 kbp. Because individual fosmids are only approximately 40 kbp in size, their assembly from random shotgun sequences is much less computationally difficult than if the sequences were derived from the entire genome. In parallel, sequencing reads derived from random shotgun sequencing were assembled into another 9.8 Gbp, which were combined with contigs and scaffolds obtained by sequencing fosmids pools. As the final step of the assembly, the products of that assembly were scaffolded using sequences derived from the jumping libraries as well as from RNA-seq paired-end reads. In total, the final first version of the assembly had approximately 12 Gbp contained within scaffolds larger than 200 bp. Larger scaffolds (>10,000 bp) represent only 4.3 Gbp of the Norway spruce genome (~22%), reflecting the challenges of generating a high-quality contiguous assembly for such a complex genome.

The quality of genome sequence assemblies is frequently measured by their N50 and NG50 values (Chapter 1, this volume), which corresponded to 4,869 bp and 721 bp for the Norway spruce assembly. The quality of the assembly was also measured by evaluating the proportion of previously sequenced full-length cDNA from a related species, Sitka spruce (*Picea sitchensis*), aligning with the draft genome. Alignment was observed for over 90% of the full-length cDNA at a 95% nucleotide identity threshold. While a significant fraction of the coding sequences (63%) aligned almost entirely to a single scaffold, the remaining fraction was distributed among several scaffolds, due to the lack of continuity in the sequence. Another approach to evaluate the quality of the assembly relied on comparing the abundance of repeats in assemblies derived from the fosmid pools and WGS, relative to that of the draft assembly of the genome. As expected, the assembly of the fosmids was shown to represent the genome better than the WGS assembly.

Annotation of the gene space of the Norway spruce genome was done primarily using an *ab initio* approach, which relied on training gene-prediction methods based on a set of well-curated genes from the species. Using this approach, a total of 70,968 proteins were predicted. These proteins were then classified as high, medium, and low confidence, based on similarity to previously described ESTs or UniProt gene models. A set of 28,354 proteins was considered to be of high confidence, as over 70% of their sequence showed similarity to previously characterized genes. For these genes, exon size appeared to be similar to that of other plants sequenced previously. However, the mean intron size of Norway spruce is significantly larger than that reported for other angiosperms such as *Arabidopsis thaliana* (1,017 bp versus 168 bp; Arabidopsis Genome Initiative, 2000), reaching up to 68 kbp. Clustering these genes into related groups identified 6,615 gene families, with a mean of 3.9 genes per family, which is modestly higher than the mean reported for previously sequenced angiosperm genomes (3.4). A reasonably high number of gene families (1,021) were not observed in other plant species and could represent novel genes that are unique to Norway spruce, conifers, or gymnosperms.

Repetitive elements have been considered the main reason for the large size of conifer genomes (Morse *et al.*, 2009; Kovach *et al.*, 2010). To uncover the repetitive sequences that are most abundant in the Norway spruce genome, a library of sequences present at high copy number was created and used to estimate the repeat content. Estimates were that approximately 70% of the Norway spruce genome is composed of high-copy-number repeats—primarily *Ty1/Copia* and *Ty3/Gypsy* elements. A phylogenetic analysis of these repeats showed that several subfamilies expanded extensively in Norway spruce and other conifer genomes, compared with angiosperms. Furthermore, these expansion events appear to have occurred over several tens of millions of years and are thus ancient (over 10 million years ago).

The loblolly pine genome

The genus *Pinus* diverged from its closest relative (genus *Picea*) approximately 85 million years ago, and comprises some of the most ecologically important woody species, such as loblolly pine. The natural range of loblolly pine extends through the south-eastern USA. In addition to its ecological value, the species is widely planted throughout the world for timber, and for pulp and paper production. Estimates of its genome size indicate that loblolly pine is approximately 10% larger than the *Picea* spp. sequenced previously. The 2C nuclear DNA content was estimated as 44.3 pg (Zonneveld, 2012), or approximately 21.7 Gbp. Loblolly pine was the third gymnosperm but the first non-spruce conifer species to be sequenced.

Instead of basing the sequencing strategy on compartmentalizing the genome into fosmid pools, the majority of the loblolly pine genome sequence was derived primarily from random shotgun sequencing (Neale *et al.*, 2014). The loblolly pine genome was sequenced using haploid and diploid DNA collected from one megagametophyte and from pine needles, respectively. Haploid DNA was used to construct a series of DNA libraries with fragments of 200–600 bp, which were sequenced from both ends (paired-end sequencing) at 64× depth of coverage. Furthermore, 48 jumping libraries with DNA inserts between 1 and 5.5 kbp were constructed and their ends sequenced (13× coverage), as well as nine fosmid libraries with larger inserts (approximately 40 kbp). The assembly began with the construction of super-reads, which are sequences derived from paired-end reads where the gap between the end reads has been filled by other sequences. Because super-reads cannot be contained within any other super-read, they are unique, even though they could occur multiple times in the genome. The purpose of generating the super-reads was primarily to significantly reduce the amount of data to be managed in the downstream assembly. In that process, the read dataset became 27-fold smaller than that available initially, making it suitable for analysis using an overlap-based assembler.

Construction of the loblolly pine genome utilized the newly developed assembler MaSuRCA, which combines a de Brujin graph-based assembly in the generation of the super-reads, followed by an overlap–layout–consensus assembly that utilizes the super-reads. The de Brujin graph-based approach was developed to handle large amounts of NGS short reads but is not flexible in managing reads of distinct length. However, the overlap–layout–consensus approach is well suited for sequences of different lengths and has been used mostly in the analysis of longer Sanger sequencing reads.

The assembly of the loblolly pine genome was validated by comparing fosmid pool sequence data with the original assembly. Fosmid pools can be used to compartmentalize the genome and simplify the assembly (see above). By providing a sequence assembly of high confidence, the sequences of these pools are also an ideal template for evaluation of the whole-genome assembly. A pool of 4,600 fosmid clones was sequenced, and all assembled fragments with over 20 kbp (109 Mbp) were used in the comparison with the genome assembly. While this high-confidence assembly represents less than 1% of the pine genome, it was shown that over 98% was covered by the assembly generated via WGS sequencing. The assembly of the loblolly pine genome had an N50 of 8.2 kbp for contiguous, complete sequences (contigs). For scaffolds, the N50 size reached 66.9 kbp. Despite the significant improvement relative to the Norway spruce assembly, the loblolly pine sequence remains dispersed over several million contigs.

The loblolly pine genome was annotated with the MAKER-P pipeline (Campbell *et al.*, 2014), which integrates the alignments in known transcripts with *ab initio* gene-prediction models to identify genes. An initial survey with this approach uncovered over 90,000 gene models, which far exceeded the number reported for most plant genomes characterized previously. The observation that almost half of these models only aligned partially to known genes suggests that many are, in fact, pseudogenes. Additional filtering of this original set of genes resulted in the selection of approximately 50,000 genes that are supported by homology to previously reported protein sequences (Wegrzyn *et al.*, 2014). As reported in the Norway spruce genome (Nystedt *et al.*, 2013), the majority of the loblolly pine genome is composed of retroelements, particularly LTR retroelements (~42%), with a predominance of *Gypsy* and *Copia* elements.

The white spruce genome

White spruce is a species native to the northern latitudes of North America and is the most widely commercially planted tree in Canada. The tree selected to be sequenced was clone PG29, an important genotype used by tree breeding programs. The genome of white spruce is estimated to contain over 20 Gbp (Murray, 1998).

Similar to the strategy used to characterize the loblolly pine genome, white spruce was also sequenced using an exclusively WGS approach based entirely on paired-end sequencing and sequencing of jumping libraries. In a significant departure from other conifer sequencing projects, sequencing of the white spruce genome did not utilize the haploid DNA from megagametophytes. Instead, all libraries were produced from diploid genomic DNA.

For sequencing, DNA libraries were constructed with 250 and 500 bp fragments. In addition, large-fragment mate-pair (jumping) libraries of sizes of 6, 8, and 12 kbp were sequenced to provide linkage information across repeat structures. Libraries with DNA fragments of 250 and 500 bp were sequenced from both ends (paired-end) to an extent that resulted in overlap between reads. In addition, a high coverage of 150-bp paired-end sequences derived by sequencing the ends of 500 bp fragments was generated. Another novel approach used in sequencing of the white spruce genome was the development and implementation of methods to allow longer reads to be produced from a high-throughput NGS platform.

The assembly strategy followed a few steps designed primarily to optimize the assembly of regions of the genome with high sequence complexity, instead of low-complexity regions that are likely composed primarily of repetitive DNA. The computational platform used for the assembly of the reads was based on the ABySS algorithm (Simpson *et al.*, 2009). In the first step, sequences derived from libraries of DNA fragments of 250 or 500 bp, and sequenced from both ends (paired-end) to the point of generating a sequence overlap, were merged. In the second step, 150-bp paired-end sequences derived by sequencing the ends of 500-bp fragments were used to link the merged reads created in the first step of the assembly. Finally, mate-pair sequences derived from the libraries with longer fragments were utilized to bridge linked merged reads.

To measure the quality of the assembly, six BACs derived from the reference genotype PG29 were aligned with the draft genome assembly. In total, the draft sequence had a 62% overlap with BAC sequences. In addition, a set of 248 ultra-conserved core eukaryotic genes (Parra *et al.*, 2007) was searched, and at least partial sequences were identified for 74% of them. A search for approximately 13,000 full-length cDNA also identified 83% of them among scaffolds. Because most merged reads can only be linked to other merged reads, if there is enough complexity in the sequence ends, it is likely that this strategy results in a diminished representation of low-complexity, repetitive regions of the genome. These regions are likely to be of lower interest to most biologists but are still essential if a detailed understanding of the genome complexity and structure is desired. In total, the first draft of the white spruce genome had a reported 20.8 Gbp distributed in 4.9 million scaffolds, with a scaffold N50 of 20.3 kbp.

The publication of the first draft of the white spruce genome was followed by later releases of improved assemblies and the first annotation for gene content (Warren *et al.*, 2015). To improve the assembly of the PG29 genome, merged reads were scaffolded using a set of additional resources, including over 27,000 white spruce cDNAs and mate-pair sequences from 3, 8, and 12 kbp fragments. The new assembly resulted in a significant improvement in the contiguity of the genome, with the scaffold NG50 increasing to 71.5 kbp (compared with 41.9 kbp in the previous version). The updated PG29 genome was accompanied by the release of a second draft sequence for genotype WS77111, a different white spruce tree that is widely used in breeding programs in eastern Canada. WS77111 was sequenced and assembled largely using the same strategy applied to the sequencing of the PG29 genome. However, a significant increase in throughput and sequence length in later generations of sequencers resulted in an improved first assembly of WS77111. This assembly was then further improved by using the PG29 draft to scaffold additional sequences, resulting in the final NG50 length of 83 kbp.

The white spruce genome was annotated using the MAKER-P pipeline, used also in the analysis of the loblolly pine genome (Campbell *et al.*, 2014). Briefly, after masking the genome for repetitive, low-complexity regions, gene regions were predicted based on training sets derived from white spruce ESTs, as well as *A. thaliana* gene models. Annotated genes were then aligned with cDNA sequences and RNA-seq data from various libraries of white spruce, as well as with all proteins in the Swiss-Prot database (www.uniprot.org/, accessed July 15, 2019). From this analysis, a high-confidence set of 16,386 coding genes was obtained.

Major conclusions and observations from the sequencing of conifer genomes

With the availability of the first conifer genome, species comparisons with data from other land plants is allowing an increase in our understanding of the features that are common to seed plants as a whole, and those that are restricted to either angiosperms or gymnosperms.

- All conifer genome sequencing projects have resulted in the annotation of a number of genes that is not significantly larger than that reported for angiosperms. Thus, the larger genomes of conifers are not the result of a larger number of genes. Furthermore, conifer genes appear to be largely similar to those from angiosperms, with relatively few conifer-specific gene families.
- Despite recent evidence that supports the hypothesis that ancient WGDs have occurred in conifer lineages, there is no evidence of such events in their recent evolutionary history that could explain their large genome sizes.
- Lack of recent WGD and the observation of a large number of LTR-RTs indicate that the expansion of the conifer genomes is likely to be associated with a proliferation of these transposable elements. The relative stability of genome size among related species of the genera *Pinus* and *Picea* suggests that this expansion was a relatively ancient event. Some evidence suggests that the lack of an efficient mechanism for eliminating LTRs is the source of the genome expansion.
- The two fundamental differences between angiosperms and other seed plants relate to reproduction and water-conducting ability, and the genes found in the *Picea* genome provide information on these systems. Norway spruce lacks *FLOWERING LOCUS T*, a member of a set of genes that are key activators of flowering in angiosperms. However, the species contains an expanded set of *FT/TFL1-like* genes, which may repress flowering. In contrast, the genetic control of water conduction is not as clear. Water transport in conifers is accomplished by cells called tracheids, but most angiosperms have more efficient conducting cells (vessels). Angiosperm-specific innovations in water conduction are controlled by a gene family (*VASCULAR NAC DOMAIN*, or *VND*) that may have originated in gymnosperms, or possibly earlier—two *VND* genes were detected in Norway spruce, compared with seven in *A. thaliana*.

Conifer Transcriptomes

The sequencing and annotation of genomes and detection of the existing genetic variation within and between species represent the first level of genomic characterization of a species, population, or individual. However, the genome and gene sequences have no functional relevance on an individual unless its genes are transcribed into mRNA and then translated into proteins. Thus, measuring transcription of the ensemble of expressed genes, or "transcriptomics," represents the next level of genomic information that needs to be studied after the genome sequence. Surprisingly, methods to measure expression of the whole transcriptome were developed before NGS became available. Transcriptome quantification using cDNA microarrays was introduced before the sequencing of the first eukaryote genome (Schena *et al.*, 1995), and was later improved by the use of oligonucleotides printed directly on two-dimensional surfaces (Chapter 1, this volume). The approach has gradually been replaced by RNA-seq studies that are based on sequencing several million reads representing the population of mRNA molecules expressed in a sample at a given time point and/or under a certain set of conditions.

Most efforts to characterize conifer transcriptomes began with the sequencing of ESTs that provided the substrate (cDNA) or gene information necessary to design microarrays. The first transcriptome study in a conifer was carried out using a microarray developed by printing approximately 3,000 PCR-amplified cDNA clones obtained from differentiating xylem (Whetten *et al.* 2001). For this study, the microarray was used to identify genes differentially expressed between different developmental stages (juvenile and mature wood) of the same loblolly pine tree, and between trees submitted to mechanical stress and control growth conditions. This initial study was soon followed by several other applications of the same or related cDNA microarray platforms. Studies included: analysis of the needle transcriptome of loblolly pine trees submitted to different levels of drought stress (Watkinson *et al.*, 2003); gene expression during adventitious root development in lodgepole pine (*Pinus contorta*; Brinker *et al.*, 2004); and during early and latewood formation (Egertsdotter *et al.*, 2004), and seasonal variation in gene expression of loblolly pine accessions from different geographical regions. Because cDNA microarrays are based on hybridization of labelled cDNA to sequences printed on the slide, and a certain level of sequence similarity between genes in members of the Pinaceae was expected, attempts were also made to use the same platform developed from a loblolly pine EST resource to analyze other conifer genera such as *Picea* (van Zyl *et al.*, 2002; Stasolla *et al.*, 2004).

The rapidly expanding EST resources that followed early studies in pines resulted in the establishment of cDNA microarrays comprised of an order of magnitude more transcripts than those used initially. These included a large cDNA microarray with 21,840 unique features that was developed for white spruce and used in one of the first studies aimed at understanding how the transcriptome of individuals in distinct populations respond to cold acclimation (Holliday *et al.*, 2008). Another large microarray composed of 26,496 cDNAs from loblolly pine was used to characterize the transcriptome response to drought stress in roots (Lorenz *et al.*, 2011).

Finally, development of NGS methods and the huge reduction in sequencing costs are leading to a dramatic proliferation of plant transcriptome studies using RNA-seq as the gene-expression measurement tool (Weber, 2015). A significant advantage of this approach is that it is "platform-free" (i.e. the development of an analytical tool such as a microarray is not required). Instead, gene-expression estimates are generated by simply counting reads derived by sequencing cDNA synthesized from mRNA, and aligning them with the species reference gene set. Furthermore, if a reference to which reads are aligned does not exist, it can be created by the assembly of all reads into a unigene set. RNA-seq studies are still in their infancy in forestry and conifer studies but have already been applied to uncover genes involved in western white pine (*Pinus monticola*) blister rust resistance (Liu *et al.*, 2013), and transcriptome changes that arise in *Picea* and *Pinus* from changes in temperature, humidity, and day length (Yeaman *et al.*, 2014).

Conifer Proteomes

Proteomics is the study of the proteins expressed in a given sample, including their identification and quantification. This field also considers all protein modifications as an area of interest. Proteomics has gone through exceptional developments in recent decades with the establishment of improved methods of extracting proteins from tissues and cells, protein separation and quantification, as well as data analysis. Nonetheless, the core methodology remains largely unchanged, based primarily on one- or two-dimensional gel electrophoresis coupled to mass spectrometry.

One of the most critical limitations of proteomics methods has been the low proteome coverage, which results in a very limited fraction of proteins being characterized. For instance, reports on most conifer studies to date describe the detection of only a few hundred to several thousand spots detected by two-dimensional gel electrophoresis, of which typically only a few dozen have been identified (reviewed by Abril *et al.*, 2011). Despite these limitations, proteomics studies were first reported in conifers, primarily in maritime pine, where it was used to study wood development (Plomion *et al.*, 2000; Gion *et al.*, 2005).

Summary

Gymnosperms are vascular, woody, perennial plants that comprise approximately

1,000 species, the majority of which are conifers. The gymnosperms separated from the angiosperms approximately 300 million years ago and have significant morphological and developmental differences in their reproductive and vascular systems that distinguish them from flowering plants. Today, they dominate the boreal forests of the Northern Hemisphere, and several species have high economic value as a source of softwood used for the production of timber, pulp, paper, and industrial reagents.

Studying conifers or any other gymnosperm is challenging because of their physical size and long generation times but also because of their large genomes. However, conifers have the advantage of producing seeds with haploid megagametophytes, which can be analyzed to generate genetic maps. Mapping studies capitalized on this feature in the early days of isozyme genotyping. Later, studies expanded to gene maps based on NGS genotyping and DNA chips, which positioned several thousand loci in dense genetic maps of *Pinus* and *Picea* spp. and other conifers.

Mapping studies showed that the genome structure of conifer species has remained relatively stable and well conserved, despite millions of years of divergence. Comparing the position of genetic elements in linkage maps showed that the order of loci is largely in agreement among species in the Pinaceae, such as those within the genera *Pinus* and *Piceae*. However, the high synteny cannot be assumed for more distant lineages, as shown in comparisons between species in the families Pinaceae and Cupressaceae.

Gymnosperms have large genomes that differ widely in size (8.1 billion–70 billion bp), although the variation in conifers is more restricted. Despite the role of WGD and polyploidization in expanding genome size, conifers do not appear to have undergone such events since their divergence from other gymnosperms. The most likely cause for the large size of conifer genomes is the proliferation of LTR-RTs, particularly those from the *Ty3/Gypsy* superfamily.

Detailed characterization of conifer genomes began with sequencing of the gene-encoding fraction in EST projects. Initial studies focused heavily on characterizing transcripts from wood-forming tissues, and later expanded to the analysis of roots, developing embryos, and beyond. Following the development of NGS methods, the sequencing of transcripts greatly expanded the breadth of species analyzed. Further, initial characterization of conifer genomes was made by the sequencing of a few BACs. To support this effort, large BAC libraries were generated for loblolly pine and other species. The sequencing of a few clones provided a first glimpse into the intergenic space, as well as gene structure.

Following the sequencing of the first large angiosperm genomes (e.g. maize), and the reduction in cost of sequencing triggered by advanced methods of NGS, initial attempts were made to sequence conifer genomes. These studies used a combination of shotgun sequencing with mate-pair libraries and sequencing of pools of BACs to simplify the genome assembly. Despite these efforts, all conifer genome sequences remain at a draft stage, composed of many contigs of relatively short size. The first conifer genome to be sequenced was that of Norway spruce, which concluded with an assembly N50 of approximately 5 kbp. Despite its limitations, this first draft showed that conifers do not have a significantly larger number of genes than angiosperms, although pseudogenes are abundant and intron length is greatly expanded. The Norway spruce genome was found to be largely composed of high-copy repeats, as expected. Sequencing of Norway spruce was soon followed by the generation of a draft genome sequence for loblolly pine and white spruce.

References

Abril, N., Gion, J.-M., Kerner, R., Müller-Starck, G., Cerrillo, R.M.N., *et al.* (2011) Proteomics research on forest trees, the most recalcitrant and orphan plant species. *Phytochemistry* 72, 1219–1242.

Acheré, V., Faivre-Rampant, P., Jeandroz, S., Besnard, G., Markussen, T., *et al.* (2004) A full saturated linkage map of *Picea abies* including AFLP, SSR, ESTP, 5S rDNA and morphological markers. *Theoretical and Applied Genetics* 108, 1602–1613.

Adams, M.D., Kelley, J.M., Gocayne, J.D., Dubnick, M., Polymeropoulos, M.H., *et al.* (1991) Complementary DNA sequencing: expressed sequence tags and human genome project. *Science* 252, 1651–1656.

Ahuja, M.R., Devey, M.E., Groover, A.T., Jermstad, K.D., and Neale, D.B. (1994) Mapped DNA probes from loblolly pine can be used for restriction fragment length polymorphism mapping in other conifers. *Theoretical and Applied Genetics* 88, 279–282.

Alexeyenko, A., Nystedt, B., Vezzi, F., Sherwood, E., Ye, R., *et al.* (2014) Efficient de novo assembly of large and complex genomes by massively parallel sequencing of Fosmid pools. *BMC Genomics* 15, 439.

Allona, I., Quinn, M., Shoop, E., Swope, K., St Cyr, S., *et al.* (1998) Analysis of xylem formation in pine by cDNA sequencing. *Proceedings of the National Academy of Sciences, USA* 95, 9693–9698.

Arabidopsis Genome Initiative (2000) Analysis of the genome sequence of the flowering plant *Arabidopsis thaliana. Nature* 408, 796–815.

Bautista, R., Villalobos, D.P., Díaz-Moreno, S., Cantón, F.R., Cánovas, F.M., and Claros, M.G. (2007) Toward a *Pinus pinaster* bacterial artificial chromosome library. *Annals of Forest Science* 64, 855–864.

Bell, C.D., Soltis, D.E., and Soltis, P.S. (2010) The age and diversification of the angiosperms re-revisited. *American Journal of Botany* 97, 1296–1303.

Brinker, M., van Zyl, L., Liu, W., Craig, D., Sederoff, R.R., *et al.* (2004) Microarray analyses of gene expression during adventitious root development in *Pinus contorta. Plant Physiology* 135, 1526–1539.

Brown, G.R., Kadel, E.E., Bassoni, D.L., Kiehne, K.L., Temesgen, B., *et al.* (2001) Anchored reference loci in loblolly pine (*Pinus taeda* L.) for integrating pine genomics. *Genetics* 159, 799–809.

Burgtorf, C., Kepper, P., Hoehe, M., Schmitt, C., Reinhardt, R., *et al.* (2003) Clone-based systematic haplotyping (CSH): a procedure for physical haplotyping of whole genomes. *Genome Research* 13, 2717–2724.

Burleigh, J.G., Barbazuk, W.B., Davis, J.M., Morse, A.M., and Soltis, P.S. (2012) Exploring diversification and genome size evolution in extant gymnosperms through phylogenetic synthesis. *Journal of Botany* 2012, 1–6.

Buschiazzo, E., Ritland, C., Bohlmann, J., and Ritland, K. (2012) Slow but not low: genomic comparisons reveal slower evolutionary rate and higher dN/dS in conifers compared to angiosperms. *BMC Evolutionary Biology* 12, 8.

Cairney, J., Zheng, L., Cowels, A., Hsiao, J., Zismann, V., *et al.* (2006) Expressed sequence tags from loblolly pine embryos reveal similarities with angiosperm embryogenesis. *Plant Molecular Biology* 62, 485–501.

Campbell, M.S., Law, M., Holt, C., Stein, J.C., Moghe, G.D., *et al.* (2014) MAKER-P: a tool kit for the rapid creation, management, and quality control of plant genome annotations. *Plant Physiology* 164, 513–524.

Canales, J., Bautista, R., Label, P., Gómez-Maldonado, J., Lesur, I., *et al.* (2014) *De novo* assembly of maritime pine transcriptome: implications for forest breeding and biotechnology. *Plant Biotechnology Journal* 12, 286–299.

Chancerel, E., Lepoittevin, C., Le Provost, G., Lin, Y.-C., Jaramillo-Correa, J.P., *et al.* (2011) Development and implementation of a highly-multiplexed SNP array for genetic mapping in maritime pine and comparative mapping with loblolly pine. *BMC Genomics* 12, 368.

Clarke, J.T., Warnock, R.C.M., and Donoghue, P.C.J. (2011) Establishing a time-scale for plant evolution. *New Phytologist* 192, 266–301.

de Miguel, M., Bartholomé, J., Ehrenmann, F., Murat, F., Moriguchi, Y., *et al.* (2015) Evidence of intense chromosomal shuffling during conifer evolution. *Genome Biology and Evolution* 7, 2799–2809.

Devey, M.E., Jermstad, K.D., Tauer, C.G., and Neale, D.B. (1991) Inheritance of RFLP loci in a loblolly pine three-generation pedigree. *Theoretical and Applied Genetics* 83, 238–242.

Devey, M.E., Fiddler, T.A., Liu, B.H., Knapp, S.J., and Neale, D.B. (1994) An RFLP linkage map for loblolly pine based on a three-generation outbred pedigree. *Theoretical and Applied Genetics* 88, 273–278.

Devey, M.E., Delfino-Mix, A., Kinloch, B.B., and Neale, D.B. (1995) Random amplified polymorphic DNA markers tightly linked to a gene for resistance to white pine blister rust in sugar pine. *Proceedings of the National Academy of Sciences USA* 92, 2066–2070.

Devey, M.E., Sewell, M.M., Uren, T.L., and Neale, D.B. (1999) Comparative mapping in loblolly and radiata pine using RFLP and microsatellite markers. *Theoretical and Applied Genetics* 99, 656–662.

Dixon, R.K., Solomon, A.M., Brown, S., Houghton, R.A., Trexier, M.C., and Wisniewski, J. (1994)

Carbon pools and flux of global forest ecosystems. *Science* 263, 185–190.

Donis-Keller, H., Green, P., Helms, C., Cartinhour, S., Weiffenbach, B., *et al.* (1987) A genetic linkage map of the human genome. *Cell* 51, 319–37.

Doudrick, R.L. (1996) Genetic recombinational and physical linkage analyses on slash pine. *Symposia of the Society for Experimental Biology* 50, 53–60.

Echt, C.S., Saha, S., Krutovsky, K.V., Wimalanathan, K., Erpelding, J.E., *et al.* (2011) An annotated genetic map of loblolly pine based on microsatellite and cDNA markers. *BMC Genetics* 12, 17.

Eckenwalder, J.E. (2009) *Conifers of the World: The Complete Reference.* Timber Press, Portland, Oregon.

Eckert, A.J., Bower, A.D., Wegrzyn, J.L., Pande, B., Jermstad, K.D., *et al.* (2009) Association genetics of coastal Douglas fir (*Pseudotsuga menziesii* var. *menziesii,* Pinaceae). I. Cold-hardiness related traits. *Genetics* 182, 1289–1302.

Eckert, A.J., van Heerwaarden, J., Wegrzyn, J.L., Nelson, C.D., Ross-Ibarra, J., *et al.* (2010) Patterns of population structure and environmental associations to aridity across the range of loblolly pine (*Pinus taeda* L., Pinaceae). *Genetics* 185, 969–982.

Egertsdotter, U., van Zyl, L.M., MacKay, J., Peter, G., Kirst, M., *et al.* (2004) Gene expression during formation of earlywood and latewood in loblolly pine: expression profiles of 350 genes. *Plant Biology* 6, 654–663.

El-Kassaby, Y.A., Sziklai, O., and Yeh, F.C. (1982) Linkage relationships among 19 polymorphic allozyme loci in coastal Douglas fir (*Pseudotsuga menziesii* var. *menziesii*). *Canadian Journal of Genetics and Cytology* 24, 101–108.

Ellegren, H. (2004) Microsatellites: simple sequences with complex evolution. *Nature Reviews Genetics* 5, 435–445.

Farjon, A. (2010) *A Handbook of the World's Conifers (two vols.).* Brill Academic Publishers, Leiden, The Netherlands.

Ferguson-Smith, M.A. and Trifonov, V. (2007) Mammalian karyotype evolution. *Nature Reviews Genetics* 8, 950–662.

Gion, J.-M., Lalanne, C., Le Provost, G., Ferry-Dumazet, H., Paiva, J., *et al.* (2005) The proteome of maritime pine wood forming tissue. *Proteomics* 5, 3731–3751.

Goncharenko, G.G., Padutov, V.E., and Silin, A.E. (1994) Construction of genetic maps for some Eurasian coniferous species using allozyme genes. *Biochemical Genetics* 32, 223–236.

Grotkopp, E., Rejmánek, M., Sanderson, M.J., and Rost, T.L. (2004) Evolution of genome size in pines (*Pinus*) and its life-history correlates: supertree analyses. *Evolution* 58, 1705–1729.

Hauge, B.M., Hanley, S.M., Cartinhour, S., Cherry, J.M., Goodman, H.M., *et al.* (1993) An integrated genetic/RFLP map of the *Arabidopsis thaliana* genome. *Plant Journal* 3, 745–754.

Hizume, M., Kondo, T., Shibata, F., and Ishizuka, R. (2001) Flow cytometric determination of genome size in the Taxodiaceae, Cupressaceae *sensu stricto* and Sciadopityaceae. *Cytologia* 66, 307–311.

Holliday, J.A., Ralph, S.G., White, R., Bohlmann, J., and Aitken, S.N. (2008) Global monitoring of autumn gene expression within and among phenotypically divergent populations of Sitka spruce (*Picea sitchensis*). *New Phytologist* 178, 103–122.

Howe, G.T., Yu, J., Knaus, B., Cronn, R., Kolpak, S., *et al.* (2013) A SNP resource for Douglas-fir: *de novo* transcriptome assembly and SNP detection and validation. *BMC Genomics* 14, 137.

Huang, H.-H., Xu, L.-L., Tong, Z.-K., Lin, E.-P., Liu, Q.-P., *et al.* (2012) *De novo* characterization of the Chinese fir (*Cunninghamia lanceolata*) transcriptome and analysis of candidate genes involved in cellulose and lignin biosynthesis. *BMC Genomics* 13, 648.

Kang, B.-Y., Mann, I.K., Major, J.E., and Rajora, O.P. (2010) Near-saturated and complete genetic linkage map of black spruce (*Picea mariana*). *BMC Genomics* 11, 515.

Khoshoo, T.N. (1961) Chromosome numbers in gymnosperms. *Silvae Genetica* 10, 1–9.

Kim, Y.-Y., Choi, H.-S., and Kang, B.-Y. (2005) An AFLP-based linkage map of Japanese red pine (*Pinus densiflora*) using haploid DNA samples of megagametophytes from a single maternal tree. *Molecules and Cells* 20, 201–209.

King, M.C. and Wilson, A.C. (1975) Evolution at two levels in humans and chimpanzees. *Science* 188, 107–116.

Kinlaw, C.S., Ho, T., Gertula, S.M., Gladstone, E., Harry, D.E., *et al.* (1996) Gene discovery in loblolly pine through DNA sequencing. In: Ahuja, M.R., Boerjan, W., and Neale, D.B. (eds) *Somatic Cell Genetics and Molecular Genetics of Trees.* Forestry Sciences series, vol. 49. Springer, Dordrecht, The Netherlands, pp. 175–182.

Kirst, M., Johnson, A.F., Baucom, C., Ulrich, E., Hubbard, K., *et al.* (2003) Apparent homology of expressed genes from wood-forming tissues of loblolly pine (*Pinus taeda* L.) with *Arabidopsis thaliana. Proceedings of the National Academy of Sciences USA* 100, 7383–7388.

Komulainen, P., Brown, G.R., Mikkonen, M., Karhu, A., García-Gil, M.R., *et al.* (2003) Comparing EST-based genetic maps between *Pinus sylvestris* and *Pinus taeda. Theoretical and Applied Genetics* 107, 667–678.

Kovach, A., Wegrzyn, J.L., Parra, G., Holt, C., Bruening, G.E., *et al.* (2010) The *Pinus taeda* genome is characterized by diverse and highly diverged repetitive sequences. *BMC Genomics* 11, 420.

Krutovsky, K.V., Troggio, M., Brown, G.R., Jermstad, K.D., and Neale, D.B. (2004) Comparative mapping in the Pinaceae. *Genetics* 168, 447–461.

Kurz, W.A., Dymond, C.C., Stinson, G., Rampley, G.J., Neilson, E.T., *et al.* (2008) Mountain pine beetle and forest carbon feedback to climate change. *Nature* 452, 987–990.

Leitch, I. (2001) Nuclear DNA C-values complete familial representation in gymnosperms. *Annals of Botany* 88, 843–849.

Li, R., Fan, W., Tian, G., Zhu, H., He, L., *et al.* (2010) The sequence and *de novo* assembly of the giant panda genome. *Nature* 463, 311–317.

Li, Z., Baniaga, A.E., Sessa, E.B., Scascitelli, M., Graham, S.W., *et al.* (2015) Early genome duplications in conifers and other seed plants. *Science Advances* 1, e1501084.

Ling, H.-Q., Zhao, S., Liu, D., Wang, J., Sun, H., *et al.* (2013) Draft genome of the wheat A-genome progenitor *Triticum urartu*. *Nature* 496, 87–90.

Liu, J.-J., Sturrock, R.N., and Benton, R. (2013) Transcriptome analysis of *Pinus monticola* primary needles by RNA-seq provides novel insight into host resistance to *Cronartium ribicola*. *BMC Genomics* 14, 884.

Liu, W., Thummasuwan, S., Sehgal, S.K., Chouvarine, P., and Peterson, D.G. (2011) Characterization of the genome of bald cypress. *BMC Genomics* 12, 553.

Lorenz, W.W., Sun, F., Liang, C., Kolychev, D., Wang, H., *et al.* (2006) Water stress-responsive genes in loblolly pine (*Pinus taeda*) roots identified by analyses of expressed sequence tag libraries. *Tree Physiol* 26, 1–16.

Lorenz, W.W., Alba, R., Yu, Y.S., Bordeaux, J.M., Simões, M., and Dean, J.F. (2011) Microarray analysis and scale-free gene networks identify candidate regulators in drought-stressed roots of loblolly pine (*P. taeda* L.). *BMC Genomics* 12, 264.

Lundkvist, K. (2009) Allozyme frequency distributions in four Swedish populations of Norway spruce (*Picea abies*). *Hereditas* 90, 127–143.

Magbanua, Z.V., Ozkan, S., Bartlett, B.D., Chouvarine, P., Saski, C.A., *et al.* (2011) Adventures in the enormous: a 1.8 million clone BAC library for the 21.7 Gb genome of loblolly pine. *PLoS One* 6, e16214.

Martínez-García, P.J., Stevens, K.A., Wegrzyn, J.L., Liechty, J., Crepeau, M., *et al.* (2013) Combination of multipoint maximum likelihood (MML) and regression mapping algorithms to construct a high-density genetic linkage map for loblolly pine (*Pinus taeda* L.). *Tree Genet Genomes* 9, 1529–1535.

Men, L., Yan, S., and Liu, G. (2013) *De novo* characterization of *Larix gmelinii* (Rupr.) Rupr. transcriptome and analysis of its gene expression induced by jasmonates. *BMC Genomics* 14, 548.

Moriguchi, Y., Ujino-Ihara, T., Uchiyama, K., Futamura, N., Saito, M., *et al.* (2012) The construction of a high-density linkage map for identifying SNP markers that are tightly linked to a nuclear-recessive major gene for male sterility in *Cryptomeria japonica* D. Don. *BMC Genomics* 13, 95.

Morse, A.M., Peterson, D.G., Islam-Faridi, M.N., Smith, K.E., Magbanua, Z., *et al.* (2009) Evolution of genome size and complexity in *Pinus*. *PLoS One* 4, e4332.

Murray, B. (1998) Nuclear DNA amounts in gymnosperms. *Annals of Botany* 82, 3–15.

Myburg, A.A., Grattapaglia, D., Tuskan, G.A., Hellsten, U., Hayes, R.D., *et al.* (2014) The genome of *Eucalyptus grandis*. *Nature* 510, 356–362.

Neale, D.B. and Adams, W.T. (1981) Inheritance of isozyme variants in seed tissue of balsam fir (*Abis balsamea*). *Canadian Journal of Botany* 59, 1285–1291.

Neale, D.B., Wegrzyn, J.L., Stevens, K.A., Zimin, A.V., Puiu, D., *et al.* (2014) Decoding the massive genome of loblolly pine using haploid DNA and novel assembly strategies. *Genome Biology* 15, R59.

Nelson, C.D., Nance, W.L., and Doudrick, R.L. (1993) A partial genetic linkage map of slash pine (*Pinus elliottii* Engelm. var. *elliottii*) based on random amplified polymorphic DNAs. *Theoretical and Applied Genetics* 87, 145–151.

Neves, L.G., Davis, J.M., Barbazuk, W.B., and Kirst, M. (2014) A high-density gene map of loblolly pine (*Pinus taeda* L.) based on exome sequence capture genotyping. *G3: Genes Genomes Genetics* 4, 29–37.

Niu, S.-H., Li, Z.-X., Yuan, H.-W., Chen, X.-Y., Li, Y., and Li, W. (2013) Transcriptome characterisation of *Pinus tabuliformis* and evolution of genes in the *Pinus* phylogeny. *BMC Genomics* 14, 263.

Nystedt, B., Street, N.R., Wetterbom, A., Zuccolo, A., Lin, Y.-C., *et al.* (2013) The Norway spruce genome sequence and conifer genome evolution. *Nature* 497, 579–584.

O'Malley, D., Grattapaglia, D., Chaparro, J., Wilcox, P., Amerson, H., *et al.* (1996) Molecular markers, forest genetics and tree breeding. In: Gustafson, J. and Flavell, R. (eds) *Genomes of Plants and Animals: 21st Stadler Genetics Symposium*. Plenum Press, New York, pp. 89–104.

Parra, G., Bradnam, K., and Korf, I. (2007) CEGMA: a pipeline to accurately annotate core genes in eukaryotic genomes. *Bioinformatics* 23, 1061–1067.

Pavy, N., Paule, C., Parsons, L., Crow, J.A., Morency, M.-J., *et al.* (2005) Generation, annotation, analysis and database integration of 16,500 white spruce EST clusters. *BMC Genomics* 6, 144.

Pavy, N., Pelgas, B., Beauseigle, S., Blais, S., Gagnon, F., *et al.* (2008) Enhancing genetic mapping of complex genomes through the design of highly-multiplexed SNP arrays: application to the large and unsequenced genomes of white spruce and black spruce. *BMC Genomics* 9, 21.

Pavy, N., Pelgas, B., Laroche, J., Rigault, P., Isabel, N., and Bousquet, J. (2012) A spruce gene map infers ancient plant genome reshuffling and subsequent slow evolution in the gymnosperm lineage leading to extant conifers. *BMC Biology* 10, 84.

Pavy, N., Gagnon, F., Rigault, P., Blais, S., Deschênes, A., *et al.* (2013) Development of high-density SNP genotyping arrays for white spruce (*Picea glauca*) and transferability to subtropical and nordic congeners. *Molecular Ecology Resources* 13, 324–336.

Pavy, N., Lamothe, M., Pelgas, B., Gagnon, F., Birol, I., *et al.* (2017) A high-resolution reference genetic map positioning 8.8 K genes for the conifer white spruce: structural genomics implications and correspondence with physical distance. *Plant Journal* 90, 189–203.

Pinosio, S., González-Martínez, S.C., Bagnoli, F., Cattonaro, F., Grivet, D., *et al.* (2014) First insights into the transcriptome and development of new genomic tools of a widespread circum-Mediterranean tree species, *Pinus halepensis* Mill. *Molecular Ecology Resources* 14, 846–856.

Plomion, C., Pionneau, C., Brach, J., Costa, P., and Baillères, H. (2000) Compression wood-responsive proteins in developing xylem of maritime pine (*Pinus pinaster* ait.). *Plant Physiology* 123, 959–969.

Plomion, C., Bartholomé, J., Lesur, I., Boury, C., Rodríguez-Quilón, I., *et al.* (2015) High-density SNP assay development for genetic analysis in maritime pine (*Pinus pinaster*). *Molecular Ecology Resources* 16, 574–587.

Ralph, S.G., Chun, H.J.E., Kolosova, N., Cooper, D., Oddy, C., *et al.* (2008) A conifer genomics resource of 200,000 spruce (*Picea* spp.) ESTs and 6,464 high-quality, sequence-finished full-length cDNAs for Sitka spruce (*Picea sitchensis*). *BMC Genomics* 9, 484.

Remington, D.L., Whetten, R.W., Liu, B.H., and O'Malley, D.M. (1999) Construction of an AFLP genetic map with nearly complete genome coverage in *Pinus taeda*. *Theoretical and Applied Genetics* 98, 1279–1292.

Rudin, D. and Ekberg, I. (1978) Linkage studies in *Pinus sylvestris* using macro-gametophyte allonyms. *Silvae Genetica* 27, 1–12.

SanMiguel, P., Tikhonov, A., Jin, Y.-K., Motchoulskaia, N., Zakharov, D., *et al.* (1996) Nested retrotransposons in the intergenic regions of the maize genome. *Science* 274, 765–768.

Sarri, V., Ceccarelli, M., and Cionini, P.G. (2011) Quantitative evolution of transposable and satellite DNA sequences in *Picea* species. *Genome* 54, 431–435.

Sax, K. and Sax, H.J. (1933) Chromosome number and morphology in the conifers. *Journal of the Arnold Arboretum* 14, 356–389.

Schena, M., Shalon, D., Davis, R.W., and Brown, P.O. (1995) Quantitative monitoring of gene expression patterns with a complementary DNA microarray. *Science* 270, 467–470.

Schnable, P.S., Ware, D., Fulton, R.S., Stein, J.C., Wei, F., *et al.* (2009) The B73 maize genome: complexity, diversity, and dynamics. *Science* 326, 1112–1115.

Seoane-Zonjic, P., Cañas, R.A., Bautista, R., Gómez-Maldonado, J., Arrillaga, I., *et al.* (2016) Establishing gene models from the *Pinus pinaster* genome using gene capture and BAC sequencing. *BMC Genomics* 17, 148.

Sewell, M.M., Sherman, B.K., and Neale, D.B. (1999) A consensus map for loblolly pine (*Pinus taeda* L.). I. Construction and integration of individual linkage maps from two outbred three-generation pedigrees. *Genetics* 151, 321–330.

Simpson, J.T., Wong, K., Jackman, S.D., Schein, J.E., Jones, S.J.M., and Birol, I. (2009) ABySS: a parallel assembler for short read sequence data. *Genome Research* 19, 1117–1123.

Soltis, D.E., Soltis, P.S., Bennett, M.D., and Leitch, I.J. (2003) Evolution of genome size in the angiosperms. *American Journal of Botany* 90, 1596–1603.

Soltis, P.S., Marchant, D.B., van de Peer, Y., and Soltis, D.E. (2015) Polyploidy and genome evolution in plants. *Current Opinion in Genetics & Development* 35, 119–125.

Stasolla, C., Bozhkov, P.V., Chu, T.-M., van Zyl, L., Egertsdotter, U., *et al.* (2004) Variation in transcript abundance during somatic embryogenesis in gymnosperms. *Tree Physiology* 24, 1073–1085.

Tang, H., Bowers, J.E., Wang, X., Ming, R., Alam, M., and Paterson, A.H. (2008) Synteny and collinearity in plant genomes. *Science* 320, 486–488.

Tuskan, G.A., Difazio, S., Jansson, S., Bohlmann, J., Grigoriev, I., *et al.* (2006) The genome of black cottonwood, *Populus trichocarpa* (Torr. & Gray). *Science* 313, 1596–1604.

van Zyl, L., von Arnold, S., Bozhkov, P., Chen, Y., Egertsdotter, U., *et al.* (2002) Heterologous array analysis in Pinaceae: hybridization of *Pinus taeda* cDNA arrays with cDNA from needles and embryogenic cultures of *P. taeda*, *P. sylvestris* or *Picea abies*. *Comparative and Functional Genomics* 3, 306–318.

Vos, P., Hogers, R., Bleeker, M., Reijans, M., van de Lee, T., *et al.* (1995) AFLP: a new technique for DNA fingerprinting. *Nucleic Acids Research* 23, 4407–4414.

Wang, X.-Q. and Ran, J.-H. (2014) Evolution and biogeography of gymnosperms. *Molecular Phylogenetics and Evolution* 75, 24–40.

Warren, R.L., Keeling, C.I., Yuen, M.M., Raymond, A., Taylor, G.A., *et al.* (2015) Improved white spruce (*Picea glauca*) genome assemblies and annotation of large gene families of conifer terpenoid and phenolic defense metabolism. *Plant Journal* 83, 189–212.

Watkinson, J.I., Sioson, A.A., Vasquez-Robinet, C., Shukla, M., Kumar, D., *et al.* (2003) Photosynthetic acclimation is reflected in specific patterns of gene expression in drought-stressed loblolly pine. *Plant Physiology* 133, 1702–1716.

Weber, A.P.M. (2015) Discovering new biology through sequencing of RNA. *Plant Physiology* 169, 1524–1531.

Wegrzyn, J.L., Lin, B.Y., Zieve, J.J., Dougherty, W.M., Martínez-García, P.J., *et al.* (2013) Insights into the loblolly pine genome: characterization of BAC and fosmid sequences. *PLoS One* 8, e72439.

Wegrzyn, J.L., Liechty, J.D., Stevens, K.A., Wu, L.-S., Loopstra, C.A., *et al.* (2014) Unique features of the loblolly pine (*Pinus taeda* L.) megagenome revealed through sequence annotation. *Genetics* 196, 891–909.

Westbrook, J.W., Chhatre, V.E., Wu, L.-S., Chamala, S., Neves, L.G., *et al.* (2015) A consensus genetic map for *Pinus taeda* and *Pinus elliottii* and extent of linkage disequilibrium in two genotype-phenotype discovery populations of *Pinus taeda*. *G3: Genes Genomes Genetics* 5, 1685–1694.

Whetten, R., Sun, Y.-H., Zhang, Y., and Sederoff, R. (2001) Functional genomics and cell wall biosynthesis in loblolly pine. *Plant Molecular Biology* 47, 275–291.

Yeaman, S., Hodgins, K.A., Suren, H., Nurkowski, K.A., Rieseberg, L.H., *et al.* (2014) Conservation and divergence of gene expression plasticity following *c.* 140 million years of evolution in lodgepole pine (*Pinus contorta*) and interior spruce (*Picea glauca×Picea engelmannii*). *New Phytologist* 203, 578–591.

Yin, T.-M., Wang, X.-R., Andersson, B., and Lerceteau-Köhler, E. (2003) Nearly complete genetic maps of *Pinus sylvestris* L. (Scots pine) constructed by AFLP marker analysis in a full-sib family. *Theoretical and Applied Genetics* 106, 1075–1083.

Zhao, Y., Thammannagowda, S., Staton, M., Tang, S., Xia, X., *et al.* (2013) An EST dataset for *Metasequoia glyptostroboides* buds: the first EST resource for molecular genomics studies in *Metasequoia*. *Planta* 237, 755–770.

Zonneveld, B.J.M. (2012) Conifer genome sizes of 172 species, covering 64 of 67 genera, range from 8 to 72 picogram. *Nordic Journal of Botany* 30, 490–502.

3 Genomics of Hardwoods

Introduction

Hardwoods are non-coniferous woody perennial trees that belong to the plant division Angiospermae, also known as the flowering plants. Angiosperms are characterized primarily by a reproductive system where ovules are enclosed in an ovary. Other important features that distinguish angiosperms from conifers and other gymnosperms are broad leaves instead of needle-like leaves and the presence of vessels for water transport. Because vessels are more efficient at water transport than coniferous tracheids (Sperry, 2003), angiosperm wood possess more fibers, resulting in the "hardwood" designation, as opposed to the use of "softwood" for conifers. Finally, while most conifers are evergreen, angiosperm trees that are native to temperate and boreal zones are deciduous. The two largest groups of plants within the Angiospermae are the monocotyledons (i.e. have a single embryonic leaf) or monocots (including grasses, such as rice and wheat), and the eudicotyledons or eudicots (Angiosperm Phylogeny Group, 2016). All hardwoods are eudicots (Fig. 3.1).

Angiosperms first appeared almost 200 million years ago (Bell *et al.*, 2010) but remained limited in their distribution until the mid- to late Cretaceous, when they expanded to become the most diverse plant division. The rapid expansion and diversification of the angiosperms has long puzzled plant biologists and was famously coined by Darwin as an "abominable mystery" (Darwin *et al.*, 1903). Phenomena such as whole-genome duplication (WGD) have long been considered a significant driver of this diversification (Levin, 1983), but specific molecular mechanisms that facilitated the expansion

of the angiosperms remain to be discovered. Some of the evolutionary innovations that appeared in the angiosperms may have made them more able to compete with the dominant division of plants until then, the conifers. In addition to a reproductive structure that can be more efficient at seed dispersal, the appearance of vessels was another significant advantage. Because of the more efficient water transport, higher photosynthetic rates and productivity were possible, fueled by the higher temperatures and abundant availability of water in the late Cretaceous (Hay and Floegel, 2012). Current estimates generated by a comprehensive review of species of vascular plants, organized by the Royal Botanic Gardens, Kew Gardens, and the Missouri Botanical Garden, have identified more than 350,000 species of angiosperms (www.theplantlist.org/, accessed July 15, 2019). Angiosperms dominate the vast majority of ecosystems on Earth, and include all our major food crops. Hardwoods are used to produce a variety of wood-based products including lumber, veneer, pulp, and paper. More recently, their use as a feedstock for renewable energy has also been highlighted, as has their value for mitigating elevated levels of carbon dioxide (CO_2) in the atmosphere.

Other than the morphological and developmental features that separate hardwoods from conifers, they also followed different paths with regard to genome evolution. The genome of hardwoods is typically one to two orders of magnitude smaller than that of conifers. Also different from conifers, hardwoods (and most angiosperms) have undergone several events of WGD, and appear to have utilized this as an approach to create genetic and phenotypic diversity, leading to the ability to adapt to a broader range of

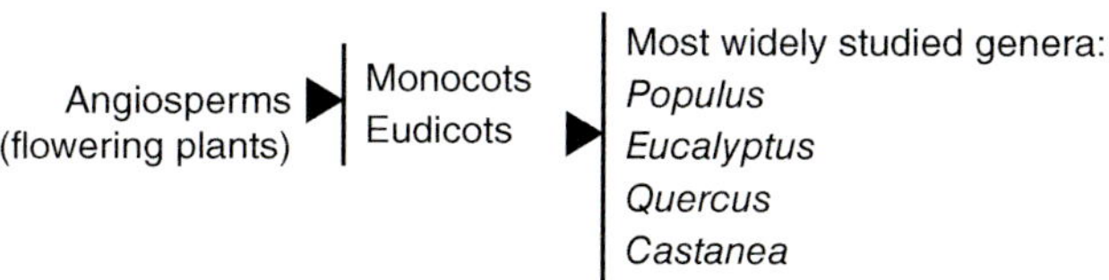

Fig. 3.1. The phylogeny of hardwoods. The flowering plants or angiosperms are separated into two groups, of which the eudicots include all the hardwood species.

environments (van de Peer *et al.*, 2017). A consequence of the occurrence of WGD across different lineages is that angiosperm genome sizes can differ by up to a 1,000-fold (Bennett and Smith, 1976). In addition, transposable elements appear to have been remarkably active in the recent history of the angiosperms. This complex evolutionary history that combines WGD events with extensive reshuffling of nuclear DNA by transposable elements has not only resulted in a wide variation in genome size across the angiosperms but also limited collinearity, except among closely related taxa. This is in sharp contrast to the extensive conservation in synteny and collinearity seen in conifers.

Why Study the Genome of Hardwoods?

Hardwoods include some of the most economically and ecologically important tree species worldwide, including those from the genera *Populus* and *Eucalyptus*. Compared with conifers and other woody gymnosperms, several hardwood species have comparatively small genomes and are relatively easy to propagate and genetically modify, making them useful models to understand the biology of wood formation, perenniality, dormancy, and other unusual properties of trees.

A model for woody species

Until the sequencing of the black cottonwood (*Populus trichocarpa*) genome (Tuskan *et al.*, 2006), tree molecular biologists had to rely on the sequence of *Arabidopsis thaliana* and other annual, herbaceous plants to elucidate gene function. Thus, a "model" tree was critical to complement the genetic resources that had been developed previously for other model plants. The phylogenetic proximity of *Populus* spp. to *A. thaliana* also offered an opportunity for comparative analysis of unique tree properties with an annual herbaceous plant, and led to the discovery that many of the molecular mechanisms are largely shared.

Understanding woody biomass productivity

Several tree species are exploited to produce a variety of wood-based products, including lumber, veneer, pulp, and paper. More recently, their use as a source of renewable energy has also been highlighted, as has their value for mitigating climate change brought on by elevated levels of atmospheric CO_2. However, woody biomass production is impacted by a broad range of environmental and genetic factors. The genomic study of short-rotation hardwood species offers an opportunity to uncover the genetic elements that contribute to biomass productivity and their interaction with the environment.

Understanding wood formation

Wood is the product of secondary, radial growth of the plant stem. In woody plants, this secondary growth is a result of the vascular cambium, which produces secondary xylem (wood), and secondary phloem (bark). Understanding the molecular mechanisms regulating wood formation is critical to be able to effectively modify its properties and tailor them for various end products, such as bioenergy or pulp and paper production. The ease with which some hardwoods can be genetically modified offers the opportunity to validate these molecular mechanisms.

Understanding dormancy and perenniality

Trees have growth habits that distinguish them from most agricultural crops, particularly with respect to the ability to survive under winter conditions by undergoing dormancy and becoming productive again the following growing season. This is in contrast to the annual status of most agricultural crops. Perenniality has often been considered a potentially valuable property for agricultural crops.

Genotyping and Genetic Mapping of Hardwoods

Genomic studies of most species began with the analysis of genetic polymorphisms, or molecular markers. Genotyping hardwood trees focused initially on genetic fingerprinting using allozymes and Restriction fragment length polymorphism (RFLP) (Rajora, 1988; Keim *et al.*, 1989). Soon after, efforts to identify linkage groups were successful in species of *Populus*, using allozymes (Muller-Starck, 1992) and a combination of allozymes and RFLP markers, which resulted in the construction of the first low-density linkage map with 57 loci (Liu and Furnier, 1993). However, it was not until polymerase chain reaction (PCR) was developed, and PCR-based markers evolved (Chapter 1, this volume), that the analysis of genome-wide polymorphisms and creation of genetic maps became possible. Table 3.1 shows the progression in the use of genetic markers in

the genus *Eucalyptus* with respect to the number of loci and regions of the genome targeted for analysis of DNA variation.

RAPD

Random amplified polymorphic DNA (RAPD) markers were first used in forestry for fingerprinting clones from several species in the genus *Populus*, where extensive genetic variation was observed (Castiglione *et al.*, 1993). Soon afterward, the first genetic map of moderate marker density was created for *Eucalyptus* (Grattapaglia and Sederoff, 1994; Grattapaglia *et al.*, 1995), based solely on RAPD markers genotyped in a pseudo-testcross mapping population. In this type of mapping population, a cross between two highly heterozygous individuals of flooded gum (*Eucalyptus grandis*) and Timor white gum (*E. urophylla*) was analyzed, and two separate maps (maternal and paternal) were created (Chapter 1, this volume). Because of the high degree of genetic diversity in the progenitors, large numbers of markers occurred in a single dose in one of the parents, while being null in the other (e.g. parent 1: A/–, parent 2: –/–). Consequently, markers segregating 1:1 in the progeny followed a testcross configuration and could be mapped in the parent that contained a single dose of the RAPD marker (Fig. 3.2). This approach resulted in over 200 markers being mapped in single-tree maps for each of the parent species, in 11 (*E. urophylla*) to 14 (*E. grandis*) linkage groups, and covering 95% of the genome. This moderate density

Table 3.1. Main advances in the genotyping of *Eucalyptus*.

Year	Marker	Marker type[a]	Detection	No. of markers	Reference
1994	RAPD	Anonymous	Gel electrophoresis	~200–300	Grattapaglia and Sederoff (1994)
2003	AFLP	Anonymous	Gel/capillary electrophoresis	803–824	Myburg *et al.* (2003)
2006	Microsatellite/ SSR	Targeted, intergenic	Gel/capillary electrophoresis	234	Brondani *et al.* (2006)
2011	SFP	Genes	Microarray	1,845	Neves *et al.* (2011)
2015	SNP	Genes	DNA chip	~60,000	Silva-Junior *et al.* (2015)

[a]Markers were classified as different types based on the amplification of unknown (anonymous) regions of the genome, targeting of specific loci (in the case of known microsatellite loci) or assaying of genetic variation in genes. SFP, single-feature polymorphism.

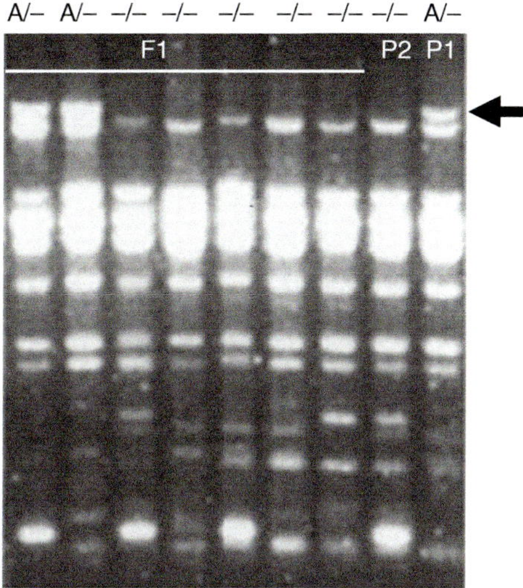

Fig. 3.2. Genetic inheritance and segregation of a pseudo-testcross RAPD marker in *Eucalyptus grandis* (P1) and *E. urophylla* (P2) and their hybrid progeny (F1). The marker (black arrow) is present in one of the parents, but segregates in the progeny, supporting that it is heterozygous (A/–) in parent P1. Individual genotypes of parents and progeny are described in the top of the gel image. (Modified from Grattapaglia *et al.*, 1995.)

genetic map was later explored extensively for mapping of quantitative trait loci (QTLs) (Fig. 3.3) (Grattapaglia *et al.*, 1995, 1996).

The first RAPD-based genetic map in *Eucalyptus* was soon followed by a similar effort in *Populus* (Bradshaw *et al.*, 1994), where additional markers based on RFLPs and sequence-tagged sites (STSs) were also employed. The mapping population used was an F_2, derived by selfing an F_1 generated by an interspecific hybridization between black cottonwood and eastern cottonwood (*P. deltoides*). In this pedigree configuration, heterozygous, dominant RAPD markers (A/–) in the F_1 are expected to segregate in a 3:1 ratio (A/A, A/– , –/A: –/–) in the progeny. Because markers that are heterozygous in the F_1 are derived from polymorphisms between black cottonwood and eastern cottonwood, the genetic map represents interspecific recombination. This is in contrast to the single-tree maps generated using the pseudo-testcross strategy of Grattapaglia and Sederoff (1994). The resulting map had 343 markers, including

215 RFLPs, 111 RAPDs, and 17 STS markers, and was later applied to QTL mapping studies (Bradshaw and Stettler, 1995; Wu *et al.*, 1997). While this effort still included a large set of restriction fragment-based makers (RFLPs and STSs), their use became less common as the speed and practicality of using PCR-based markers became more evident.

Amplified Fragment-length Polymorphisms (AFLPs)

While RAPDs popularized the use of genetic markers in plants, their application was plagued by a lack of repeatability, as well the relatively limited number of polymorphisms detected (Chapter 1, this volume). These AFLP markers were introduced to partially address these limitations (Vos *et al.*, 1995). By detecting variation at restriction sites and flanking regions, AFLPs provided a genotyping platform that could capture more genetic variation, while also being relatively robust.

In hardwoods, AFLP markers were first used in the analysis of a pseudo-testcross of forest red gum (*E. tereticornis*) and Tasmanian bluegum (*E. globulus*), and resulted in the mapping of over 200 loci in the two parental individuals (Marques *et al.*, 1998). The study also included a few markers segregating 3:1, which could be mapped in both parental species and resulted in the first detection of homeologous linkage groups. The use of AFLPs was later extended to populations of other hardwood species, including *Populus* (Wu *et al.*, 2000; Cervera *et al.*, 2001) and *Salix* spp. (Tsarouhas *et al.*, 2002).

A major limitation of genetic markers that rely on fragment separation in a gel matrix is the low throughput when identifying and analyzing the segregation of bands that correspond to amplified fragments. To partially address this limitation, Myburg *et al.* (2001) introduced the use of an AFLP detection protocol based on a PCR amplification step that used two distinct infrared dye-labeled primer combinations. Fragments amplified by distinct primers could then be multiplexed and separated on equipment designed to detect fluorescently labelled fragments (Fig. 3.4). The use of this approach, in combination

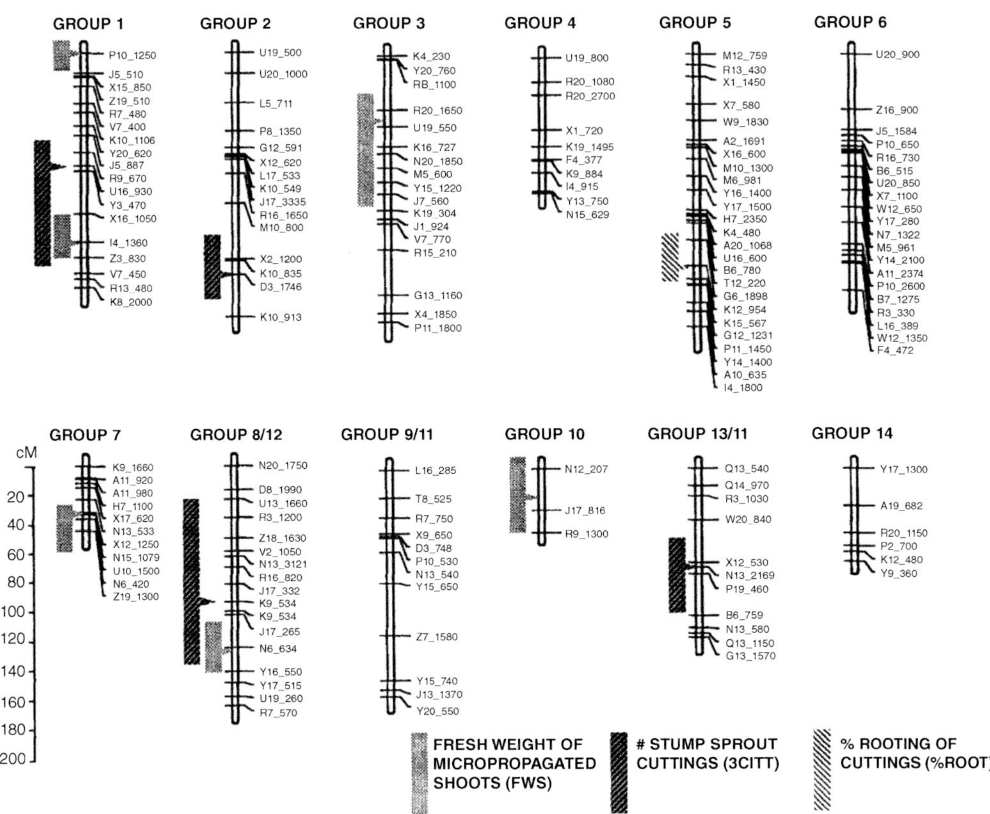

Fig. 3.3. Linkage groups and quantitative trait loci detected in an RAPD-based genetic map of *Eucalyptus grandis*. Genetic markers mapped in the linkage map are identified on the right of each linkage group. Bars on the left show the support intervals for the location of QTLs for traits related to vegetative propagation, and arrows indicate the most likely position. (From Grattapaglia *et al.*, 1995.)

with methods of digital imaging, allowed a significant expansion of the number of loci characterized. For example, it led to the genotyping of over 800 loci in a large population of *Eucalyptus* (Myburg *et al.*, 2003).

Microsatellites

The use of RAPD and AFLP markers revolutionized the genetic analysis of plants because of the ease in applying them to any species and population with no prior information. However, in addition to the relative lack of reproducibility, it is difficult to transfer RAPD and AFLP marker information derived from one species to another, or even between populations within the same species.

Marker transferability is critical for the comparative analysis of genomes, including the investigation of genome synteny among species. Finally, both RAPD and AFLP markers have relatively low information content (Chapter 1, this volume), and most loci are monomorphic in a population. Microsatellites, or simple sequence repeats (SSRs), address several of the limitations of RAPDs and AFLPs. Microsatellites are sequences of 2–6 nucleotides (nt) that occur as tandem repeats. Such repeats are caused by slippage of the DNA polymerase during genomic DNA replication (Chapter 1, this volume). The mutation rate at these loci is significantly higher than that detected in individual nucleotides, and the variable number of repeats results in multi-allelic genetic markers that typically have

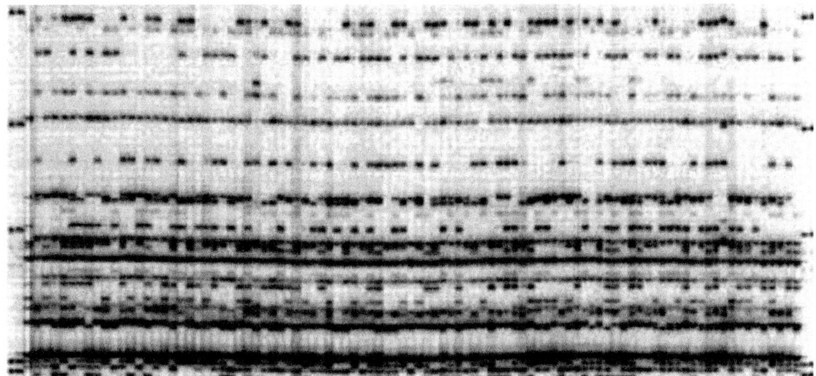

Fig. 3.4. Detection of AFLP fragments in *Eucalyptus* using infrared dye-labeled primers. The two lanes at each end of the gel contain molecular weight standard markers. The first two lanes on the left show the AFLP fragments of the parents of a full-sib family followed by the full-sib progeny of 94 individuals. (From Myburg *et al.*, 2001.)

higher information content than RAPDs and AFLPs. Consequently, despite the fact that SSR loci are normally analyzed individually and are consequently of low throughput, their high level of genetic variation often results in up to four segregating alleles being detected in a bi-parental population, and mapping of the locus in both parental maps. Furthermore, because the analysis of these loci relies on PCR amplification based on primers that anneal to conserved flanking regions, they are often transferable across populations of a given species, or often even among related species. This property results in the possibility of aligning linkage groups derived from different segregating populations (and even of different species) to evaluate chromosome synteny and collinearity, and to compare the position of genome features, such as QTLs (Fig. 3.5).

In hardwoods, the first microsatellite loci were identified in shining gum (*E. nitens*), by screening a genomic DNA library using $(CA)_n$ and $(GA)_n$ oligonucleotide probes (Byrne *et al.*, 1996). While only four loci were described, they showed extremely high levels of polymorphisms (9.5 alleles observed on average among 20 individuals) and extensive conservation across related *Eucalyptus* spp. The high level of variation at microsatellites led to the development of additional loci for *Eucalyptus* spp., including *E. grandis*, *E. urophylla*, *E. globulus*, and *E. sieberi* (Glaubitz *et al.*, 2001; Steane

et al., 2001; Brondani *et al.*, 2002) and the construction of a microsatellite-only consensus map with 234 microsatellites covering over 90% of the *Eucalyptus* genome (Brondani *et al.*, 2006). Microsatellite transferability was also assessed across related species and even related genera. For instance, transferability among *Eucalyptus* spp. of the subgenus *Symphyomyrtus* can be higher than 80%, 50–60% for species of different subgenera, and even 25% in the related genus *Corymbia* (Grattapaglia and Kirst, 2008). Such a high level of conservation across species allowed the development of comparative maps for verification of genome homology in multiple species through marker collinearity (see below).

The value of microsatellite markers for mapping and population genetics diversity studies, as well as the ease in identifying these markers from increasingly abundant gene and genome sequences, has rapidly expanded their availability in other hardwoods. Since the early development of microsatellites in *Eucalyptus*, large numbers of this type of marker have been identified and established for genotyping species of commercial value and ecological importance, such as those from the genera *Swietenia* (Lemes *et al.*, 2002); *Acacia* (Omondi *et al.*, 2010); and *Cedrela* (Soldati *et al.*, 2013). However, the development of microsatellites based on the enrichment of genomic DNA libraries for specific repeat sequences was still a major hurdle for

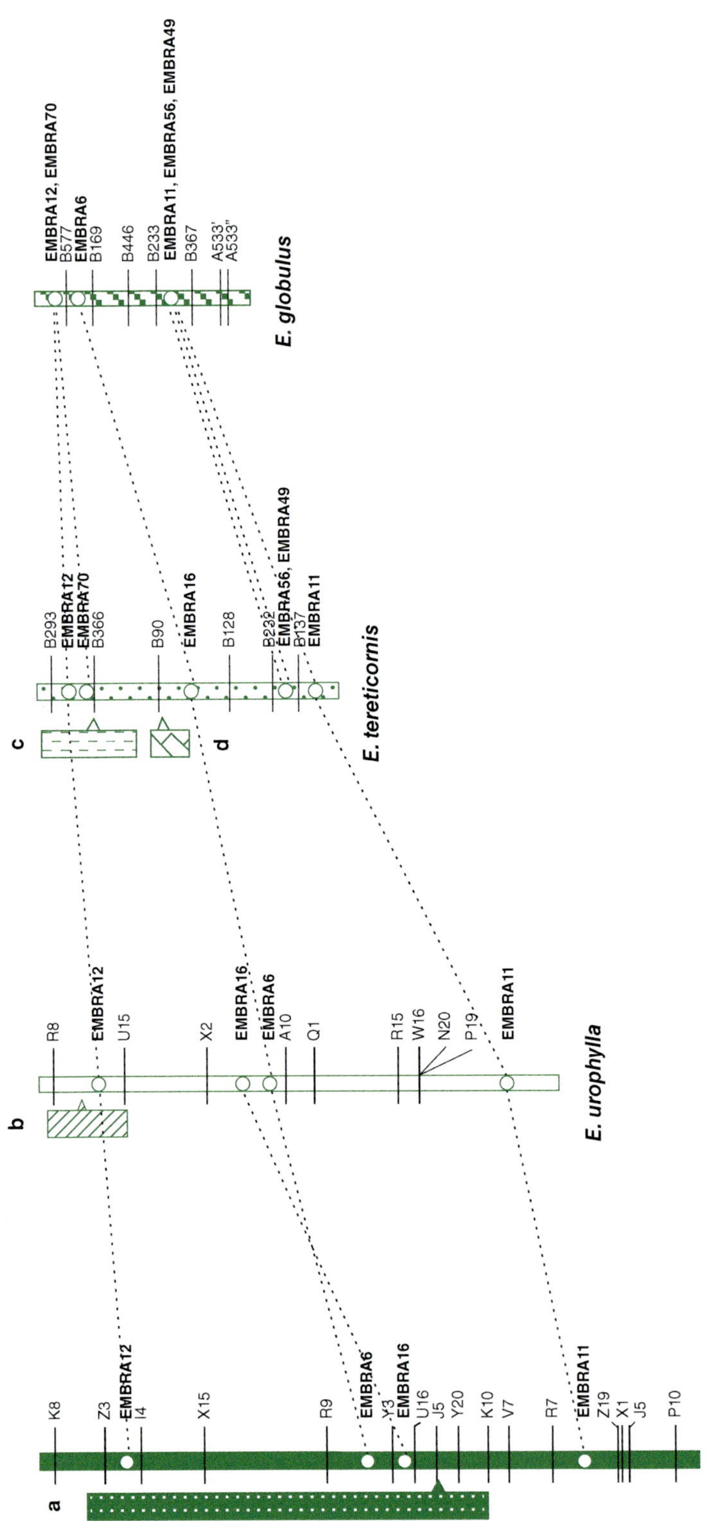

Fig. 3.5. Homeologous linkage groups in four *Eucalyptus* spp. genetic maps, identified based on shared microsatellites (EMBRA (*Eucalyptus* microsatellites from Brazil) markers in bold, linked by dashed lines). The position of QTLs for vegetative traits (bars on left side of linkage group) can be compared based on the conservation of microsatellite markers across the species. (From Marques *et al.*, 2002.)

the wider discovery and adoption of these markers. This obstacle was resolved in part by the development of next-generation sequencing (NGS) methods, which allowed the identification of new microsatellite loci from randomly sequenced genomic regions, or from transcriptome sequences. Computational tools for identification of these loci from these sources were also developed (Faircloth, 2008), resulting in microsatellites being developed for several additional hardwood species.

Single-feature polymorphisms (SFPs)

These genetic markers are derived from the analysis of signals detected in microarrays, which have typically been used for gene-expression studies. Hybridization of complementary DNA (cDNA) derived from mRNA or from genomic DNA to a microarray may fail if the sequence of the oligonucleotide is not perfectly complementary to the sample being hybridized. The resulting presence/absence of signal can be treated as a dominant marker. This genotyping approach was used extensively in *A. thaliana* (Borevitz *et al.*, 2003, 2007) and to a limited extent in *Populus* (Drost *et al.*, 2009) and *Eucalyptus* (Neves *et al.*, 2011). In both cases where the method was applied to tree species, it resulted in the detection and mapping of several hundred to a few thousand genetic markers.

SNPs

With the development of genomic applications that require genome-wide analysis of genetic variation (e.g. genome-wide association studies and genome-wide selection), the limitations of genetic markers such as RAPDs, AFLPs, and microsatellites soon became evident. The low throughput of the genotyping assays required to analyze these loci, and their relatively low abundance in the genome, resulted in the need for development of a different type of genetic marker that could easily be assayed. The most abundant form of sequence variation in the genome of most organisms are SNPs. While bi-allelic

SNPs are not as informative genetically as multi-allelic microsatellites (Chapter 1, this volume), their frequency in the genome and the ease of genotyping has led to their selection as the preferred type of polymorphism for genomic studies. SNPs can be assayed on a genome-wide scale using essentially two distinct strategies (Chapter 1, this volume). The first involves the direct sequencing of genomic DNA, using NGS platforms. In this strategy, sequencing can target the whole genome, or a fraction of it in the case where complexity reduction methods are used. Alternatively, SNPs can be genotyped by platforms that assay variation at a specific nucleotide, usually using DNA chips. A detailed description of both strategies for genotyping with SNPs, including their variations, and advantages or disadvantages, is provided in Chapter 1 (this volume).

SNP genotyping by sequencing

The dramatic reduction in the cost of DNA sequencing and extremely high throughput of existing sequencing platforms has created the opportunity to analyze the entire set of polymorphisms that occur in one or more individuals by sequencing their genomes. The first demonstration of this approach in forestry was in the analysis of a population of black cottonwood (Slavov *et al.*, 2012; Evans *et al.*, 2014), which was genotyped based on resequencing the entire genome of 544 individuals. The study identified and genotyped almost 18 million SNP loci, as well as over a quarter of a million insertions and deletions. Despite the value of this approach, the complexity of analyzing and uncovering information that is biologically relevant (such as DNA polymorphisms) from whole-genome sequences in many individuals is still daunting. Nonetheless, as the computation methods for SNP detection and genotyping improve, and computational power for such analysis becomes widely available, it is likely that future genotyping projects in most species will migrate toward a whole-genome sequencing approach (Chapter 1, this volume).

Until whole-genome sequencing becomes a mainstream approach to genotyping, detection of DNA variants in hardwood species using NGS remains based primarily on the analysis of a limited fraction of the genome, using genome-complexity-reduction strategies. Genotyping SNPs based on NGS of a reduced fraction of the genome has relied on two approaches: (i) the use of restriction enzymes to cut and clone genomic fragments that represent only part of the genome and (ii) the use of oligonucleotide probes that capture regions of interest by hybridization. Detailed descriptions of these methods, and their respective advantages and disadvantages, are presented in Chapter 1 (this volume). Different variations of the restriction enzyme-based methods of complexity reduction have been developed (Davey *et al.*, 2011) and applied to genotyping of hardwood species in the genera *Eucalyptus* (Petroli *et al.*, 2012) and *Populus* (Schilling *et al.*, 2014).

The use of methods that capture specific regions of the genome has also been applied to the analysis of a population of black cottonwood (Zhou *et al.*, 2014; Holliday *et al.*, 2016) and eastern cottonwood (Fahrenkrog *et al.*, 2016). Both studies focused on the detection and genotyping of polymorphisms in a majority of coding sequences and a set of intergenic regions, and characterized large natural populations for detecting signatures of natural selection and adaptation (Chapter 4, this volume) and bioenergy (Chapter 5, this volume).

SNP genotyping by chip

The development of methods to synthesize large numbers of oligonucleotides on a two-dimensional surface allowed the creation of novel approaches to massive parallel genotyping (Chapter 1, this volume). In tree species, these novel genotyping platforms (commonly referred to as "genotyping DNA chips") were first used in a conifer (Pavy *et al.*, 2008). The successful use in such a complex genome led to the realization that the development of similar tools for hardwood species would be feasible.

The first generation of genotyping DNA chips developed for hardwoods was designed for species in the genus *Eucalyptus* (Grattapaglia *et al.*, 2011). While it was used to assay only 768 SNP loci, the study demonstrated that this genotyping method was applicable to highly genetically diverse species. *Eucalyptus* spp. typically carry extremely high levels of genetic diversity, with nucleotide diversity rates exceeding 1%. Thus, there was significant concern that polymorphisms in the flanking region of the target loci could interfere with the assay performance. In order to minimize the negative effect of flanking DNA variants, stringent filtering procedures were adopted to eliminate loci that could be unreliable for genotyping. This included removing all SNPs flanked by any other variants located 60 nt upstream and downstream of the target region, and resulted in an initial pool of 66,254 loci being reduced to only 1,329 possible targets. While this improves the likelihood of success of the genotyping assay design, stringent filtering also severely reduced the diversity and range of loci that could be analyzed. For many applications, severely constraining the number of genotyped loci may still result in an adequate number of polymorphisms for analysis. However, if genotyping all or most variants within a genomic region is necessary, stringent filtering may be unsuitable. The final design of the first *Eucalyptus* DNA chip contained 768 loci, of which the vast majority (>90%) could be successfully genotyped. Thus, the analysis supported the expectation that, following stringent filtering, this genotyping platform was suitable for analysis of highly genetically diverse species.

Despite this early success, analysis of several individuals of *E. grandis* (the species that had been used for the DNA chip design), as well as individuals from related species, only approximately half or fewer loci were observed to be polymorphic. The limited variation detected in the loci represented in this first-generation *Eucalyptus* DNA chip demonstrated the need to define a list of variable loci with high confidence. To address this need, a second-generation genotyping chip for *Eucalyptus* was preceded by resequencing the genomes of 240 trees from 12 different species in the genus (Silva-Junior *et al.*, 2015). This preliminary study resulted in the detection of almost 50 million putative SNPs, of which approximately 60,000 were selected based on significant evidence of the presence of DNA variants and high information content within species. Furthermore, several SNPs were shared across species and are likely to have originated from mutations that arose prior to the divergence of different lineages. Consequently, this genotyping resource is not only useful for analysis of individual species but also for comparative genome analysis among different species of *Eucalyptus*. In total, the second-generation *Eucalyptus* chip (referred to as EUChip60k) contained 59,222 loci that could be effectively genotyped, of which 86% were shown to be variable in at least one of 14 different *Eucalyptus* spp tested. The selection of widely distributed loci also resulted in a consistent coverage of almost the entire genome, with one variable locus located every 10–20 kbp.

A similar genotyping resource was also developed for another model hardwood species, black cottonwood (Geraldes *et al.*, 2013). In this case, the identification of suitable SNPs to be included in the genotyping chip relied on the selection of 34,131 loci detected in a set of 34 black cottonwood individuals that had been fully resequenced. Instead of targeting widely distributed loci, as was done with *Eucalyptus*, this chip was designed primarily to assay variants in a set of 3,543 candidate genes previously identified as potentially involved in the control of complex traits. As with the *Eucalyptus* chip, the design was highly successful, with over 95% of the loci producing reliable genotyping calls, and a large number of these loci being polymorphic in the analysis of even relatively small populations. The use of this genotyping platform, developed based on the genome sequence of black cottonwood, showed reasonably high transferability across the genus. For example, over 70% of the SNP loci could be genotyped in other *Populus* spp.

Comparative Genetic Maps of Hardwoods

Comparative mapping refers to the analysis of the organization and position of loci in linkage maps across different segregating populations. Comparative maps have many uses, including the assessment of the conservation of QTLs across populations and uncovering the cause of hybrid inviability associated with genome differences between species (Myburg *et al.*, 2003). Comparing maps from different species also enables the identification of syntenic linkage groups and evaluation of the extent of collinearity of loci between genomes. In conifers, comparative mapping uncovered extensive synteny and collinearity among the genomes of different species (Chapter 2, this volume; Devey *et al.*, 1999; Brown *et al.*, 2001), leading to the conclusion that their genomes evolved slowly and have undergone limited shuffling since divergence. In contrast to conifers, the genome of hardwoods (and angiosperms in general) underwent a complex evolution, with numerous WGD events and extensive reshuffling by mobile elements. These events have often been considered a major driver of the diversification of angiosperms and their adaptation to most terrestrial habitats (Soltis *et al.*, 2008). However, despite a much more complex genome evolution, relatively high levels of synteny and collinearity have been observed between species of the most intensively studied genera, including *Eucalyptus* and *Populus*.

For two maps to be compared, they need to share a common set of orthologous loci. Thus, genetic markers that are broadly transferable across species, such as microsatellites, are particularly valuable. Microsatellites were used to construct linkage maps for eastern cottonwood, European black poplar (*P. nigra*), and black cottonwood to identify homoeologous groups among these species (Cervera *et al.*, 2001). In *Eucalyptus*, the availability of a large set of microsatellite markers identified syntenic linkage groups between maps from *E. grandis* and *E. urophylla* (Brondani *et al.*, 2002). Genotyping of 46 microsatellite markers in the two populations allowed the alignment of 10 linkage groups and demonstrated that almost all loci were in a similar order (collinear) between the two parental maps (Fig. 3.6). The use of microsatellites that could be transferred across species of the genus *Eucalyptus* was then applied to compare the positions of QTLs for traits related to vegetative propagation in *E. grandis*, *E. urophylla*, *E. tereticornis*, and *E. globulus* (Marques *et al.*, 2002).

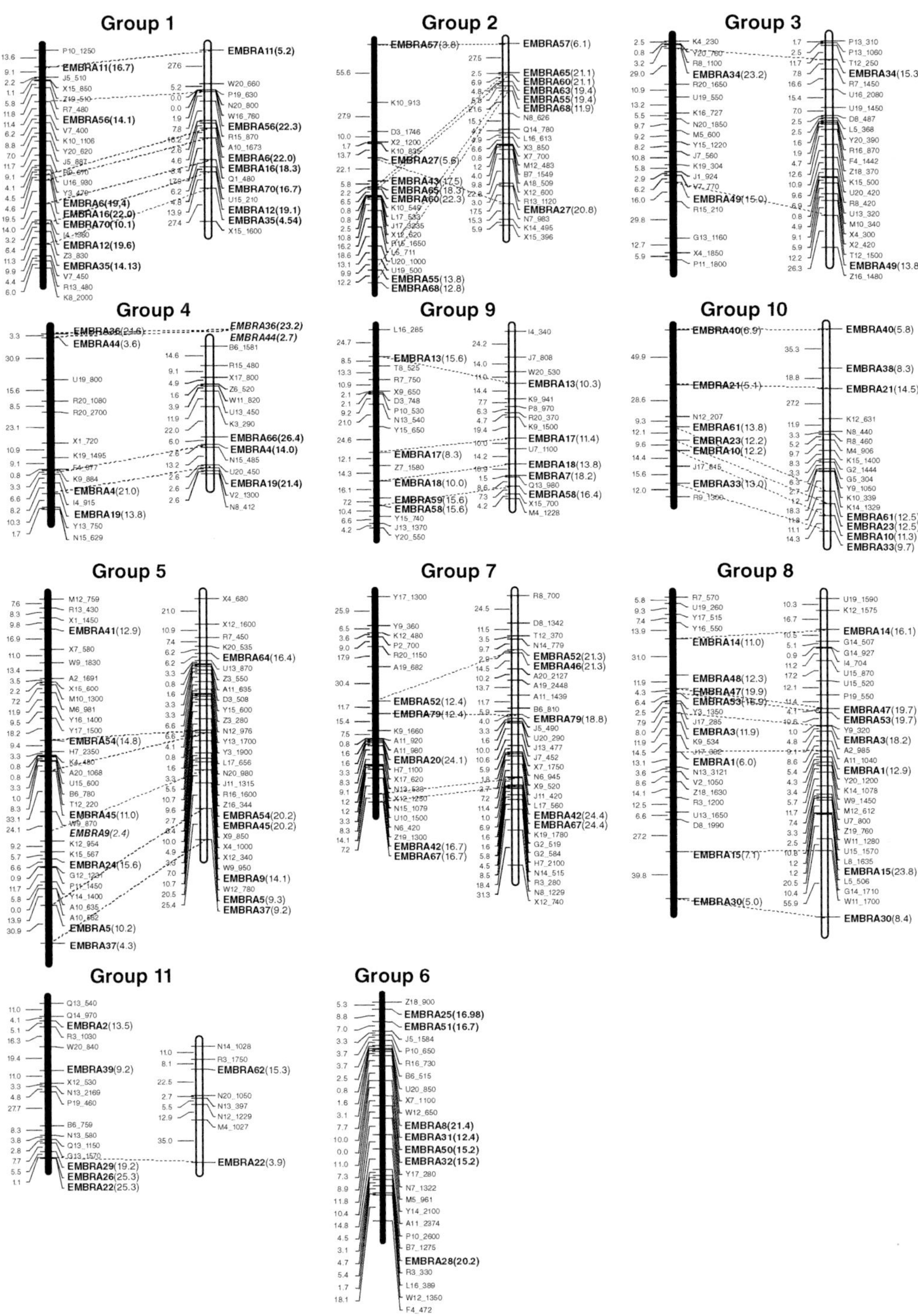

Fig. 3.6. Genetic mapping of conserved microsatellite loci across *Eucalyptus grandis* and *E. urophylla*. Shared microsatellites allowed the detection of homeologous linkage groups across species. (From Brondani *et al.*, 2002.)

Following these initial reports, more detailed maps have been created expanding the range of species compared. In the genus *Populus*, parental maps were developed for Chinese aspen (*P. adenopoda*) and silver poplar (*P. alba*) (*Populus* section), based on microsatellites identified on the black cottonwood (*Tacamahaca* section) genome (Wang et al., 2011). The same loci were also used in the construction of a map of eastern cottonwood and Canadian poplar (*P. euramericana*) (*Aigeiros* section). Thus, the study allowed a detailed comparison of the level of synteny and collinearity across different species belonging to the same and distinct sections of the genus. The alignment showed a high degree of marker synteny and collinearity and a closer relationship between the *Aigeiros* and *Tacamahaca* sections than between the *Populus* and *Tacamahaca* sections. Moreover, there was evidence for the chromosomal duplication and interchromosomal reorganization involving some poplar linkage groups, suggesting a complicated course of fission or fusion in one of the lineages.

The release of the black cottonwood genome (Tuskan et al., 2006) also facilitated the detection and mapping of orthologous loci in genomes of related species, even from different genera. For instance, genetic maps were developed for species of the genus *Salix* using orthologous loci developed from the genome of black cottonwood (Berlin et al., 2010). Both genera diverged approximately 45 million years ago. Two linkage maps were constructed based on these loci and showed a high degree of synteny and gene order conservation between *Salix* and *Populus*. Exceptions were the observation of two extensive interchromosomal rearrangements involving two *Populus* chromosomes and one chromosome of *Salix*, suggesting chromosome fission or fusion events, and a few potential chromosome inversions.

ESTs

These are generated by sequencing cDNA, synthesized from mRNA (Chapter 1, this volume). An EST collection obtained from a biological sample can be viewed as a representation of the diversity of genes expressed in that sample. Before sequencing entire tree genomes became feasible and affordable, genome studies in hardwoods and conifers focused primarily on generating EST sequences. In trees, large EST collections were first constructed for Eurasian aspen (*P. tremula*) and black cottonwood, where 5,692 sequences derived from mRNA collected from stem tissue (developing xylem, cambium, and phloem) were characterized (Sterky et al., 1998). This study provided for the first time a broad overview of the genes that are implicated in wood formation, and detected 3,719 unique transcripts, of which 4% were previously known to be involved in cell-wall formation. Interestingly, a relatively large proportion of transcripts had no similarity to genes characterized previously in herbaceous annual plants, which initially suggested that they could have a specific role in wood formation. Later studies indicated, however, that the vast majority of genes appear to be shared between woody and non-woody plants (Kirst et al., 2003). In the years that followed the development of the first EST resource for species of the genus *Populus*, other studies rapidly expanded this resource, including the generation of over 100,000 ESTs from 18 tissues of *Populus*, which created for the first time a transcriptome atlas of genes expressed across a broad range of woody plant tissues (Sterky et al., 2004). This and other EST resources are particularly valuable in identifying genes that participate in the development of specific cell types or organs. Efforts were also dedicated to the sequencing of full-length ESTs (Ralph et al., 2008). Full-length ESTs can be very useful for the annotation of genomes because they not only identify the occurrence of coding sequences but also define the position of untranslated regions and exon–intron boundaries.

While EST resources developed very rapidly for poplars, sequencing of ESTs in other hardwood species advanced at a much slower pace. Even for well-studied species of the genus *Eucalyptus*, there were only a few thousand sequences available in public databases, derived from small-scale EST sequencing projects (Kirst *et al.*, 2004; Paux *et al.*, 2004; Foucart *et al.*, 2006), which were later expanded to almost 10,000 sequences (Rengel *et al.*, 2009). These pioneering EST sequencing studies of *Populus*, *Eucalyptus*, and other hardwoods created a preliminary overview of the genes expressed in a number of tissues that are unique to woody plants and essential for wood formation. However, it was the development of NGS methods that dramatically accelerated the pace of EST sequencing and the creation of detailed and extensive catalogues of genes expressed in tree species. The dramatic progress in EST sequencing that resulted from NGS approaches can be illustrated by the first sequencing of ESTs in *Eucalyptus* using advanced technologies (Novaes *et al.*, 2008). While only a few thousand EST sequences were available for *Eucalyptus* prior to this report, the size of this database increased 20-fold with a single study that generated over 1 million sequences, resulting in an assembly with 71,384 contigs. The large number of contigs and the relative small size (247 bp) of this initial assembly implied that the sequences of transcripts were largely incomplete. However, as this technology matured and its use expanded, larger EST sequencing projects that increased the number and size of sequencing reads led to more complete transcript resources for a large number of hardwood species.

Since the first sequencing of ESTs from a tree species were developed using NGS technology, this type of resource has expanded tremendously to other plant species, including a large number of hardwoods of ecological and economic value. In addition to these EST sequencing efforts targeted to individual species, facile and low-cost generation of comprehensive EST collections using NGS has led to the development of studies that sought to survey the ensemble of genes that occur in broad taxa. For instance, the onekp project (www.onekp.org, accessed July 15, 2019; Matasci *et al.*, 2014) was created from an international consortium that initially aimed to carry out gene sequencing for 1,000 plant species. The database now includes ESTs from over 1,000 species, including 830 angiosperms (as of June 2016), including many hardwoods species such as kenaf (*Hibiscus cannabinus*), white cedar (*Melia azedarach*), yellow trumpet tree (*Tabebuia umbellata*), Peruvian nutmeg (*Laurelia sempervirens*), camphor tree (*Cinnamomum camphora*), and sassafras (*Sassafras albidum*), to name a few.

Transcriptome Analysis of Hardwoods

The analysis of transcript abundance from large numbers of genes advanced rapidly with the development of cDNA and oligonucleotide microarrays (Chapter 1, this volume), and became widely adopted in forestry, and for hardwoods in particular. The early availability of the reference genome of black cottonwood led to the development of the first microarrays that allowed the analysis of expression of all known genes in the species. For example, a poplar GeneChip (Affymetrix) containing short (25-mer) probe sets that represented 56,055 transcripts was developed based on the genes annotated in the first draft of the black cottonwood genome, as well as publically available ESTs from related *Populus* spp. (Wilkins *et al.*, 2009a). A related microarray analysis platform was developed simultaneously, based on the analysis of longer (60-mer) probes (Drost *et al.*, 2009). Development of this platform began with the screening of six to seven probes representing each of 45,555 predicted gene models reported in the black cottonwood genome, as well as 10,238 ESTs and less-supported gene models for which there was transcriptional evidence (Tuskan *et al.*, 2006). The screening of large numbers of probes was carried out to eliminate those that might provide biased estimates of expression due to the occurrence of polymorphisms between the genome of a sample and the reference used for probe design (Kirst *et al.*, 2006). After this screening step, the microarray contained individual probes optimized for the expression analysis of 46,001 genes. While these microarrays were utilized extensively, the development of high-throughput and low-cost NGS methods has led to the steady migration from microarrays to RNA-seq for transcriptomic analyses (Chapter 1, this volume). It is expected that the shift from microarrays to RNA-seq will continue and that most future studies will utilize primarily a sequencing approach.

As transcriptome data from increasingly comprehensive and diverse studies were unraveled, databases were created to represent transcriptomic information in an easily searchable format. Valuable tools have surfaced for mining these resources. For instance, the *Populus* Genome Integrative Explorer (PopGenIE, http://popgenie.org, accessed July 15, 2019), which integrates tools for analysis of the *Populus* genome and transcriptome, is now widely used for visualization of gene expression across a range of organs and tissues (Sjödin *et al.*, 2009). This resource is particularly valuable to those who lack expertise in raw transcriptome data analysis and wish to avoid the need to carry out such analysis to interpret their data. Alternatively, large transcriptome datasets produced for hardwood species by microarrays and RNA-seq are currently available in the public databases maintained by government organizations such as the National Institutes of Health (NIH). For example, the Gene Expression Omnibus (GEO) from the NIH (www.ncbi.nlm.nih.gov/geo/, accessed July 15, 2019) is a data repository that supports microarray and sequence-based transcriptome data. While the GEO does not offer the visualization and search tools that PopGenIE provides, it allows users to download data from well-curated experiments and gene-expression profiles for further analysis. Similar databases have been developed and are maintained by other governmental institutions, such as Array-Express (www.ebi.ac.uk/arrayexpress/, accessed July 15, 2019) at the European Bioinformatics Institute (Brazma *et al.*, 2003).

Since the first microarray for a hardwood species was reported, transcriptome analyses have been conducted to address numerous biological questions, ranging from uncovering the genes involved in wood formation to insect herbivory. The main areas of research and discoveries are described below.

Transcriptome of wood formation

The development of a cell from the cambial meristem into wood follows a series of stages that initiate with its division, expansion, secondary wall formation, lignification, and programmed cell death. Wood chemical and physical properties are determined by genes expressed during these stages, and their discovery is leading toward an understanding of this molecular pathway and its manipulation (Demura and Fukuda, 2007). Much of the early efforts to apply transcriptome analysis tools to hardwoods were directed at uncovering the genes that are involved in wood formation and differences in its properties. A pioneering study characterized transcript abundance in tangential sections collected from the different zones of differentiating xylem of *Populus* (Hertzberg *et al.*, 2001). While this study relied on a microarray designed from ESTs and contained only 2,995 putative genes, it showed that genes known to encode enzymes in the lignin and cellulose biosynthetic pathways, as well as regulators of xylogenesis, are expressed in defined cell types under strict transcriptional programming. This preliminary work was later expanded to include the majority of genes expressed during the stages of xylogenesis.

This early survey was followed by other cDNA microarray-based studies in *Populus* and *Eucalyptus* that characterized gene expression in developing xylem of plants submitted to gravitational stimulus (Paux *et al.*, 2004; Andersson-Gunnerås *et al.*, 2006). These plants form tension wood, which is distinct from normal wood in that fibers contain a cell-wall layer composed mainly of crystalline cellulose, and deficient in lignin and hemicelluloses. These studies reported increases in sucrose synthase transcripts, in agreement with a higher flux of carbon to cellulose, and differential expression of cellulose synthases. Also reported was a decrease in transcription of genes involved in pathways that lead to synthesis of hemicelluloses, lignin, and the cell-wall matrix. Other studies followed that evaluated the contribution of plant hormones and their interaction with the transcriptome during wood formation (Björklund *et al.*, 2007;

Nilsson *et al.*, 2008), as well as a comparison of primary and secondary growth (Dharmawardhana *et al.*, 2010).

Interest in understanding wood formation in hardwoods at the level of the transcriptome led to the early development of microarray tools for gene-expression analysis in poplars and eucalypts. However, the rapid development of NGS methods and their use for RNA-seq allowed new types of studies that previously were not feasible using microarrays. For instance, RNA-seq was used to uncover the entire diversity of the transcriptome and alternative splicing in the developing xylem of *Populus* (Bao *et al.*, 2013). RNA-seq was also used to compare expressed genes across distinct populations (Thavamanikumar *et al.*, 2014) and species (Salazar *et al.*, 2013), in order to identify characteristics that make them unique. Later, the approach was even applied to uncover the aspects that distinguish wood formation in species of different genera, when a differentiating xylem transcriptome was contrasted between species of *Populus* and *Eucalyptus* (Hefer *et al.*, 2015). These studies would not have been possible with microarrays, which rely on the use of specific oligonucleotides and their hybridization to complementary RNA sequences. Instead, by direct sequencing of the mRNA and the use of abundance of reads to quantify transcripts and isoforms, a much more comprehensive inventory of the diverse transcripts involved became possible.

Transcriptome response to abiotic stress

In parallel with the transcriptome analysis of wood formation, similar efforts were undertaken to understand the molecular response of trees to sources of abiotic stress. Early transcriptomic studies focused primarily on identifying genes differentially regulated in response to elevated levels of atmospheric CO_2 (Taylor *et al.*, 2005). This was an attempt to understand the effects of long-term exposure of trees to greenhouse gases, and the consequences for forest survival and biomass productivity. Surprisingly, relatively few genes showed differential regulation as a result of long-term (6 years) exposure, suggesting a plastic response to this source of stress. However, an increase in CO_2 levels in the atmosphere has been associated with a delay in autumnal senescence. Thus, a subsequent transcriptome study sought to characterize the *Populus* transcriptome during leaf senescence, with altered levels of CO_2 (Tallis *et al.*, 2010). The study showed an increase in autumnal leaf sugar accumulation, as well as the upregulation of genes involved in anthocyanin biosynthesis. This increase was associated with an increase in leaf longevity and a consequent delay in autumnal senescence.

Drought is another significant source of abiotic stress in several hardwoods. There has been a keen interest in developing drought-tolerant germplasm, particularly for species within the genus *Populus*, which are often riparian species. Several studies compared the transcriptomic changes in trees with distinct drought-tolerance profiles. One of the first of such studies analyzed a segregating population of *P. trichocarpa* × *P. deltoides* to identify individuals with obvious responses to drought (Street *et al.*, 2006). In this study, individuals that showed extreme sensitivity and insensitivity to drought were grouped and analyzed by microarrays. An unusual experimental design was used, in that cDNA obtained from individuals from each group was pooled for microarray analysis. The approach, which attempted to find transcriptional responses that are common among pooled individuals, resulted in the detection of a few hundred genes that responded differently to drought stress between the two groups. Considering the relevance of drought tolerance to plant productivity, several other recent studies have explored the relationship between gene expression and drought response, primarily in *Populus* (Watkinson *et al.*, 2003; Wilkins *et al.*, 2009b; Cohen *et al.*, 2010; Hamanishi *et al.*, 2010, 2015; Raj *et al.*, 2011; Plavcová *et al.*, 2013; Viger *et al.*, 2016), but also in *Eucalyptus* (Villar *et al.*, 2011).

In addition to elevated CO_2 and drought, salt stress is viewed as a potentially significant threat to forest plantations, considering the rise in sea level and continuous depletion of groundwater. Thus, transcriptome analysis of plants exposed to high salinity has been conducted to understand how tolerance to this stress can be introgressed into tree species to expand their range into areas that currently are not considered suitable for agriculture, or to tolerate transient increases in salinity. Certain hardwood species, such as desert poplar (*P. euphratica*), are well adapted to areas of high salinity, and exhibit high salt tolerance (Chen and Polle, 2010). Thus, desert poplar has been used as a model to better understand the genetic mechanisms of salt tolerance in tree species, leading to the sequencing of its genome (Ma *et al.*, 2013). While several studies have compared expression patterns of desert poplar with that of other *Populus* spp. exposed to high salinity, results from these studies have often been contradictory. A comparison of transcriptome and metabolite abundance in desert poplar with other species, under salt stress, indicated that the former does not seem to overexpress genes involved in stress pathways. Instead, desert poplar seems to have permanently active control mechanisms for osmotic adjustment, ion compartmentalization, and detoxification of reactive oxygen species (Janz *et al.*, 2010). In sharp contrast to that initial observation, another transcriptome analysis of desert poplar subjected to salt stress identified thousands of genes differentially expressed in salt-stressed calli, leaves, and roots of seedlings (Ma *et al.*, 2013). The conflicting results often found in transcriptome studies could be caused by subtle differences in experimental conditions and analytical methods, which may greatly impact transcriptome profiles and the inferences made.

Finally, transcriptome studies have also been used to uncover individual or species-specific responses to drought and toxic levels of soil-derived elements, such as aluminum (Grisel *et al.*, 2010) and cadmium (He *et al.*, 2013; Yang *et al.*, 2015).

Transcriptome responses to biotic stress and plant biotic interactions

In contrast to agricultural crops, the time between planting and harvesting trees can span decades. Consequently, tree species are exposed to regular and prolonged attacks by insects and diseases. While chemical control is feasible in the early years, the long growth cycle and large size of forest plantations has resulted in a significant interest in identifying genes that may confer resistance to the most important pests. Another area of interest is the understanding of which genes are involved in plant biotic interactions, some of which may be advantageous for plant development. As in other areas where transcriptome studies have been done, most of the effort has been focused on species within the genus *Populus* and their main pathogens. Among these, the poplar leaf rust caused by *Melampsora* spp. is one of the primary threats. Transcriptome changes in poplar leaves during infection by *Melamspora* have been characterized using cDNA and oligonucleotide microarrays (Miranda *et al.*, 2007; Rinaldi *et al.*, 2007; Azaiez *et al.*, 2009), and, more recently, with RNA-seq (Petre *et al.*, 2012). Transcriptome analysis of the pathogen has also been carried out to uncover genes that are involved in the infection process (Duplessis *et al.*, 2011). Overall, these studies have revealed several thousand genes that are differentially regulated during pathogen infection and the induction of well-known pathogen defense-related genes. To a lesser extent, studies have also evaluated the transcriptional response of hardwoods to insect herbivory (Philippe *et al.*, 2010).

Genomes of Hardwoods

The organization of the genome of hardwoods and other angiosperms reflects a complex evolution, with numerous WGD events and intensive reshuffling by mobile elements. At least one triplication event (γ) is likely to have been shared by all core eudicots (Bowers *et al.*, 2003). Additional and more ancient WGD events appear to have occurred even earlier in angiosperm evolution (Soltis *et al.*, 2008). A consequence of the occurrence of WGD across different lineages is that angiosperm genome sizes may differ more than 1,000-fold (Bennett and Smith, 1976). In addition, transposable elements have been very active in the recent history of the angiosperms. This complex pattern of genome evolution has not only resulted in a wide variation in genome size across the angiosperms, but also a very limited occurrence of collinearity, except among closely related taxa.

Evidence of a complex evolutionary history of hardwoods began to emerge with the sequencing of the black cottonwood genome, which contains large syntenic blocks (Tuskan *et al.*, 2006). This seminal work demonstrated the challenges of assembling such genomes, but was largely successful because it primarily utilized sequencing reads derived from Sanger technology (Chapter 1, this volume). While Sanger sequencing has lower throughput and is more costly per nucleotide than NGS methods, it generates longer sequences and these are typically of higher quality. The same technology was largely used later, to sequence the *E. grandis* genome, another hardwood species with high economic value worldwide. With the dramatic reduction in costs and increase in throughput of NGS, characterization of most genomes sequenced after black cottonwood and flooded gum has utilized more advanced technology. As a result, the most recently sequenced tree genomes are based primarily on short reads (Chapter 1, this volume). While these approaches will produce a reasonable assembly in regions of the genome that have high complexity and low abundance of repetitive DNA, assemblies have remained largely fragmented. This outcome reflects the difficulty in using short reads to produce a high-quality genome sequence that spans regions of low and high complexity. Nonetheless, it is expected that as next-generation technologies improve, the quality and length of the sequence generated will become equivalent or better than that of Sanger technology.

The main features of the black cottonwood and *E. grandis* genomes are described below. They represent the most detailed genome resources currently available for hardwoods. Both reference genomes have become a template for the sequencing and assembly of other genomes from related species, such as desert poplar (Ma *et al.*, 2013) and Tasmanian bluegum (Myburg *et al.*, 2014), but produced using NGS methods.

The *P. trichocarpa* genome

The genus *Populus* diverged from the model plant *A. thaliana* 100 million years ago. Approximately 60 million years ago, *Populus* diverged from its sister genus, *Salix*. The genus *Populus* now includes 29 species that have been grouped into six sections, although there are disagreements about the definitive phylogeny. Among the various species, black cottonwood is one of the most broadly distributed poplars in North America, and together with other species in the genus, it has been utilized extensively as a model tree for the study of perenniality and secondary growth (Taylor, 2002; Wullschleger *et al.*, 2002; Jansson and Douglas, 2007). As a consequence of its widespread use as model tree species and potential source of renewable energy, black cottonwood was selected to be the first tree to have its genome sequenced (Tuskan *et al.*, 2006). It also has a relatively small genome compared with other widely planted woody species.

The black cottonwood genome was sequenced and assembled using a strategy that combined WGS sequencing, with BAC-end sequencing and high-density genetic mapping. For WGS sequencing, DNA libraries made of inserts of 3 and 8 kbp were generated and characterized by Sanger sequencing, for an average depth of 7.5×. In addition, a BAC library with genome depth of coverage of 9.5× was generated. Fingerprinted BACs were paired-end sequenced and linked to the WGS assembly, which allowed positioning of additional sequences into linkage groups. The assembly of these sequences resulted in 2,447 scaffolds representing approximately 410 Mbp, the equivalent to ~85% of the total genome sequence. In order to identify the relative position of each scaffold, microsatellites identified in individual scaffolds were mapped, resulting in the assignment of 155, representing 335 Mbp of sequence. Posterior studies uncovered the genetic position of additional scaffolds based on genetic mapping (Drost *et al.*, 2009). The quality and completeness of the black cottonwood genome sequence was assessed by comparing the assembly results with previous genomic data. Regions of similarity were identified for more than 95% of cDNA that had previously been reported for the species, suggesting that a vast majority of the coding fraction of the genome is represented in the initial assembly. Furthermore, 91% of the microsatellites were collinear with the assembly, indicating that the order and orientation of scaffolds was highly reliable.

For identification of gene-coding sequences, a set of full-length cDNAs was utilized to train gene-finding programs, resulting in the identification of 45,555 genes in the nuclear genome. Thus, black cottonwood contains significantly more coding sequences than *A. thaliana*, possibly a consequence of a recent genome duplication and slow evolution. In fact, analysis of the *Populus* genome sequence showed a large proportion of paralogous genes, of similar age. Together with cytological evidence, this indicates that the poplar genome underwent at least one WGD event following the divergence from the *A. thaliana* lineage. A consequence of this relatively recent duplication event and the slow evolution of the *Populus* genome is the consistent observation that genome segments contain equivalent (syntenic) regions somewhere else in the genome. Another possible reason for the larger number of coding sequences is the need for additional types of transcripts for survival and adaptation. In contrast to herbaceous annual plants, trees need to be able to adapt to regular and diverse sources of biotic and abiotic stress for many years. Not surprisingly, an abundance of genes related to disease and insect resistance, meristem development, and nutrient transport were observed in the genome.

Sequencing of the first woody perennial resulted in a number of discoveries that are relevant to understanding genome evolution of other woody perennial species. For instance, the sequence divergence observed between paralogous genes derived from the most recent duplication of the *Populus* genome was estimated to have occurred less than 15 million years ago, based on parameters commonly used for analysis of annual plants. However, the fact that this duplication event was shared with the *Salix* lineage indicates that it was, in fact, far more ancient, as it predates the separation of both lineages. The most plausible explanation for this discrepancy is a slower evolutionary clock in *Populus*. Similar evidence for a slower clock in trees has also been reported for other species, such as those within *Eucalyptus* and *Pinus*. A slower clock could be due to the fact that long-lived perennial species may contribute to gametes for several generations; thus, the "ancient" alleles may be passed to future generations over long periods of time. This explanation does not consider that gametes are likely to undergo mutations over time, and has yet to be confirmed biologically.

The *E. grandis* genome

The genus *Eucalyptus* includes more than 700 species from the family Myrtaceae, mostly native to Australia. Several species in the genus display broad adaptability and fast growth, which made it a favorite for forest plantations in tropical and temperate regions. Currently, eucalypts are the most cultivated hardwood species worldwide, with over 20 million ha of forest plantations (Myburg *et al.*, 2014). *Eucalyptus* spp. are mostly diploid, with a haploid chromosome number of 11, and estimated genome sizes in the range of 370–700 million bp (Grattapaglia and Bradshaw, 1994). Of all the species, *E. grandis* is the most widely cultivated and is the foundational species of most tree-breeding programs. For this reason, a genotype of *E. grandis* was selected to be the second hardwood tree to have its genome sequenced, following sequencing of the black cottonwood genome almost a decade earlier (Tuskan *et al.*, 2006).

The *E. grandis* genome, with approximately 640 million bp, was sequenced using a whole-genome random shotgun strategy, in combination with sequencing of the ends of BACs. Genome libraries were prepared for fragments of approximately 2.6, 6, and 36–40 kb, and were paired-end sequenced to 6.7× coverage. End sequences from BAC libraries with insert sizes of 127–155 kbp were also generated for long-range linking of contigs generated by assembling reads from smaller insert libraries. The assembly resulted in approximately 6000 scaffolds with an L50 size of 4.9 Mbp and a total scaffold size of ~700 Mbp. To integrate scaffolds into chromosomes, previously constructed dense genetic maps were used. Based on the mapping of contigs and scaffolds, a final assembly was generated with 4,952 scaffolds with L50 of 67.2 kbp (contig) and 53.9 Mbp (scaffolds).

Predicted protein-coding sequences in the genome were identified initially based on similarity to a database of peptides described previously in other plant species, along with assemblies generated from 260,000 ESTs from *E. grandis* and related species. Genomic regions of similarity were extended and submitted to gene prediction tools, resulting in the identification of 36,376 coding sequences. Completeness of the assembly was evaluated by aligning over a million ESTs sequenced from the reference genotype, which resulted in an alignment of ~99%. This suggests that the assembly represented the vast majority of the expressed genes that had been identified. Finally, repetitive regions of the genome were identified and major repetitive units showed a preponderance of retrotransposons (44.5%), the majority of which were LTR-RTs.

The *E. grandis* genome revealed some unique properties that have not previously been reported for plant species. The first striking observation was that 34% of its genes appear in tandem duplications. Tandemly duplicated genes are often related to stress response, suggesting that they may be related to the adaptation of the species to diverse environments. Interestingly, the pattern of tandem duplications appears to be highly variable across different *Eucalyptus* spp., pointing to a dynamic genome. The *Eucalyptus* genome also contains a number of extended gene families that relate to a characteristic property of the species: the extreme diversity and concentration of essential oils, which are composed of mixtures of mono- and sesquiterpenes. Not surprisingly, the *E. grandis* genome contains the largest number of terpene synthase genes ever reported.

Summary

Hardwoods are non-coniferous woody perennial trees that belong to the plant division Angiospermae. Angiosperms appeared over 200 million years ago and today have the highest plant diversity, with more than 350,000 species that dominate the vast majority of terrestrial ecosystems. Hardwoods include some of the most important commercial tree species planted worldwide, including those from the genera *Eucalyptus* and *Populus*, which are used for production of lumber, veneer, pulp, and paper, and are increasingly considered ideal feedstocks for renewable energy.

Whole-genome duplications (WGDs) have been considered a significant driver for the diversification of angiosperms, although their genomes are generally much smaller than those of conifers. The complex evolutionary history of their genomes has resulted in limited synteny and collinearity across taxa, in contrast to the conifers. Due to their relatively small genome and ease of transformation and vegetative propagation, hardwoods from the genus *Populus* have become a favorite model to understand the biology of wood formation, perenniality, dormancy, and other unique properties of trees.

As with conifers, genomic studies of hardwoods began with genetic maps created using isozymes and progressed toward denser maps with PCR-based markers, DNA chips, and genotyping based on sequencing. Characterization of coding regions, through sequencing of ESTs, was generated first for hardwoods, particularly from the genus *Populus*. These early EST sequencing studies showed that the vast majority of genes appear to be shared between hardwood and herbaceous angiosperms, and such studies have expanded to other hardwood species following the development of NGS.

A significant event in the genomics of hardwoods was the sequencing of the first tree genome, *P. trichocarpa* cv. Nisqually-1, in 2006. *Populus trichocarpa* is one of the most widely distributed hardwood species in North America. Because of its widespread use as model tree species and in commercial plantations, and as a potential source of renewable energy, it was selected to be the first tree to have its genome sequenced. Sequencing of Nisqually-1 allowed the cataloguing of most genes in the species, and enabled the first full transcriptome studies in a tree to be conducted. Soon after the *Populus* genome sequence became available, several microarray platforms were developed. These allowed the analysis of expression of all genes in the genome, triggering studies that characterized transcriptome changes associated with each stage of wood formation, and responses to sources of biotic and abiotic stress. With the development of NGS and RNA-seq for gene-expression analysis, transcriptome studies have rapidly expanded to other hardwoods, particularly those from the genera *Eucalyptus* and *Quercus*, as well as other species in the Salicaceae.

The sequencing of the *P. trichocarpa* genome was carried out using the Sanger sequencing approach, based on combining data from genomic DNA libraries of different sizes, and resulted in an assembly that contained 410 Mbp in its scaffolds. The annotation of the genome uncovered 45,555 genes in the nuclear genome, significantly more than *A. thaliana*, possibly as a consequence of a recent genome duplication and slow evolution. Sequencing of *Populus* was followed several years later by the sequencing of *E. grandis*, the most cultivated hardwood species worldwide. The final assembly based on NGS resulted in scaffolds in the order of megabases and uncovered approximately 36,000 genes. A striking property of the *E. grandis* genome was the observation that 34% of its genes appear in tandem duplications. Other hardwood genomes, of willow, oak, and chestnut, have recently been published or are currently in their final stages of assembly and annotation.

References

Andersson-Gunnerås, S., Mellerowicz, E.J., Love, J., Segerman, B., Ohmiya, Y., *et al.* (2006) Biosynthesis of cellulose-enriched tension wood in *Populus*: global analysis of transcripts and metabolites identifies biochemical and developmental regulators in secondary wall biosynthesis. *Plant Journal* 45, 144–165.

Angiosperm Phylogeny Group (2016) An update of the Angiosperm Phylogeny Group classification for the orders and families of flowering plants: APG IV. *Botanical Journal of the Linnean Society* 181, 1–20.

Azaiez, A., Boyle, B., Levée, V., and Séguin, A. (2009) Transcriptome profiling in hybrid poplar following interactions with *Melampsora* rust fungi. *Molecular Plant–Microbe Interactions* 22, 190–200.

Bao, H., Li, E., Mansfield, S.D., Cronk, Q.C.B., El-Kassaby, Y.A., and Douglas, C.J. (2013) The developing xylem transcriptome and genome-wide analysis of alternative splicing in *Populus trichocarpa* (black cottonwood) populations. *BMC Genomics* 14, 359.

Bell, C.D., Soltis, D.E., and Soltis, P.S. (2010) The age and diversification of the angiosperms re-revisited. *American Journal of Botany* 97, 1296–1303.

Bennett, M.D. and Smith, J.B. (1976) Nuclear DNA amounts in angiosperms. *Philosophical Transactions of the Royal Society B: Biological Sciences* 274, 227–274.

Berlin, S., Lagercrantz, U., von Arnold, S., Öst, T., Rönnberg-Wästljung, A., *et al.* (2010) High-density linkage mapping and evolution of paralogs and orthologs in *Salix* and *Populus*. *BMC Genomics* 11, 129.

Björklund, S., Antti, H., Uddestrand, I., Moritz, T., and Sundberg, B. (2007) Cross-talk between gibberellin and auxin in development of *Populus* wood: gibberellin stimulates polar auxin transport and has a common transcriptome with auxin. *Plant Journal* 52, 499–511.

Borevitz, J.O., Liang, D., Plouffe, D., Chang, H.-S., Zhu, T., *et al.* (2003) Large-scale identification of single-feature polymorphisms in complex genomes. *Genome Research* 13, 513–523.

Borevitz, J.O., Hazen, S.P., Michael, T.P., Morris, G.P., Baxter, I.R., *et al.* (2007) Genome-wide patterns of single-feature polymorphism in *Arabidopsis thaliana*. *Proceedings of the National Academy of Sciences USA* 104, 12057–12062.

Bowers, J.E., Chapman, B.A., Rong, J., and Paterson, A.H. (2003) Unravelling angiosperm genome evolution by phylogenetic analysis of chromosomal duplication events. *Nature* 422, 433–438.

Bradshaw, H.D. and Stettler, R.F. (1995) Molecular genetics of growth and development in populus. IV. Mapping QTLs with large effects on growth, form, and phenology traits in a forest tree. *Genetics* 139, 963–973.

Bradshaw, H.D., Villar, M., Watson, B.D., Otto, K.G., Stewart, S., and Stettler, R.F. (1994) Molecular genetics of growth and development in *Populus*.

III. A genetic linkage map of a hybrid poplar composed of RFLP, STS, and RAPD markers. *Theoretical and Applied Genetics* 89, 167–178.

Brazma, A., Parkinson, H., Sarkans, U., Shojatalab, M., Vilo, J., *et al.* (2003) ArrayExpress – a public repository for microarray gene expression data at the EBI. *Nucleic Acids Research* 31, 68–71.

Brondani, R.P.V., Brondani, C., and Grattapaglia, D. (2002) Towards a genus-wide reference linkage map for *Eucalyptus* based exclusively on highly informative microsatellite markers. *Molecular Genetics and Genomics* 267, 338–347.

Brondani, R.P.V., Williams, E.R., Brondani, C., and Grattapaglia, D. (2006) A microsatellite-based consensus linkage map for species of *Eucalyptus* and a novel set of 230 microsatellite markers for the genus. *BMC Plant Biology* 6, 20.

Brown, G.R., Kadel, E.E., Bassoni, D.L., Kiehne, K.L., Temesgen, B., *et al.* (2001) Anchored reference loci in loblolly pine (*Pinus taeda* L.) for integrating pine genomics. *Genetics* 159, 799–809.

Byrne, M., Marquezgarcia, M., Uren, T., Smith, D., and Moran, G. (1996) Conservation and genetic diversity of *Microsatellite loci* in the genus *Eucalyptus*. *Australian Journal of Botany* 44, 331–341.

Castiglione, S., Wang, G., Damiani, G., Bandi, C., Bisoffi, S., and Sala F. (1993) RAPD fingerprints for identification and for taxonomic studies of elite poplar (*Populus* spp.) clones. *Theoretical and Applied Genetics* 87, 54–59.

Cervera, M.T., Storme, V., Ivens, B., Gusmão, J., Liu, B.H., *et al.* (2001) Dense genetic linkage maps of three *Populus* species (*Populus deltoides*, *P. nigra* and *P. trichocarpa*) based on AFLP and microsatellite markers. *Genetics* 158, 787–809.

Chen, S. and Polle, A. (2010) Salinity tolerance of *Populus*. *Plant Biology* 12, 317–333.

Cohen, D., Bogeat-Triboulot, M.-B., Tisserant, E., Balzergue, S., Martin-Magniette, M.-L., *et al.* (2010) Comparative transcriptomics of drought responses in *Populus*: a meta-analysis of genome-wide expression profiling in mature leaves and root apices across two genotypes. *BMC Genomics* 11, 630.

Darwin, C., Darwin, F., and Seward, A.C. (eds) (1903) *More Letters of Charles Darwin. A Record of his Work in a Series of Hitherto Unpublished Letters*. J. Murray, London.

Davey, J.W., Hohenlohe, P.A., Etter, P.D., Boone, J.Q., Catchen, J.M., and Blaxter, M.L. (2011) Genome-wide genetic marker discovery and genotyping using next-generation sequencing. *Nature Reviews Genetics* 12, 499–510.

Demura, T. and Fukuda, H. (2007) Transcriptional regulation in wood formation. *Trends in Plant Science* 12, 64–70.

Devey, M.E., Sewell, M.M., Uren, T.L., and Neale, D.B. (1999) Comparative mapping in loblolly and radiata pine using RFLP and microsatellite markers. *Theoretical and Applied Genetics* 99, 656–662.

Dharmawardhana, P., Brunner, A.M., and Strauss, S.H. (2010) Genome-wide transcriptome analysis of the transition from primary to secondary stem development in *Populus trichocarpa*. *BMC Genomics* 11, 150.

Drost, D.R., Novaes, E., Boaventura-Novaes, C., Benedict, C.I., Brown, R.S., *et al.* (2009) A microarray-based genotyping and genetic mapping approach for highly heterozygous outcrossing species enables localization of a large fraction of the unassembled *Populus trichocarpa* genome sequence. *Plant Journal* 58, 1054–1067.

Duplessis, S., Hacquard, S., Delaruelle, C., Tisserant, E., Frey, P., *et al.* (2011) *Melampsora larici-populina* transcript profiling during germination and time-course infection of poplar leaves reveals dynamic expression patterns associated with virulence and biotrophy. *Molecular Plant–Microbe Interactions* 24, 808–818.

Evans, L.M., Slavov, G.T., Rodgers-Melnick, E., Martin, J., Ranjan, P., *et al.* (2014) Population genomics of *Populus trichocarpa* identifies signatures of selection and adaptive trait associations. *Nature Genetics* 46, 1089–1096.

Fahrenkrog, A., Neves, L., Resende, M.F.R., Vasquez, A., de Los Campos, G., *et al.* (2016) Genome-wide association study reveals putative regulators of bioenergy traits in *Populus deltoides*. *New Phytologist* 213, 799–811.

Faircloth, B.C. (2008) MSATCOMMANDER: detection of microsatellite repeat arrays and automated, locus-specific primer design. *Molecular Ecology Resources* 8, 92–94.

Foucart, C., Paux, E., Ladouce, N., San-Clemente, H., Grima-Pettenati, J., and Sivadon, P. (2006) Transcript profiling of a xylem vs phloem cDNA subtractive library identifies new genes expressed during xylogenesis in *Eucalyptus*. *New Phytologist* 170, 739–752.

Geraldes, A., DiFazio, S.P., Slavov, G.T., Ranjan, P., Muchero, W., *et al.* (2013) A 34K SNP genotyping array for *Populus trichocarpa*: design, application to the study of natural populations and transferability to other *Populus* species. *Molecular Ecology Resources* 13, 306–323.

Glaubitz, J.C., Emebiri, L.C., and Moran, G.F. (2001) Dinucleotide microsatellites from *Eucalyptus sieberi*: inheritance, diversity, and improved scoring of single-base differences. *Genome* 44, 1041–1045.

Grattapaglia, D. and Bradshaw, H.D. Jr. (1994) Nuclear DNA content of commercially important *Eucalyptus* species and hybrids. *Canadian Journal of Forest Research* 24, 1074–1078.

Grattapaglia, D. and Kirst, M. (2008) *Eucalyptus* applied genomics: from gene sequences to breeding tools. *New Phytologist* 179, 911–929.

Grattapaglia, D. and Sederoff, R. (1994) Genetic linkage maps of *Eucalyptus grandis* and *Eucalyptus urophylla* using a pseudo-testcross: mapping strategy and RAPD markers. *Genetics* 137, 1121–1137.

Grattapaglia, D., Bertolucci, F.L., and Sederoff, R.R. (1995) Genetic mapping of QTLs controlling vegetative propagation in *Eucalyptus grandis* and *E. urophylla* using a pseudo-testcross strategy and RAPD markers. *Theoretical and Applied Genetics* 90, 933–947.

Grattapaglia, D., Bertolucci, F.L., Penchel, R., and Sederoff, R.R. (1996) Genetic mapping of quantitative trait loci controlling growth and wood quality traits in *Eucalyptus grandis* using a maternal half-sib family and RAPD markers. *Genetics* 144, 1205–1214.

Grattapaglia, D., Silva-Junior, O.B., Kirst, M., de Lima, B.M., Faria, D.A., and Pappas, G.J. (2011) High-throughput SNP genotyping in the highly heterozygous genome of *Eucalyptus*: assay success, polymorphism and transferability across species. *BMC Plant Biology* 11, 65.

Grisel, N., Zoller, S., Künzli-Gontarczyk, M., Lampart, T., Münsterkötter, M., *et al.* (2010) Transcriptome responses to aluminum stress in roots of aspen (*Populus tremula*). *BMC Plant Biology* 10, 185.

Hamanishi, E.T., Raj, S., Wilkins, O., Thomas, B.R., Mansfield, S.D., *et al.* (2010) Intraspecific variation in the *Populus balsamifera* drought transcriptome. *Plant, Cell & Environment* 33, 1742–1755.

Hamanishi, E.T., Barchet, G.L.H., Dauwe, R., Mansfield, S.D., and Campbell, M.M. (2015) Poplar trees reconfigure the transcriptome and metabolome in response to drought in a genotype- and time-of-day-dependent manner. *BMC Genomics* 16, 329.

Hay, W.W. and Floegel, S. (2012) New thoughts about the Cretaceous climate and oceans. *Earth-Science Reviews* 115, 262–272.

He, J., Li, H., Luo, J., Ma, C., Li, S., *et al.* (2013) A transcriptomic network underlies microstructural and physiological responses to cadmium in *Populus × canescens*. *Plant Physiology* 162, 424–439.

Hefer, C.A., Mizrachi, E., Myburg, A.A., Douglas, C.J., and Mansfield, S.D. (2015) Comparative interrogation of the developing xylem transcriptomes of two wood-forming species: *Populus trichocarpa* and *Eucalyptus grandis*. *New Phytologist* 206, 1391–1405.

Hertzberg, M., Aspeborg, H., Schrader, J., Andersson, A., Erlandsson, R., *et al.* (2001) A transcriptional roadmap to wood formation. *Proceedings of the National Academy of Sciences USA* 98, 14732–14737.

Holliday, J.A., Zhou, L., Bawa, R., Zhang, M., and Oubida, R.W. (2016) Evidence for extensive parallelism but divergent genomic architecture of adaptation along altitudinal and latitudinal gradients in *Populus trichocarpa*. *New Phytologist* 209, 1240–1251.

Jansson, S. and Douglas, C.J. (2007) *Populus*: a model system for plant biology. *Annual Review of Plant Biology* 58, 435–458.

Janz, D., Behnke, K., Schnitzler, J.-P., Kanawati, B., Schmitt-Kopplin, P., and Polle, A. (2010) Pathway analysis of the transcriptome and metabolome of salt sensitive and tolerant poplar species reveals evolutionary adaption of stress tolerance mechanisms. *BMC Plant Biology* 10, 150.

Keim, P., Paige, K.N., Whitham, T.G., and Lark, K.G. (1989) Genetic analysis of an interspecific hybrid swarm of *Populus*: occurrence of unidirectional introgression. *Genetics* 123, 557–565.

Kirst, M., Johnson, A.F., Baucom, C., Ulrich, E., Hubbard, K., *et al.* (2003) Apparent homology of expressed genes from wood-forming tissues of loblolly pine (*Pinus taeda* L.) with *Arabidopsis thaliana*. *Proceedings of the National Academy of Sciences USA* 100, 7383–8.

Kirst, M., Myburg, A.A., De León, J.P.G., Kirst, M.E., Scott, J., and Sederoff, R. (2004) Coordinated genetic regulation of growth and lignin revealed by quantitative trait locus analysis of cDNA microarray data in an interspecific backcross of eucalyptus. *Plant Physiology* 135, 2368–2378.

Kirst, M., Caldo, R., Casati, P., Tanimoto, G., Walbot, V., *et al.* (2006) Genetic diversity contribution to errors in short oligonucleotide microarray analysis. *Plant Biotechnology Journal* 4, 489–498.

Lemes, M.R., Brondani, R.P.V., and Grattapaglia, D. (2002) Multiplexed systems of microsatellite markers for genetic analysis of mahogany, *Swietenia macrophylla* King (Meliaceae), a threatened neotropical timber species. *Journal of Heredity* 93, 287–290.

Levin, D.A. (1983) Polyploidy and novelty in flowering plants. *American Naturalist* 122, 1–25.

Liu, Z. and Furnier, G.R. (1993) Inheritance and linkage of allozymes and restriction fragment length polymorphisms in trembling aspen. *Journal of Heredity* 84, 419–424.

Ma, T., Wang, J., Zhou, G., Yue, Z., Hu, Q., *et al.* (2013) Genomic insights into salt adaptation in a desert poplar. *Nature Communications* 4, 2797.

Marques, C.M., Araújo, J.A., Ferreira, J.G., Whetten, R., O'Malley, D.M., *et al.* (1998) AFLP genetic maps of *Eucalyptus globulus* and *E. tereticornis*. *Theoretical and Applied Genetics* 96, 727–737.

Marques, M., Brondani, V., Grattapaglia, D., and Sederoff, R. (2002) Conservation and synteny of SSR loci and QTLs for vegetative propagation in four *Eucalyptus* species. *Theoretical and Applied Genetics* 105, 474–478.

Matasci, N., Hung, L.-H., Yan, Z., Carpenter, E.J., Wickett, N.J., *et al.* (2014) Data access for the 1,000 Plants (1KP) project. *Gigascience* 3, 17.

Miranda, M., Ralph, S.G., Mellway, R., White, R., Heath, M.C., *et al.* (2007) The transcriptional response of hybrid poplar (*Populus trichocarpa* × *P. deltoides*) to infection by *Melampsora medusae* leaf rust involves induction of flavonoid pathway genes leading to the accumulation of proanthocyanidins. *Molecular Plant–Microbe Interactions* 20, 816–831.

Muller-Starck, G. (1992) Genetic control and inheritance of isoenzymes in poplars of the Tacamahaca section and hybrids. *Silvae Genetica* 41, 87–95.

Myburg, A.A., Remington, D.L., O'Malley, D.M., Sederoff, R.R., and Whetten, R.W. (2001) High-throughput AFLP analysis using infrared dye-labeled primers and an automated DNA sequencer. *Biotechniques* 30, 348–357.

Myburg, A.A., Griffin, A.R., Sederoff, R.R., and Whetten, R.W. (2003) Comparative genetic linkage maps of *Eucalyptus grandis*, *Eucalyptus globulus* and their F_1 hybrid based on a double pseudo-backcross mapping approach. *Theoretical and Applied Genetics* 107, 1028–1042.

Myburg, A.A., Grattapaglia, D., Tuskan, G.A., Hellsten, U., Hayes, R.D., *et al.* (2014) The genome of *Eucalyptus grandis*. *Nature* 510, 356–62.

Neves, L.G., McMamani, E., Alfenas, A.C., Kirst, M., and Grattapaglia, D. (2011) A high-density transcript linkage map with 1,845 expressed genes positioned by microarray-based Single Feature Polymorphisms (SFP) in *Eucalyptus*. *BMC Genomics* 12, 189.

Nilsson, J., Karlberg, A., Antti, H., Lopez-Vernaza, M., Mellerowicz, E., *et al.* (2008) Dissecting the molecular basis of the regulation of wood formation by auxin in hybrid aspen. *Plant Cell* 20, 843–855.

Novaes, E., Drost, D.R., Farmerie, W.G., Pappas, G.J., Grattapaglia, D., *et al.* (2008) High-throughput gene and SNP discovery in *Eucalyptus grandis*, an uncharacterized genome. *BMC Genomics* 9, 312.

Omondi, S.F., Kireger, E., Dangasuk, O.G., Chikamai, B., Odee, D.W., *et al.* (2010) Genetic diversity and population structure of *Acacia senegal* (L) Willd. in Kenya. *Tropical Plant Biology* 3, 59–70.

Paux, E., Tamasloukht, M., Ladouce, N., Sivadon, P. and Grima-Pettenati, J. (2004) Identification of genes preferentially expressed during wood formation in *Eucalyptus*. *Plant Molecular Biology* 55, 263–280.

Pavy, N., Pelgas, B., Beauseigle, S., Blais, S., Gagnon, F., *et al.* (2008) Enhancing genetic mapping of complex genomes through the design of highly-multiplexed SNP arrays: application to the large and unsequenced genomes of white spruce and black spruce. *BMC Genomics* 9, 21.

Petre, B., Morin, E., Tisserant, E., Hacquard, S., Da Silva, C., *et al.* (2012) RNA-Seq of early-infected poplar leaves by the rust pathogen *Melampsora larici-populina* uncovers *PtSultr3;5*, a fungal-induced host sulfate transporter. *PLoS One* 7, e44408.

Petroli, C.D., Sansaloni, C.P., Carling, J., Steane, D.A., Vaillancourt, R.E., *et al.* (2012) Genomic characterization of DArT markers based on high-density linkage analysis and physical mapping to the *Eucalyptus* genome. *PLoS One* 7, e44684.

Philippe, R.N., Ralph, S.G., Mansfield, S.D., and Bohlmann, J. (2010) Transcriptome profiles of hybrid poplar (*Populus trichocarpa* × *deltoides*) reveal rapid changes in undamaged, systemic sink leaves after simulated feeding by forest tent caterpillar (*Malacosoma disstria*). *New Phytologist* 188, 787–802.

Plavcová, L., Hacke, U.G., Almeida-Rodriguez, A.M., Li, E., and Douglas, C.J. (2013) Gene expression patterns underlying changes in xylem structure and function in response to increased nitrogen availability in hybrid poplar. *Plant, Cell & Environment* 36, 186–199.

Raj, S., Bräutigam, K., Hamanishi, E.T., Wilkins, O., Thomas, B.R., *et al.* (2011) Clone history shapes *Populus* drought responses. *Proceedings of the National Academy of Sciences USA* 108, 12521–12526.

Rajora, O.P. (1988) Allozymes as aids for identification and differentiation of some *Populus maximowiczii* Henry clonal varieties. *Biochemical Systematics and Ecology* 16, 635–640.

Ralph, S.G., Chun, H.J.E., Cooper, D., Kirkpatrick, R., Kolosova, N., *et al.* (2008) Analysis of 4,664 high-quality sequence-finished poplar full-length cDNA clones and their utility for the discovery of genes responding to insect feeding. *BMC Genomics* 9, 57.

Rengel, D., San Clemente, H., Servant, F., Ladouce, N., Paux, E., *et al.* (2009) A new genomic resource dedicated to wood formation in *Eucalyptus*. *BMC Plant Biology* 9, 36.

Rinaldi, C., Kohler, A., Frey, P., Duchaussoy, F., Ningre, N., *et al.* (2007) Transcript profiling of poplar leaves upon infection with compatible and incompatible strains of the foliar rust *Melampsora larici-populina*. *Plant Physiology* 144, 347–366.

Salazar, M.M., Nascimento, L.C., Camargo, E.L.O., Gonçalves, D.C., Lepikson Neto, J., *et al.* (2013) Xylem transcription profiles indicate potential metabolic responses for economically relevant characteristics of *Eucalyptus* species. *BMC Genomics* 14, 201.

Schilling, M.P., Wolf, P.G., Duffy, A.M., Rai, H.S., Rowe, C.A., *et al.* (2014) Genotyping-by-sequencing for *Populus* population genomics: an assessment of genome sampling patterns and filtering approaches. *PLoS One* 9, e95292.

Silva-Junior, O.B., Faria, D.A., and Grattapaglia, D. (2015) A flexible multi-species genome-wide 60K SNP chip developed from pooled resequencing of 240 *Eucalyptus* tree genomes across 12 species. *New Phytologist* 206, 1527–1540.

Sjödin, A., Street, N.R., Sandberg, G., Gustafsson, P., and Jansson, S. (2009) The *Populus* Genome Integrative Explorer (PopGenIE): a new resource for exploring the *Populus* genome. *New Phytologist* 182, 1013–1025.

Slavov, G.T., DiFazio, S.P., Martin, J., Schackwitz, W., Muchero, W., *et al.* (2012) Genome resequencing reveals multiscale geographic structure and extensive linkage disequilibrium in the forest tree *Populus trichocarpa*. *New Phytologist* 196, 713–725.

Soldati, M.C., Fornes, L., van Zonneveld, M., Thomas, E., and Zelener, N. (2013) An assessment of the genetic diversity of *Cedrela balansae* C. D.C. (Meliaceae) in Northwestern Argentina by means of combined use of SSR and AFLP molecular markers. *Biochemical Systematics and Ecology* 47, 45–55.

Soltis, D.E., Bell, C.D., Kim, S., and Soltis, P.S. (2008) Origin and early evolution of angiosperms. *Annals of the New York Academy of Sciences* 1133, 3–25.

Sperry, J.S. (2003) Evolution of water transport and xylem structure. *International Journal of Plant Sciences* 164, S115–S127.

Steane, D., Vaillancourt, R., Russell, J., Powell, W., Marshall, D., and Potts, B. (2001) Development and characterisation of microsatellite loci in *Eucalyptus globulus* (Myrtaceae). *Silvae Genetica* 50, 89–91.

Sterky, F., Bhalerao, R.R., Unneberg, P., Segerman, B., Nilsson, P., *et al.* (2004) A *Populus* EST resource for plant functional genomics. *Proceedings of the National Academy of Sciences USA* 101, 13951–13956.

Sterky, F., Regan, S., Karlsson, J., Hertzberg, M., Rohde, A., *et al.* (1998) Gene discovery in the wood-forming tissues of poplar: analysis of 5,692 expressed sequence tags. *Proceedings of the National Academy of Sciences USA* 95, 13330–13335.

Street, N.R., Skogström, O., Sjödin, A., Tucker, J., Rodríguez-Acosta, M., *et al.* (2006) The genetics and genomics of the drought response in *Populus*. *Plant Journal* 48, 321–341.

Tallis, M.J., Lin, Y., Rogers, A., Zhang, J., Street, N.R., *et al.* (2010) The transcriptome of *Populus* in elevated CO reveals increased anthocyanin biosynthesis during delayed autumnal senescence. *New Phytologist* 186, 415–428.

Taylor, G. (2002) *Populus*: arabidopsis for forestry. Do we need a model tree? *Annals of Botany* 90, 681–689.

Taylor, G., Street, N.R., Tricker, P.J., Sjödin, A., Graham, L., *et al.* (2005) The transcriptome of *Populus* in elevated CO_2. *New Phytologist* 167, 143–154.

Thavamanikumar, S., Southerton, S., and Thumma, B. (2014) RNA-Seq using two populations reveals genes and alleles controlling wood traits and growth in *Eucalyptus nitens*. *PLoS One* 9, e101104.

Tsarouhas, V., Gullberg, U., and Lagercrantz, U. (2002) An AFLP and RFLP linkage map and quantitative trait locus (QTL) analysis of growth traits in *Salix*. *Theoretical and Applied Genetics* 105, 277–288.

Tuskan, G.A., Difazio, S., Jansson, S., Bohlmann, J., Grigoriev, I., *et al.* (2006) The genome of black cottonwood, *Populus trichocarpa* (Torr. & Gray). *Science* 313, 1596–1604.

van de Peer, Y., Mizrachi, E., and Marchal, K. (2017) The evolutionary significance of polyploidy. *Nature Reviews Genetics* 18, 411–424.

Viger, M., Smith, H.K., Cohen, D., Dewoody, J., Trewin, H., *et al.* (2016) Adaptive mechanisms and genomic plasticity for drought tolerance identified in European black poplar (*Populus nigra* L.). *Tree Physiology* 36, 909–928.

Villar, E., Klopp, C., Noirot, C., Novaes, E., Kirst, M., *et al.* (2011) RNA-Seq reveals genotype-specific molecular responses to water deficit in eucalyptus. *BMC Genomics* 12, 538.

Vos, P., Hogers, R., Bleeker, M., Reijans, M., van de Lee, T., *et al.* (1995) AFLP: a new technique for DNA fingerprinting. *Nucleic Acids Research* 23, 4407–4414.

Wang, Y., Zhang, B., Sun, X., Tan, B., Xu, L.-A., *et al.* (2011) Comparative genome mapping among *Populus adenopoda*, *P. alba*, *P. deltoides*, *P. euramericana* and *P. trichocarpa*. *Genes & Genetic Systems* 86, 257–268.

Watkinson, J.I., Sioson, A.A., Vasquez-Robinet, C., Shukla, M., Kumar, D., *et al.* (2003) Photosynthetic acclimation is reflected in specific patterns of gene expression in drought-stressed loblolly pine. *Plant Physiology* 133, 1702–1716.

Wilkins, O., Nahal, H., Foong, J., Provart, N.J., and Campbell, M.M. (2009a) Expansion and diversification of the *Populus* R2R3-MYB family of transcription factors. *Plant Physiology* 149, 981–993.

Wilkins, O., Waldron, L., Nahal, H., Provart, N.J., and Campbell, M.M. (2009b). Genotype and time of day shape the *Populus* drought response. *Plant Journal* 60, 703–715.

Wu, R., Bradshaw, H. and Stettler, R. (1997) Molecular genetics of growth and development in *Populus* (Salicaceae). V. Mapping quantitative trait loci affecting leaf variation. *American Journal of Botany* 84, 143.

Wu, R.L., Han, Y.F., Hu, J.J., Fang, J.J., Li, L., *et al.* (2000) An integrated genetic map of *Populus deltoides* based on amplified fragment length polymorphisms. *Theoretical and Applied Genetics* 100, 1249–1256.

Wullschleger, S.D., Jansson, S., and Taylor, G. (2002) Genomics and forest biology: *Populus* emerges as the perennial favorite. *Plant Cell* 14, 2651–2655.

Yang, J., Li, K., Zheng, W., Zhang, H., Cao, X., *et al.* (2015) Characterization of early transcriptional responses to cadmium in the root and leaf of Cd-resistant *Salix matsudana* Koidz. *BMC Genomics* 16, 705.

Zhou, L., Bawa, R., and Holliday, J.A. (2014) Exome resequencing reveals signatures of demographic and adaptive processes across the genome and range of black cottonwood (*Populus trichocarpa*). *Molecular Ecology* 23, 2486–2499.

Forest-tree Population Genomics

Introduction

Population genetics is the field of biology that studies how allele frequencies change within and among populations, and the factors that can contribute to these changes. In order to study and quantify the contribution of these factors, population geneticists first defined what is necessary for the allele frequencies in a population to remain stable from generation to generation. These requirements were outlined by the German physician Wilhelm Weinberg and the British mathematician Godfrey Harold Hardy, early in the 20th century. Both scientists demonstrated, within a few months of each other, that the frequency of alleles in a population should remain stable from generation to generation in the absence of evolutionary forces such as selection, mutation, migration, and genetic drift (Hardy, 1908).

While the Hardy–Weinberg principle provides a static model, the allele frequencies in natural and breeding populations fluctuate from generation to generation. Several evolutionary forces contribute to allele frequency changes. Mutations are the ultimate source of novel genetic variation, and the mutation rate is defined as the frequency of base substitutions per site per generation. The feasibility of sequencing multiple genomes now allows the precise quantification of mutation accumulation. This strategy has been used to show that in *Arabidopsis thaliana* the mutation rate is approximately 7×10^{-9} base substitutions per site per generation, and that most new variants represent G:C→A:T transitions (Ossowski *et al.*, 2010). This provides an estimate of the general mutation rate in plants, but factors such as genome size also contribute to differences in allele frequency. In fact, species with larger genomes have higher mutation rates (Lynch, 2010).

While mutations introduce new DNA variation in populations, evolutionary forces such as random genetic drift contribute to their removal. Random genetic drift arises from the effect of allele sampling from generation to generation. At each generation, alleles are transmitted from parent to offspring. By chance, over multiple generations, one allele at a given locus will be transmitted more often than others, resulting in the long-term loss of the alternative alleles (Hartl and Clark, 2007). Random genetic drift is a consequence of the finite number of individuals in a population and becomes increasingly significant as the size of populations decrease. Because allele fixation and loss occur randomly, different populations are likely to become differentiated from others. Mutations and random genetic drift act over the entire genome, and the balance between these two forces (the former contributing genetic variation and the latter removing it) has been considered one of the main reasons for the maintenance of genetic variation in natural populations, as explained in the neutral theory of molecular evolution (Kimura, 1968).

While mutation and random genetic drift are forces that impact genetic variation in the entire genome, forces such as selection act primarily on individual loci, although the effect may be observable in large genome segments due to selective sweeps. Selection occurs when the presence of an allele results in the individual that carries it contributing fewer (negative selection) or more (positive selection) individuals to the next generation. Selection can occur at many stages of an individual's life, from gamete selection prior to fertilization to mate choice and fertility.

The coefficient of selection is an estimate of the frequency in a subpopulation and determines the number of generations required for an allele to become fixed. For instance, a lethal allele will be lost within a single generation, but one that causes only a minor loss in fertility will require a larger number of generations to be removed. Other factors, such as the dominance relationship between alleles in the population will also determine the number of generations needed for a favorable allele to become fixed. Finally, factors such as migration (movement of alleles from population to population) can also result in changes in allele frequencies, due to the introduction of alleles from other populations.

The study of the genetics of populations started when Gregor Mendel (1822–1884) discovered the laws of segregation and independent assortment. His early genetic studies focused on the analysis of bi-parental families and simple traits regulated by one or a few genetic loci. However, the study of genetics soon expanded to the analysis of genetically unstructured populations and traits of higher complexity. These studies were pioneered by scientists such as Francis Galton (1822–1911), who introduced the statistical analysis of traits with continuous distributions in populations, and over multiple generations. Galton's early studies were followed by work by Sewall Wright (1889–1988) and Ronald Fisher (1880–1962), who are recognized as pioneers in the study of how allele frequencies change in populations as a result of mutation, migration, natural selection, and random genetic drift. Both scientists also made numerous contributions to the application of statistics to biological sciences. These approaches are still widely used in the design and analysis of results from genetic experiments in many fields of biology.

During the last century, the general principles of transmission genetics first identified by Mendel were rediscovered, and the theoretical foundation of population genetics was established. However, until recently, population genetics studies were confined largely to the development of theories to describe how evolutionary forces contribute to changes in gene and allele frequency. The lack of tools to characterize the genomic diversity in a large number of samples was a key limitation. As a consequence, population genetics studies have been restricted to the analysis of a few loci, allowing only limited inferences to be made. At the same time, computational resources required for complex types of analyses were also lacking. However, in the past decade, the field of genetics has been revolutionized by the development of high-throughput, low-cost genome sequencing methods (Chapter 1, this volume). This development led to the genome-wide analysis of genetic variants in increasingly larger populations.

In parallel to the emergence of the genomic sciences, new computational capabilities led to the development and application of novel methods of data analysis that previously were impossible. The result of this theoretical, analytical, and computational revolution was that population genetics moved from being largely theoretical to becoming more applied. Most importantly, it expanded from being concentrated on the analysis of a few loci to including the whole genome. This chapter focuses on the use of genomic information to characterize forest-tree population diversity, the relationship among populations, and the detection of loci that contribute to their adaptation. Other aspects related to population genetics of tree species, such as geographic variation, population evolution, and gene conservation have been reviewed in detail elsewhere (e.g. White *et al.*, 2007).

Why Study the Genomics of Forest Populations?

Tree populations are a fundamental component of terrestrial ecosystems. Forests influence climate and the hydrological and carbon cycles, protect soil resources, and harbor biodiversity. Approximately one-third of the land surface is covered by forests that store ~45% of all carbon present on Earth (Bonan, 2008). Natural forest populations are under increasing biotic and abiotic stress from the rapid changes in climate (Kurz *et al.*, 2008). Some of these stresses also impact tree plantations, reducing their productivity and increasing the need for more intensive silvicultural treatments, with higher inputs and costs. While affected by climate change, planted and natural forests can also be part of a global solution for mitigating climate change. For this to be possible, an understanding of the factors that impact the survival and adaptation of forest populations, as well as their productivity, will require characterization of the genomic composition of individuals within these populations. Most genetics studies of forest populations have focused on the topics described in the following sections.

Gene management and conservation in existing forests

As forests become increasingly threatened by existing and new sources of biotic and abiotic stresses, it is critical to quantify the genetic diversity of populations to measure the extent of gene loss and allele frequency changes, and the evolutionary consequences. This information is necessary to predict the performance of populations under existing and new threats, as lower genetic diversity is expected to hamper the ability to respond favorably to them. An understanding of the changes in population allele frequencies is also useful in helping breeders choose the most appropriate germplasm to generate clones better adapted to specific environmental influences.

Tree breeding

Genetic improvement programs are based on the principle that, from generation to generation, there is an increase in the frequency of favorable alleles, resulting in an overall improvement in trait characteristics. The application of genomics to breeding is commonly confined to the idea that genetic markers that regulate commercial traits can be identified and used for marker-assisted or genomic selection (Chapter 5, this volume). However, population genomic analysis is

also applied to breeding for purposes such as uncovering novel sources of germplasm, as well as the quantification of inbreeding and genomic structure in populations.

Identifying genes involved in adaptation and assisting migration

Climate change is threatening the performance and survival of existing tree populations. Accurate prediction of the fate of forests is needed to develop management and conservation strategies aimed at climate-change mitigation. This requires knowledge of adaptation and migration, two possible ways for tree populations to survive changing environmental conditions (Aitken *et al.*, 2008). The ability of a species to undergo phenotypic modification in response to environmental changes (phenotypic plasticity), and local adaptation of subpopulations within a species, affect survival capacity and influence predictions of the species distribution under future climatic scenarios (Valladares *et al.*, 2014). Many forest-tree species show high levels of phenotypic plasticity and local adaptation (Alberto *et al.*, 2013), highlighting the importance of considering intraspecific variation when modeling the effects of climate change on species' ranges (Valladares *et al.*, 2014).

Population Genomic Parameters

In order to characterize a population genetically, DNA polymorphisms have to be detected, and the frequency of alleles needs to be quantified (Chapter 1, this volume; Hartl and Clark, 2007; White *et al.*, 2007). The number of alleles detected at a locus and their frequency in a population are the most fundamental measures of genomic diversity but constitute only a crude representation of that diversity. In this section, commonly used genomic parameters and examples of their use in the analysis of tree populations are described.

Heterozygosity

The expected heterozygosity (H_E) corresponds to the frequency of individuals in a population that are expected have alternative alleles at a given locus, assuming a population in Hardy–Weinberg equilibrium (HWE). An alternative interpretation is that H_E is the probability that two alleles or haplotypes drawn randomly from a population will be different. Assuming a locus with two alleles, H_E can be estimated by:

$$H_E = 2 \times p_i \times q_j \qquad (4.1)$$

where p and q represent the observed frequency of alleles i and j, respectively, at a given locus. For loci with multiple alleles, H_E is calculated by summing the expected frequency of all possible heterozygotes (Hartl and Clark, 2007).

Another measure related to H_E is the observed heterozygosity (H_O), which is the frequency of heterozygote individuals that are identified (not estimated from allele frequencies, as with H_E) in a population. H_O is expected to be identical to H_E in a population that is in HWE. However, most populations deviate from HWE, resulting in H_O values that are often different from those expected. As a consequence, the relationship between H_E and H_O can be used to infer the presence of evolutionary forces that are acting on a population. For example, H_O is often lower than H_E in smaller populations, pointing to the occurrence of inbreeding.

Nucleotide polymorphisms and diversity

Several measures have been proposed to describe the variation in DNA sequence obtained from a population. A simple measure of diversity in DNA is the number of segregating sites that differ when sequences from several individuals are aligned. In this case, a count of the number of segregating sites is sufficient. However, this is not a very useful measure because it does not take into consideration the length of the sequence, the frequency of the alternative alleles and the number of sequences aligned to generate that estimate. All of these factors impact the likelihood that variable sites will be observed.

An alternative measure is to estimate the pairwise segregating sites (Π) (i.e. the number of sites that are variable between any two haplotypes randomly sampled from the population). This measure takes into consideration the frequency of the alleles at the segregating site (Box 4.1). Furthermore, by dividing the number of pairwise segregating sites by the length of the sequences (L), it is possible to estimate the nucleotide diversity, π (Nei and Li, 1979), as described in Eqn. 4.2. This provides an estimate of the frequency of mismatches that are expected to occur, per nucleotide, in a given genomic region and for a given population.

$$\pi = \Pi / L \qquad (4.2)$$

Another common measure of population nucleotide diversity is Watterson's estimator of θ (θ_W; Watterson, 1975), which is proportional to the mutation rate (μ) and the effective population size (N_e):

$$\theta = 4 N_e \mu \qquad (4.3)$$

Because μ and N_e are often unknown, the parameter θ can also be estimated by θ_W:

$$\theta_W = S / \alpha \qquad (4.4)$$

θ_W is estimated by counting the number of segregating sites in a DNA segment (S) and dividing it by a correction factor (α), which takes into consideration the number of haploid samples (n) used to identify these sites:

$$\alpha = \sum_{i=1}^{n-1} \frac{1}{i} = 1 + \frac{1}{2} + \frac{1}{3} + \ldots \frac{1}{n} \qquad (4.5)$$

The value of S is expected to increase as more haplotypes are sampled, justifying the need to apply the correction. The estimation of θ_W is based on the assumption that there is an infinite number of sites that may vary and that the sample size is much smaller than the effective size of the population. Under these assumptions, θ_W represents an unbiased estimator of nucleotide diversity.

Box 4.1. Measuring and Representing Nucleotide Variation in a Population

In the genomic analysis of individuals in a population, segments of homologous regions are generally sequenced and aligned, and the variation identified. Below are four sequences derived from a random sample of individuals in a population. For simplicity, it is assumed that samples are haploid. The sequence contains three segregating sites (S) in positions 1, 4, and 6:

$$
\begin{array}{ll}
\text{Sample A} & \text{A CTG}\mathbf{T}\text{AGGTC} \\
\text{Sample B} & \text{A }\mathbf{A}\text{TGCAGATC} \\
\text{Sample C} & \text{A CTGCA}\mathbf{C}\text{ATC} \\
\text{Sample D} & \text{A CTG}\mathbf{T}\text{AGATC}
\end{array}
$$

Position: 0 1 2 3 4 5 6 7 8 9

The number of pairwise segregating sites (π) can calculated by the number of pairwise comparisons that have a different nucleotide, relative to the number of all possible pairwise comparisons. For instance, at position 0, there are no pairwise nucleotide differences, but at position 1, there are three such differences (sample B is different from samples A, C, and D). In addition to the three pairwise differences at position 1, there are four pairwise differences in position 4 (sample A is different from samples B and C, and sample D is different from samples B and C) and three pairwise differences in position 6 (sample C is different from samples A, B, and D). Considering all segregating sites, there are ten pairwise differences in total. The value of π is calculated by dividing it by the number of possible pairwise differences, which is six in this case (sample A can be compared to B, C, and D; sample B can be compared to C and D; sample C can be compared to D; sample D has been compared to all other samples previously). Therefore, π is equal to:

$$
\Pi = \frac{\textit{Number of pairwise differences}}{\textit{Number of possible pairwise comparisons}} = \frac{10}{6} = 1.67
$$

The nucleotide diversity (π) can be estimated by dividing π by the length of the sequence (L; in this case, 10 nucleotides).

$$
\pi = \frac{\Pi}{L} = \frac{1.67}{10} = 0.167
$$

In summary, on average, 16.7% of sites sampled from the same genomic region, are expected to be different when any two samples from this population are characterized.

Forest-tree species are generally long lived, out-crossing, and disperse pollen and seeds over long distances. Forest populations also often occupy a large continuous natural range. These factors contribute to a limited incidence of random genetic drift and the consequent loss of neutral alleles. Even tree-breeding populations typically display high levels of nucleotide diversity, similar to natural populations, because genetic improvement programs have been through a limited number of breeding and selection cycles. While exceptions exist, perennial woody species have generally been found to have levels of nucleotide diversity (or other measures of genetic diversity) that exceed that of annual herbaceous plants. An early survey of plant studies based on isozyme markers found this to be so (Hamrick and Godt, 1996).

While the early studies using isozymes showed a consistent pattern of high genetic diversity in forest species, a very limited number of loci was surveyed. The limited sampling of genomic regions may bias estimates of genetic diversity. As a consequence, they may not be informative for estimating nucleotide polymorphisms and all the parameters that are derived during population genomic studies. This limitation started being addressed with the reduction in sequencing costs and development of high-throughput sequencing methods. Initially, studies focused on the analysis of candidate genes identified previously using expressed sequence tags (ESTs) (reviewed in conifers by Neale and Savolainen, 2004). In this work, candidate genes were amplified by polymerase chain reaction (PCR), followed by Sanger sequencing and alignment for detection of polymorphisms. Despite the technical difficulties in obtaining a consistent amplification of target loci across a diverse sample of individuals, these studies succeeded in characterizing the sequence of multiple genic regions and reporting their nucleotide diversity. Reports in the angiosperm European aspen (*Populus tremula*) (five genes; Ingvarsson, 2005) and the conifers Japanese cedar (*Cryptomeria japonica*) (seven genes; Kado *et al.*, 2003); loblolly pine (*Pinus taeda*) (19 genes; Brown *et al.*, 2004); and Douglas-fir (*Pseudotsuga menziesii*) (19 genes; Krutovsky and Neale, 2005) showed that estimates of nucleotide diversity (π and θ_{W}) ranged between 0.004 and 0.016. The highest diversity was detected in European aspen and Douglas-fir but the high level of variation observed among the genes sampled clearly indicated the need for broader surveys of the genome space to assess the variance in these estimates.

From the initial survey of fewer than a dozen genes, the use of PCR amplification of genic regions was later expanded to almost 8,000 loci when the approach was used in the analysis of 18 samples of *P. taeda* to estimate nucleotide diversity parameters over a wide range of genes (Eckert *et al.*, 2013). This expanded analysis offered the first glimpse of how nucleotide diversity varies genome wide and depends on the gene function. While overall levels of nucleotide diversity were similar to those reported previously in *P. taeda* (Brown *et al.*, 2004), genes related to disease resistance were shown to carry much higher (2×) levels of diversity.

While PCR amplification of genic regions, sequencing, and polymorphism detection were successful in *Pinus*, the approach is difficult to implement in many species because it is complex and potentially biased. Identifying genomic regions that are sufficiently conserved for PCR primer design is particularly difficult in tree species because they often have high level of genetic diversity. Successful design of primers that perform well across many individuals requires them to anneal to invariable regions. By their very nature, conserved invariable regions contain few genetic variants, which may be so because they are under selection and, therefore, biasing posterior population genetic estimates. These limitations were partially addressed by the introduction of sequence capture for genome analysis (Chapter 1, this volume). Sequence capture uses the hybridization of genomic DNA that is complementary to oligonucleotide probes that correspond to the regions of interest to retrieve the sequence of these regions in various individuals. Hybridization is carried out under conditions that allow a moderate level of mismatch (5–15%) between the genomic DNA and the oligonucleotide probes. Thus, oligonucleotide probes developed to capture specific genomic regions of an individual in the population can be used even if there are some differences between their DNA sequences. The use of the approach was first demonstrated in the conifers *P. taeda* and *P. elliottii* when it was used to partially capture the sequence of 14,729 genes in 24 haploid samples of each species (Neves *et al.*, 2013), and characterize their nucleotide diversity. The use of this approach was later expanded to capture approximately 200,000 exons from the *P. taeda* genome (~49 Mbp), resulting in the detection of almost 1,000,000 single-nucleotide polymorphisms (SNPs) in 375 individuals collected across the natural range of the species (Lu *et al.*, 2016).

The focus on the use of sequence capture for population genomics studies of conifers demonstrates the challenges of characterizing genome-wide nucleotide diversity in species with megagenomes, such as pines. Despite the rapid progress in high-throughput sequencing methods, it is still computationally and economically unfeasible to sequence the genomes of many samples of conifers to achieve an estimate of genome-wide nucleotide diversity. In contrast, for woody angiosperms, such as those in the genus *Populus*, this type of study has become a more common approach to characterization of genetic variation of populations. Pioneering work that resequenced 16 genomes of black cottonwood (*Populus trichocarpa*) (Slavov *et al.*, 2012) was soon followed by a landmark study that characterized 544 unrelated individuals from the same species and identified approximately 18 million SNPs. As expected, the study showed that nucleotide diversity is significantly higher in intergenic regions (π = 0.0064) than in the genic space (π = 0.003), consistent with purifying selection acting on coding regions.

Recombination and linkage disequilibrium

Characterizing nucleotide diversity in a population is a first step toward understanding its evolutionary history. However, nucleotide variation at a locus can also be evaluated relative to its correlation with variation at other loci, in order to obtain an estimate of linkage disequilibrium (LD) (Box. 4.2). The LD depends on the shared ancestry of the alleles at different loci, and occurs when the relationship between alleles in different loci is not random (i.e. they do not segregate independently in the population). The LD is generally described by plotting measures of LD (such as the coefficient of linkage disequilibrium, D, or r^2) relative to the distance between the two loci for which it was measured, for all pairwise measurements of LD. A non-linear regression is then used to fit the decay of LD relative to the physical distance between loci (Fig. 4.1A). Alternatively, LD of a genomic region can be depicted with a heat map that uses different shades or colors to represent the levels of disequilibrium, relative to the position of each pair of loci along a genome axis (Fig. 4.1B).

The LD measured in a population is dependent on a number of factors that impact specific, local regions in the DNA sequence, or the genome as a whole. At the local level, LD is dependent on factors such as the rate of recombination between the two loci under consideration. This rate can be highly variable in different regions of the genome, resulting in significant differences in the level of observed LD. As expected, a longer physical distance between loci generally results in a lower LD between them, as the probability of a recombination event increases. Selection

acting on a specific allele can also result in a higher LD in the genomic region surrounding it, in what is commonly known as "genetic hitchhiking." However, LD in the whole genome is impacted primarily by the mating system employed and LD is higher in selfing than in out-crossing species (Nordborg, 2000). Another significant factor is genetic drift, which can lead to an excess of a given haplotype in a population, raising overall LD. Random genetic drift is dependent on the effective size of the population; therefore, smaller populations, or populations derived from fewer recent founders, are likely to have higher levels of LD.

Knowing the extent of LD in the genome is relevant for a number of applications. First, LD can be used as a proxy for several population genetic parameters, such as for estimation of the effective population size (Waples, 2006). Second, as mentioned above, significant deviations in the genome pattern of LD in a population can be indicative of a locus under selection. Several studies have used the information on LD to identify loci that are related to adaptation in natural populations or domestication in plant breeding (Zhou *et al.*, 2015). Finally, knowing the extent of LD in a population also provides information on how much resolution can be achieved in association mapping studies (Goddard and Hayes, 2009). If LD decays very rapidly, detection of a DNA variant in association with a trait is an indication that the locus is in close proximity to the causative variant.

Most forest species are out-crossing and, consequently, have very low levels of LD. An early report based on analysis of a few genes in several conifers showed that LD decayed below an r^2 threshold of 0.2 (an arbitrary threshold often used in the literature) within 1–2 kbp (Neale and Savolainen, 2004). These findings were corroborated in an analysis of the angiosperm European aspen (*Populus tremula*), which showed even faster LD decay (Ingvarsson, 2005), and in multiple subsequent studies that greatly expanded the number of genes sampled, including in *Pinus taeda* (Lu *et al.*, 2016). In contrast to these early discoveries, some recent genome-wide studies have suggested that LD undergoes

Box 4.2. Measuring Linkage Disequilibrium

The relationship between alleles at two loci can be estimated by the linkage disequilibrium between them. For the purposes of this example, we will assume there are two loci (A and B) with two alleles and the following frequencies:

Locus	Allele	Frequency	Locus	Allele	Frequency
A	A1	$p1$	B	B1	$q1$
	A2	$p2$		B2	$q2$

If the segregation of these loci is independent of each other, then the frequency of occurrence of each haplotype is expected to be the product of their respective allele frequencies:

$f(A1B1) = p1 \times q1$
$f(A1B2) = p1 \times q2$
$f(A2B1) = p2 \times q1$
$f(A2B2) = p2 \times q2$

and

$f(A1B1) \times f(A2B2) = f(A1B2) \times f(A2B1)$
$(f(A1B1) \times f(A2B2)) - (f(A1B2) \times f(A2B1)) = 0.$

The estimate above is the coefficient of linkage disequilibrium, D. When D is equal to 0, the alleles at the two loci are independently associated, or in linkage equilibrium. If the estimate of D departs from 0, it is an indication that the alleles are not independent, and there is linkage disequilibrium.

For instance, let us assume that the following alleles were observed in two loci (A and B) in four samples collected in a population. Then the frequencies of the allelic combination are:

Sample 1   A1   B1 $f(A1B1) = 0.25$
Sample 2   A1   B2 $f(A1B2) = 0.25$
Sample 3   A2   B1 $f(A2B1) = 0.25$
Sample 4   A2   B2 $f(A2B2) = 0.25$

As can be seen, there is no correlation between the alleles detected at locus A, and those detected at locus B. As a consequence, the estimate of linkage disequilibrium, D, is zero.

$D = (f(A1B1) \times f(A2B2)) - (f(A1B2) \times f(A2B1)) = (0.25 \times 0.25) - (0.25 \times 0.25) = 0$

Consider another example:

Sample 1   A1   B1 $f(A1B1) = 0.5$
Sample 2   A1   B1 $f(A1B2) = 0.0$
Sample 3   A2   B2 $f(A2B1) = 0.0$
Sample 4   A2   B2 $f(A2B2) = 0.5$

In this case, there is a complete correlation between the alleles detected at locus A, and those detected at locus B. For instance, when allele A1 is present at locus A, allele B1 is always detected at B. Hence, the estimate of linkage disequilibrium is different from 0:

$D = (f(A1B1) \times f(A2B2)) - (f(A1B2) \times f(A2B1)) = (0.5 \times 0.5) - (0.0 \times 0.0) = 0.25$

The upper and lower limits of the value of D depend on the frequency of the alleles in the population. The measure r^2 of linkage disequilibrium is a standardized alternative to D, where the upper and lower limits are 0 (linkage equilibrium) and 1 (linkage disequilibrium), respectively. The r^2 is calculated by dividing D^2 by the product of the individual allele frequencies:

$r^2 = D^2/(f(A1) \times f(A2) \times f(B1) \times f(B2))$

For the example above, r^2 is equal to:

$r^2 = (0.25)^2/(0.5 \times 0.5 \times 0.5 \times 0.5) = 0.0625/0.0625 = 1.$

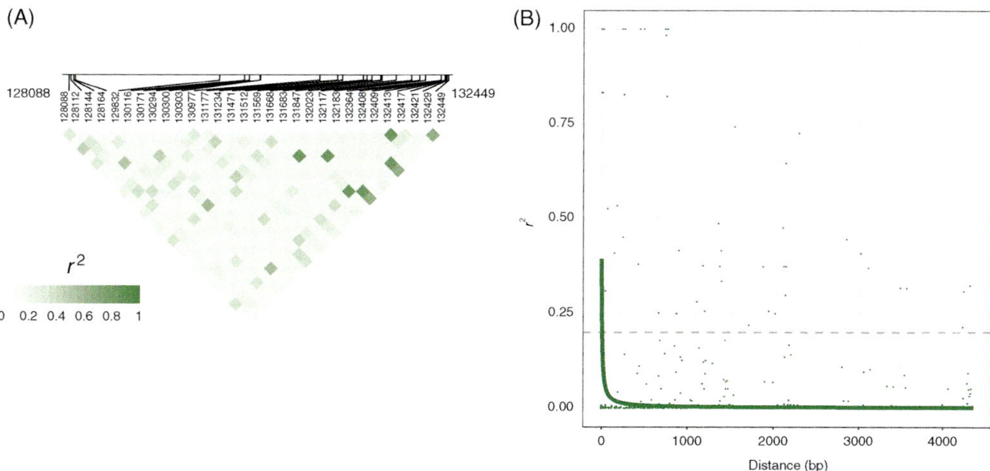

Fig. 4.1. Linkage disequilibrium (measured as r^2) between DNA polymorphisms discovered in the *Populus deltoides* homologue of the gene *POTR_0001s00260*, characterized in a natural population. (A) The upper part of the figure displays the polymorphic sites along the gene *POTR_0001s00260*, and the location of these sites between positions 128088 and 132449. The lower part (heat map) shows the value of r^2 between each pair of polymorphic sites. (B) The plot displays r^2 measured between pairs of polymorphic sites (*y*-axis) observed in gene *POTR_0001s00260* relative to their distance in base pairs between these sites (*x*-axis). The line represents the fitted non-linear regression. (Data from Fahrenkrog *et al.*, 2017.)

a significantly slower decline in some tree species, such as black cottonwood, extending to 3–6 kbp (Slavov *et al.*, 2012). Another study in *Eucalyptus* also supported a similarly slow decline in LD, which decayed within 4–6 kbp (Silva-Junior and Grattapaglia, 2015). While there is significant variation in LD along the genome, these contrasting results suggest bias in the selection of loci and samples. This bias may be explained at least in part by the difference in approach of selecting DNA variants for LD analysis. Most early studies that focused on specific genes considered all genetic variants in the estimation of LD, including those that are rare in the population (<1% minor allele frequency). In contrast, studies reporting higher LD generally considered loci where the alternative variant is at a frequency higher than 5–10%. The relationship between LD and the minor allele frequency is well known—previous population genomic studies demonstrated that a slower LD decay was observed when higher minor allele frequency thresholds were used (Yan *et al.*, 2009).

Population Genomic Structure

Populations are traditionally defined as groups of individuals that can interbreed and are, to some extent, reproductively isolated from other potential mates. The separation of individuals into isolated groups can result in distinct allele frequencies relative to other populations, a phenomenon referred to as genomic structure. Genomic structure can arise because of random genetic drift, which results in the fixation and loss of random alleles in different populations. The magnitude of random genetic drift is dependent on factors such as the size of each population and number of generations since they became isolated. Populations may also become differentiated from each other due to mutations that occur in only certain groups of individuals, and natural selection favoring specific alleles that provide a fitness advantage in certain environments.

Knowledge about genomic structure is important to inform management and conservation strategies (Boyce, 1992; Waples and Gaggiotti, 2006). In genome-wide association mapping studies, population structure can lead to the detection of spurious associations between traits and genetic markers (Chapter 5, this volume; Pritchard and Rosenberg, 1999). The same is true for environmental association studies, where the objective is to identify a correlation between markers and environmental variables (Rellstab *et al.*, 2015).

Quantifying genomic differentiation among known populations

In certain instances, the existence of genomic differentiation among groups of individuals (referred to hereafter as subpopulations) is known. In this case, population genomic studies primarily aim to uncover the extent of the differentiation among these subpopulations.

Population genomic structure results in the reduction of heterozygosity within groups of isolated individuals, relative to the heterozygosity resulting from combining all individuals in a single population. This occurs because random genetic drift and inbreeding resulting from mating between relatives increases in subpopulations, relative to a population that combines all individuals in a single, random mating group. This phenomenon was recognized by Sewall Wright, who proposed a simple measure of population differentiation based on the reduction of heterozygosity in subpopulations, relative to the heterozygosity that would result from individuals from all subpopulations being combined into a single population (Box 4.3). If all subpopulations have identical allele and heterozygote frequencies, they can be considered genetically identical. In contrast, if subpopulations have lost or fixed alternative alleles for a given locus or have different allele frequencies, then genetic structure is present. This principle is summarized in the fixation index, or F_{ST}, which quantifies the reduction in heterozygosity in a group of individuals (subpopulation), relative to the entire or total population (described in Eqn. 4.6 below) (Wright, 1943). As initially described by Wright, F_{ST} can be estimated at different levels in the hierarchy of a species

population. F_{ST} can be estimated for each locus individually; this approach is often used to identify genes that are under natural or artificial selection (e.g. F_{ST} outlier tests; see below). However, if general inferences are to be made about the existence of genetic structure among subpopulations, then multiple loci need to be used in the calculation of F_{ST}. Furthermore, to avoid bias in the detection of population structure, neutral loci (i.e. loci that are not under selection) are used. Microsatellites and SNPs in intergenic regions are often preferred for these analyses. F_{ST} is calculated as follows:

$$F_{ST} = \frac{H_T - H_S}{H_T} \tag{4.6}$$

where H_T is the expected heterozygosity of the total population, and H_S is the heterozygosity of the subpopulations (Box 4.3). There is no simple threshold to define whether an F_{ST} measure is significant (i.e. whether there is genetic differentiation between two populations). However, Wright proposed that F_{ST} values less than 0.05 indicate low genetic differentiation, values between 0.05 and 0.25 indicate moderate to high differentiation, and values greater than 0.25 suggest very high differentiation.

Uncovering unknown population structure

An estimate of F_{ST} implies that the subpopulation membership of each individual is known. However, population genomic analysis often involves samples of unknown origin, or from populations with an unknown genetic relationship. In this case, population structure can be determined in various ways. Distance-based methods rely on developing a pairwise distance matrix using genetic markers. For example, Nei's genetic distance is based on the probability that two populations share the same allele at a given locus (Nei, 1972). Individuals are then grouped based on the genetic distance between them: those with shorter distances are clustered together, while those that are more distant are clustered at a higher level. The data are

---

Box 4.3. Wright's Fixation Index

Assuming that a founder population is divided into subpopulations (A, B, and C), alleles are expected to become fixed or lost in subsequent generations due to random genetic drift, leading to a reduction in heterozygosity. For simplicity, two alleles of a hypothetical locus are represented by the color (*white* or *green*) of each individual in the population.

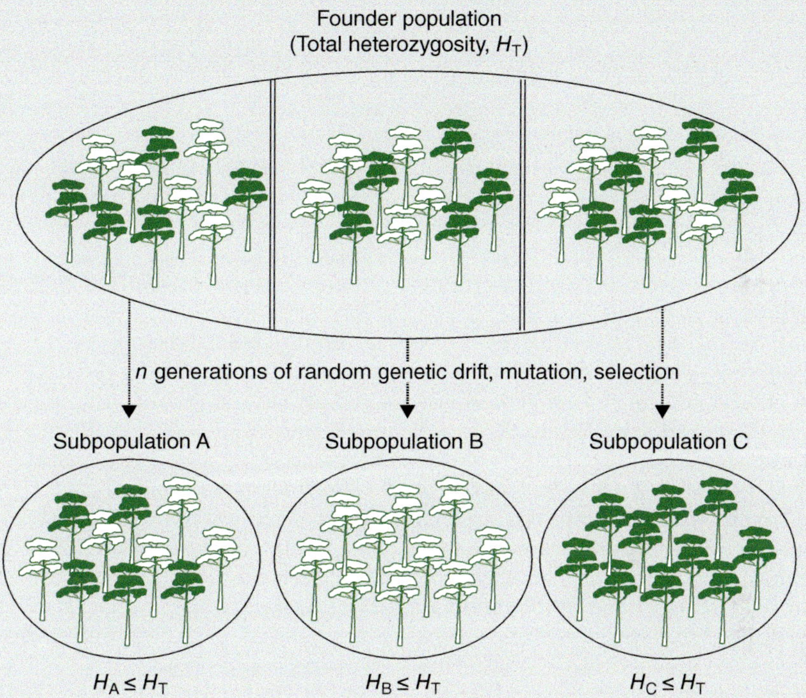

The expected heterozygosity of the subpopulations (H_A, H_B, and H_C) can be estimated based on the observed allele frequency. For the subpopulations B and C, the expected heterozygosity is zero because the alleles *white* and *green* became fixed, respectively. In subpopulation A, the frequency of both alleles, and the expected frequency of the heterozygote, is 0.5.

Subpopulation A		Subpopulation B		Subpopulation C	
Allele	Frequency	Allele	Frequency	Allele	Frequency
white	$f(w) = 0.5$	*white*	$f(w) = 1.0$	*white*	$f(w) = 0.0$
green	$f(g) = 0.5$	*green*	$f(g) = 0.0$	*green*	$f(g)=1.0$

$H_A = 2 \times f(w) \times f(g) = 2 \times 0.5 \times 0.5 = 0.5$
$H_B = 2 \times f(w) \times f(g) = 2 \times 1.0 \times 0.0 = 0$
$H_C = 2 \times f(w) \times f(g) = 2 \times 0.0 \times 1.0 = 0$

The expected heterozygosity of the subpopulations (H_S) can be calculated from the average of the subpopulation heterozygosities:

$H_S = (H_A + H_B + H_C)/3 = (0.5 + 0 + 0)/3 = 0.167$

Continued

Box 4.3. Continued.

While the founder population is not evaluated, its frequency of *white* and *green* alleles can be inferred from their frequency in the subpopulations A, B, and C:

Founder population $f(w) = (f(w_A) + f(w_B) + f(w_C))/3 = = (0.5 + 1 + 0)/3 = 0.5$

The expected heterozygosity of the founder population can be estimated as:

$H_T = 2 \times f(w) \times f(g) = 2 \times 0.5 \times (1 - 0.5) = 0.5$

In conclusion, the F_{ST} is:

$F_{ST} = (H_T - H_S)/H_T = (0.5 - 0.167)/0.5 = 0.67$

---

typically represented in a graphical form (e.g. a tree) that needs to be visually inspected to identify individuals that are likely to belong to the same subpopulation.

More recently, distance-based methods have largely been replaced by model-based approaches. In this case, individuals are considered a random sample derived from their respective subpopulations. Using methods such as Bayesian statistics, individuals are clustered iteratively into groups. At each iteration, the probability of each cluster is evaluated based on various criteria, such as departure from HWE. The process is repeated until reshuffling of individuals among clusters no longer results in an improvement in the criterion used to determine the optimal separation of individuals into groups. Possibly the most widely used approach that applies model-based clustering was described by Jonathan K. Pritchard (Pritchard *et al.*, 2000; Gilbert, 2016), which uses a Bayesian model to group individuals into subpopulations (Fig. 4.2A). This method uses genetic markers in linkage equilibrium to group individuals into clusters and simultaneously estimates population allele frequencies to obtain subpopulations in HWE. This process is run for a user-defined number of subpopulations (K) to be tested. The optimal number of subpopulations present in the sample under study can be inferred from the K value, where the log probability of the data stops increasing and reaches a plateau (Pritchard *et al.*, 2000). Another popular method to select the most likely number of subpopulations is the ΔK statistic. In this case, the magnitude of the change in the probability of the data is compared when successive K values are tested (Evanno *et al.*, 2005). Due to the uncertainty of the clustering process, both methods provide only an approximate estimate of the true number of subpopulations. Analysis of all K values with a plausible biological explanation has been recommended (Meirmans, 2015), and other statistically rigorous approaches to define the optimal number of subpopulations have been proposed more recently (Verity and Nichols, 2016).

Another popular method for the assessment of population structure is principal component analysis (PCA). This is a statistical method that converts the molecular data derived from each individual into uncorrelated variables, referred to as the principal components (Fig. 4.2B). The first component represents the one that captures the highest amount of the variance in the data and may consequently separate individuals into the most distinct subpopulations. Additional components can also be evaluated and correspond to finer subpopulation differences that may not be detected in the first principal component. However, PCA does not assign individuals to specific groups. Instead, it identifies the main axes of genetic variation present in a sample and provides the coordinates along those axes for each individual (Patterson *et al.*, 2006).

Finally, it is worth noting that population structure is normally determined based on genomic variation in the nuclear DNA, but additional sources of data from organelle DNA may be available and can provide added information about population differentiation. While the nuclear genome informs the effects of gene flow through pollen and seed

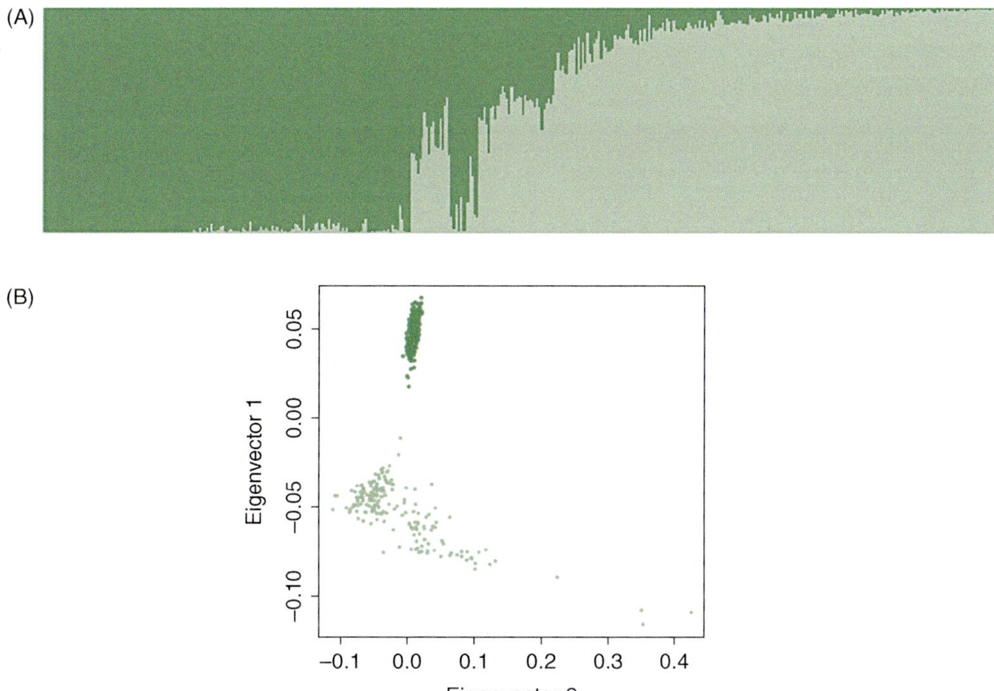

Fig. 4.2. Population genomic structure of *Populus deltoides*. (A) Ancestry coefficient bar plots obtained by model-based clustering of 425 individuals of *P. deltoides* in two subpopulations ($K = 2$). Individual columns represent the membership coefficients of each individual evaluated (indicated by dark and light bars), a measure of the probabilities of belonging to each of the subpopulations. (B) Subpopulation structure revealed by principal component analysis of the 425 individuals of the *P. deltoides* population. The analysis identified two major clusters of individuals (dark and light dots), separated in the first principal component (eigenvector 1). (Data from Fahrenkrog *et al.*, 2017.)

dispersal on genetic structure, organelle genomes (haploid and mostly maternally inherited in flowering plants) reflect only gene flow through seed dispersal (Ennos, 1994).

Population structure in forest species

Forest-tree species often have low genetic differentiation among populations. As a consequence, the genetic diversity within populations is often larger than among populations of the same species, although exceptions exist.

The genomic structure of natural forest populations has been studied extensively with isozymes since their use became widespread in the 1980s. An extensive survey of studies published up to 1990 reported data on 213 woody species and showed that those that are long lived have a significantly lower G_{ST} (a measure of population genetic differentiation analogous to F_{ST} but for multiple alleles; Nei, 1973) than annual species (0.08 and 0.36, respectively) (Hamrick and Godt, 1996). Of significance also was the geographic range of the species: a G_{ST} of 0.12–0.14 was detected for species with a narrow or endemic range, but this was lowered to 0.03 in species that were distributed more widely. These studies later expanded to PCR-based genetic markers and more informative loci, such as microsatellites, and largely confirmed the trends reported previously using isozymes, but still remained largely restricted to relatively few genomic regions.

The development of methods to reduce the complexity of genomes and the dramatic decrease in the cost and increase in throughput of sequencing (Chapter 1, this volume) greatly expanded population genetic structure studies to larger populations and a greater number of loci. One of the first genome-wide surveys of genetic diversity and the subsequent population structure analysis was carried out for black alder (*Alnus glutinosa*) by de Kort *et al.* (2014), using a genotyping-by-sequencing approach, which is based on restriction digestion (Chapter 1, this volume). In this case, the loci analyzed were not selected *a priori* and, due to the lack of a genome sequence for *Alnus* spp., there was no knowledge about their distribution in the genome. Nevertheless, the detection of almost 2,000 polymorphic loci suggested that a large proportion of the genome was represented. Black alder is a widely distributed tree species in Europe that is self-incompatible and wind pollinated, all factors that contribute to a very low F_{ST} (<0.044) being detected in all population pairwise comparisons. In parallel, the first study to resequence large populations of black cottonwood resulted in the detection of 17.9 million SNPs, a fraction of which were used to cluster the population into four genetic groups, and several subgroups by PCA (Evans *et al.*, 2014).

Subsequently, several genome-wide studies ensued for species within the genera *Populus*, *Eucalyptus*, and *Pinus*. A study with eastern cottonwood (*Populus deltoides*) assessed population structure in a sample of 425 unrelated individuals using 8,664 intergenic SNPs and detected four geographically distinct subpopulations (Fahrenkrog *et al.*, 2017). Two western subpopulations showed low genetic differentiation (F_{ST} = 0.022), although higher differentiation was uncovered between more distantly located subpopulations (pairwise F_{ST} = 0.07–0.10). In a study with *Pinus taeda*, where several hundred samples were genotyped for over 30,000 SNPs in coding and non-coding regions, model-based clustering identified two main subpopulations (average F_{ST} = 0.026), located east and west of the Mississippi River (Lu *et al.*, 2016). While these studies suggest that the genomic structure of tree populations is generally limited, there is clear variation among species, which is heavily dependent on characteristics such as the range of their natural distribution. For example, a survey of F_{ST} measured for 600 loci in natural populations of *Eucalyptus grandis*, *E. camaldulensis*, and *E. globulus* clearly showed a pattern of higher F_{ST} in the latter (*E. grandis* and *E. camaldulensis*, F_{ST} <0.05; *E. globulus* F_{ST} >0.05). *Eucalyptus grandis* and *E. camaldulensis* are widely distributed geographically along the east coast of Australia. In contrast, *E. globulus* has a much narrower geographic distribution that is restricted to Tasmania, southern Victoria, and a few islands in Bass Strait. Smaller, isolated populations likely explain the higher F_{ST} detected among these populations.

Identification of Genes and Genomic Regions that Regulate Fitness and Adaptation

The identification of genes involved in fitness and adaptation has been a major focus of population genomics studies. Knowing which genes contribute to adaptation helps us understand how populations respond to selection forces. For a breeder, the identification of genes implicated in adaptation can be used to select trees that are more suitable for introduction in certain environments, or support transfer of germplasm to a new location. Knowledge of the genes implicated in adaptation, and their individual effects, also helps predict the performance of populations in a changing environment (Sork *et al.*, 2013).

Adaptive evolution can occur through development of new mutations that provide higher fitness, or through selection acting on polymorphisms that are already present in a population. These genetic variants can be identified by several strategies (Fig. 4.3). The first approach relies on the detection of a correlation between DNA variation

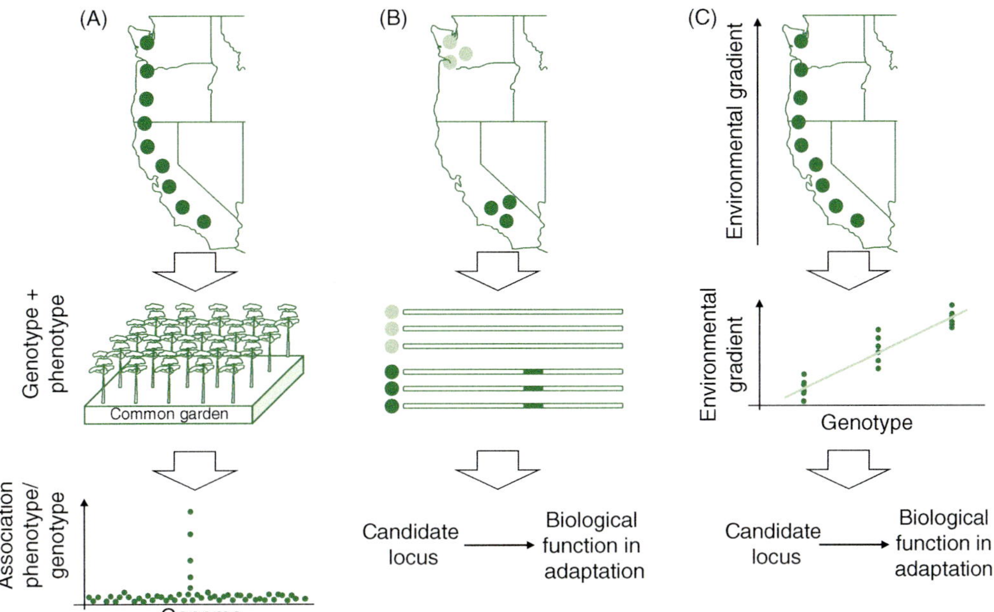

Fig. 4.3. Principles of identification of loci that contribute to fitness and adaptation. (A) Linkage and association mapping relies on genotyping individuals collected in populations adapted to different environments, and phenotyping them in common gardens for identification of loci associated with adaptive traits. (B) Outlier tests compare genetic parameters between contrasting populations to identify loci that depart from the expectation of neutral loci. (C) Environmental association analysis evaluates the correlation between allele frequency and environmental variables, while correcting for population structure. Both outlier tests and environmental association analyses are followed by the evaluation of candidate loci for potential roles in adaptive traits.

and traits related to fitness and adaptation, using methods such as linkage and association mapping. The second approach seeks to identify loci that show high genetic differentiation among populations. These loci are detected through outlier tests, by comparing them with neutral loci. A third approach tests for association between loci and the environment. The method uses environmental association analysis and is often referred to as landscape genomics because it seeks to relate DNA variants with differences in climate and other environmental variables. All these methods have significant limitations and are often used in combination so that there are multiple lines of evidence to support the relevance of specific genes in adaptation. These methods are reviewed in the following sections, including examples of their application to studies of forest-tree populations.

Linkage and association mapping

Linkage and association mapping approaches are based on identifying correlations between alleles and phenotypes in genetically structured (e.g. pseudo-backcross) or unstructured (e.g. unrelated individuals collected in the wild) populations (Fig. 4.3A). For such an analysis, individuals from the population are genotyped for genetic markers and phenotyped for the adaptive trait. Next, genetic markers are tested statistically to determine whether different alleles segregate with the trait. If there is co-segregation, this suggests that the marker is at or in close proximity to the genetic locus that regulates the trait. Details about the statistical methods of detecting association between genetic markers and traits are described in Chapter 5 (this volume).

Linkage and association mapping were initially developed for the identification of genes that regulate quantitative traits, for breeding and selection of elite individuals. However, it was recognized that this type of analysis could be expanded beyond traits related to productivity (e.g. volume growth and disease resistance) to include those related to environmental adaptations, such as phenology (e.g. bud set). For this type of analysis, populations typically consist of unrelated individuals or progeny from parents that originate from (i.e. are adapted to) different environments. These populations are then grown in common gardens, where they are exposed to the same environmental conditions, to be phenotyped. In some cases, several common gardens, established under different environments, are used.

Linkage and association mapping of traits related to evolution and adaptation has the advantage of being based on many years of established theory and statistical methods, previously developed for model systems (Mackay *et al.*, 2009). Methods to uncover markers correlated to phenotypic traits have advanced dramatically since Karl Sax used these principles for studying qualitative traits co-segregating with seed weight (Sax, 1923). For instance, the challenge of controlling for false positives is well documented, and approaches to overcome this difficulty have been proposed (Churchill and Doerge, 1994). Finally, applying linkage and association mapping to uncover genes that regulate adaptive traits offers a way to clearly demonstrate a relationship between a locus and a phenotype. Thus, if the adaptive trait under analysis is flowering time, a locus associated with it can be categorically assigned to the regulation of that trait.

A drawback of the use of linkage and association mapping approaches is that traits that are critical for adaptation of populations need to be known. Traits such as time to bud set or flowering are clearly adaptive for forest populations growing along a gradient of daylength. In contrast, forest populations growing under what appears to be similar environmental conditions may be under distinct selective pressure from less evident sources of stress (e.g. diseases).

Linkage and association mapping approaches also require growing the population in common gardens, which can be logistically difficult to establish and maintain. In addition, phenotypes may only be evident after an extended period of time because many important adaptive traits (e.g. flowering) are expressed later in the plant's life cycle. Finally, while there is a clear understanding of many of the challenges in linkage and association mapping studies, there is no obvious solution to address some of them. For example, many adaptive traits are likely to be highly complex (i.e. regulated by many genes of small effect). The statistical power to identify these loci is limited when small experimental populations are used, as is often the case in forestry. These and other challenges of linkage and association mapping methods have been extensively documented and remain, at least partially, unresolved (Strauss *et al.*, 1992; Beavis, 1998; Mackay *et al.*, 2009).

Despite the challenges of applying linkage and association mapping to uncover genes involved in the regulation of adaptive traits, genetic mapping of these traits was widely used before genome-wide methods of genotyping became available. Linkage mapping of adaptive traits was pioneered in a study to uncover loci regulating timing of growth initiation and cessation in Douglas-fir, in a progeny of 460 individuals that were genetically mapped and phenotyped in a controlled environment (Jermstad *et al.*, 2003). As the availability of genomic data and high-throughput genotyping methods evolved, linkage and association mapping studies for adaptive traits were expanded to other species and mapping populations, including in white spruce (*Picea glauca*) (Pelgas, 2011), black spruce (*Picea mariana*) (Prunier *et al.*, 2013), and black cottonwood (Evans *et al.*, 2014).

Outlier tests

Outlier tests rely on identifying loci that show unique genetic properties, or signatures of selection, that distinguish them from the rest of the genome (Fig. 4.3B).

For instance, the fixation of an allele or haplotype in a given population suggests that selection has acted on it. Following the detection of outlier loci, the next step is to relate them to an adaptive role.

Approaches to detect signatures of selection can be separated into those based on analysis of genetic diversity and site frequency spectrum to identify selection signatures across the genome (e.g. Tajima's D, see below) or the identification of loci under divergent selection among populations (e.g. F_{ST} outliers). By comparing the allele frequency spectrum of these alternative parameters with the frequency of neutral loci, it is possible to infer the occurrence of demographic pressures such as natural selection.

The identification of signatures of selection is an unbiased approach to detect genes related to adaptation, as long as it is based on a genome-wide survey. However, just like linkage and association mapping approaches, it may be difficult to identify significant loci if the adaptive trait is highly complex and controlled by many genes of small effect. Furthermore, adaptation to local conditions may cause only subtle changes in allele frequencies that are difficult to detect. These tests also have the drawback that, when an outlier is found, it is often unclear how it relates to a given evolutionary process because the trait regulated by the outlying adaptive locus is unknown.

Site frequency spectrum

An allele or site frequency spectrum is the distribution of allele frequencies in a population. Many different parameters derived from the allele frequency spectrum can be estimated, including the nucleotide diversity (θ) and the number of pairwise nucleotide differences (Π). The relationship between θ and Π is the basis for estimating Tajima's D (Tajima, 1989), which is used to identify loci that are evolving under some evolutionary force, such as natural selection, relative to neutral loci. In a population of constant size, neutral loci are expected to vary depending on the mutation rate and random genetic drift. Under these conditions, estimates of θ and Π are expected to be the same. However,

if other evolutionary forces are acting on a population, deviations are expected. These deviations occur because θ increases when the number of low-frequency variants increases, while Π is based on the average of pairwise sequence differences and is not greatly affected by low-frequency polymorphisms. Tajima's D is based on the difference between the two: $\Pi - \theta$. A negative Tajima's D indicates a selective sweep because, in this case, there is reduced variation in the nucleotides near a favored mutation in DNA. A reduction in variation results in a decrease in Π (fewer pairwise sequence differences), while θ remains largely constant. A positive Tajima's D is expected at loci undergoing balancing selection because, under these circumstances, frequencies of alternative alleles at a locus are expected to be similar. Therefore, pairwise sequence differences will be detected often between any two samples. As expected, Tajima's D is negative at a genome-wide scale when examining genic regions, where non-synonymous mutations are commonly subjected to the action of purifying selection (Subramanian, 2016).

F_{ST} outliers

When selection is acting on a locus, the frequency of a favored allele is expected to increase. As a result, there is a reduction in heterozygosity among individuals in the population under selection. When compared with other populations that are not under the same selection pressure, expected differences in the level of heterozygosity can be detected by an F_{ST} outlier test. Briefly, loci that display F_{ST} values that indicate a higher genetic distance between populations than observed for neutral loci are candidates for divergent selection. Detection of F_{ST} outliers can be hampered by evolutionary processes that affect the entire population, or the presence of genetic structure. Some of these challenges are partially addressed by the use of methods that simulate the distribution of neutral loci and that account for population genetic structure to establish better-defined significance thresholds (Beaumont and Balding, 2004; Excoffier *et al.*, 2009).

Outlier tests can easily be applied to analysis of populations because they do not require any information other than genotypic data and knowledge about the population from which the individuals came. Most importantly, the application of outlier tests avoids the need to establish common gardens and phenotype the adaptive traits. As the cost of genome-wide genotyping decreased and that of establishing field trials and phenotyping remained constant, studies have increasingly emphasized the use of outlier tests, as can be seen from the rapid expansion in their use in forest-tree species with moderate genome sizes (Evans *et al.*, 2014; Zhou *et al.*, 2014; Fahrenkrog *et al.*, 2017).

Environment association analysis

Environmental association analysis seeks to identify loci related to adaptation, based on their variation relative to environmental influences (Fig. 4.3C) (Nielsen, 2005; Siol *et al.*, 2010; Rellstab *et al.*, 2015). Identifying correlations between genetic and environmental variation is based on statistical principles that are well established. However, applying these principles to the identification of loci related to environmental variation can be challenging. First, there are numerous environmental variables that can be considered. For example, bioclimatic data are available from the WorldClim database (www.worldclim.org, accessed July 15, 2019) for 19 variables (e.g. temperature and precipitation). Because many of these variables are highly correlated, a PCA is often conducted to summarize the data into its main components, which are then analyzed. Another significant challenge related to these analyses is the fact that environmental association methods are affected by population structure (Siol *et al.*, 2010). To address this limitation, improved algorithms have been developed in an attempt to correct for confounding factors. For instance, BAYENV2 (Günther and Coop, 2013) uses a variance–covariance matrix of allele frequencies among (sub)populations (Ω) to correct for evolutionary history. In another approach, tests for correlation between environmental and genetic variation simultaneously correct for the effect of population structure (latent factors) (Frichot *et al.*, 2013). Much like outlier tests, the use of environmental association analysis methods has been expanded to include studies targeting candidate genes (Prunier *et al.*, 2012); reduced representations of genomes based on sequence capture (Fahrenkrog *et al.*, 2017); and genome-wide analysis (Evans *et al.*, 2014).

Combining strategies and species convergence to support the identification of loci related to adaptation

Each of the approaches described above adopts independent strategies for identification of loci involved in adaptation, but each has its own strengths and weaknesses. Combining results derived from multiple strategies to identify common candidate loci can reduce the likelihood that they are false positives. For example, environmental association analyses and outlier tests are often used in concert (Fig. 4.4) (Evans *et al.*, 2014; Fahrenkrog *et al.*, 2017).

As more population genomic studies are performed for related species, another approach, which is based on the detection of convergent local adaptation, has recently emerged to support the identification of loci related to adaptation. Yeaman *et al.* (2016) evaluated DNA variants uncovered in distantly related conifers species (*Pinus contorta* and species within *Picea*) for convergent evolution. These species are naturally distributed along a similar environmental gradient in North America, creating the possibility of evaluating orthologous loci for their involvement in adaptation and convergent evolution. To evaluate this hypothesis, individuals from a large number of populations from across the native range were genotyped for DNA polymorphisms in 23,000 genes. The analysis uncovered a set of 47 genes associated with spatial variation in temperature or cold hardiness in these species. While conservative, this approach offered a heightened level of confidence in the relevance of these genes for adaptation.

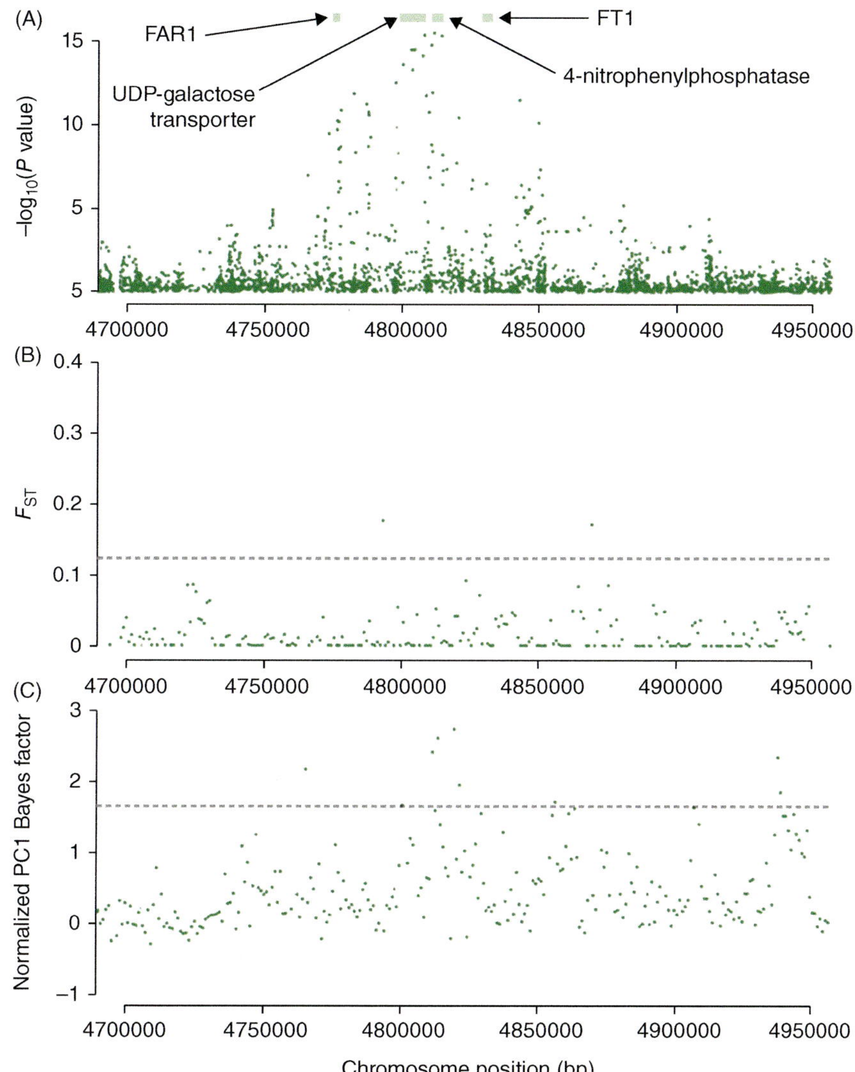

Fig. 4.4. Identification of a genomic region associated with bud flush in *Populus trichocarpa* using different strategies. (A) A genome-wide analysis identified several loci significantly associated with bud flush in chromosome 8. (B, C) An F_{ST} outlier (B) and several BAYENV outliers (C) were identified in the same region. In combination, these analyses support the hypothesis that a locus in chromosome 8 is involved in the regulation of bud flush and adaptation. FAR1, FAR-RED IMPAIRED RESPONSE 1; FT1, FLOWERING LOCUS T 1. (From Evans *et al.*, 2014.)

Summary

Evolutionary forces such as mutation, migration, and random genetic drift can cause changes to allele frequencies in populations. Population genomics aims to characterize these changes and understand their causes and consequences, increasingly using advanced next-generation sequencing (NGS) methods.

Most often, the primary parameter to be estimated in populations is their genetic diversity. This includes the number and frequency of segregating sites, and other parameters derived from it, such as nucleotide polymorphism and diversity. Because forest-tree species are generally long lived, out-crossing, and disperse pollen and seeds over long distances, there is often a low incidence of random genetic drift. This results in high levels of genetic diversity within populations.

Populations are also often characterized for the extent of LD, a measure of the non-random relationship between alleles at different loci. Because LD depends on the shared ancestry of the alleles at different loci, it reflects several population and species life-history traits, such as their mating system, and selection acting on specific loci. Because most forest-tree species are out-crossing and, consequently, have very low levels of LD. As LD also often decays much faster in tree species than in most crops, because of the higher effective size of these populations. Knowing the extent of LD in the genome of individuals from a population is relevant for estimation of several population genetic parameters, identification of loci under selection, and informing the resolution of genome-wide association studies.

Populations are often isolated from each other, resulting in distinct allele frequencies or genomic structure. Knowledge about genomic structure is relevant to inform management and conservation strategies, and in genome-wide association mapping studies, where population structure can lead to spurious associations. When the existence of differentiation among groups of individuals or populations is known, it can be quantified by the fixation index (F_{ST}). When it is unknown and there is population structure in a sample of individuals, this can be tested by model-based methods that apply Bayesian statistics, or by PCA. Because forest-tree species are generally long lived, out-crossing, and disperse pollen and seeds over long distances, there is typically low genetic differentiation among populations.

The development of methods to characterize whole-genome sequence variation has led to the pursuit of genes and genetic variants involved in fitness and adaptation. These variants can be identified by detecting correlations between DNA variation and traits related to fitness and adaptation.

Alternatively, loci of interest are detected based on their genetic differentiation among populations, through outlier tests, by comparing them with neutral loci. Finally, these loci can also be detected based on the association between their variation and environmental stimuli, in landscape genomics. These approaches represent distinct strategies for identification of loci involved in adaptation, and have unique strengths and weaknesses. Thus, the results from multiple approaches are often used in parallel to identify common loci, and reduce the likelihood of false positives.

References

Aitken, S.N., Yeaman, S., Holliday, J.A., Wang, T., and Curtis-McLane, S. (2008) Adaptation, migration or extirpation: climate change outcomes for tree populations. *Evolutionary Applications* 1, 95–111.

Alberto, F.J., Aitken, S.N., Alía, R., González-Martínez, S.C., Hänninen, H., *et al.* (2013) Potential for evolutionary responses to climate change – evidence from tree populations. *Global Change Biology* 19, 1645–1661.

Beaumont, M.A. and Balding, D.J. (2004) Identifying adaptive genetic divergence among populations from genome scans. *Molecular Ecology* 13, 969–80.

Beavis, W. (1998) Limitations of QTL mapping. In: Paterson, A.H. (ed.) *Molecular Dissection of Complex Traits*. CRC Press, Boca Raton, Florida, pp. 145–162.

Bonan, G.B. (2008) Forests and climate change: forcings, feedbacks, and the climate benefits of forests. *Science* 320, 1444–1449.

Boyce, M.S. (1992) Population viability analysis. *Annual Review of Ecology and Systematics* 23, 481–506.

Brown, G.R., Gill, G.P., Kuntz, R.J., Langley, C.H., and Neale, D.B. (2004) Nucleotide diversity and linkage disequilibrium in loblolly pine. *Proceedings of the National Academy of Sciences USA* 101, 15255–15260.

Churchill, G.A. and Doerge, R.W. (1994) Empirical threshold values for quantitative trait mapping. *Genetics* 138, 963–971.

de Kort, H., Vandepitte, K., Bruun, H.H., Closset-Kopp, D., Honnay, O., and Mergeay, J. (2014) Landscape genomics and a common garden trial reveal adaptive differentiation to temperature across Europe in the tree species *Alnus glutinosa*. *Molecular Ecology* 23, 4709–4721.

Eckert, A.J., Wegrzyn, J.L., Liechty, J.D., Lee, J.M., Cumbie, W.P., *et al.* (2013) The evolutionary genetics of the genes underlying phenotypic associations for loblolly pine (*Pinus taeda*, Pinaceae). *Genetics* 195, 1353–1372.

Ennos, R.A. (1994) Estimating the relative rates of pollen and seed migration among plant populations. *Heredity* 72, 250–259.

Evanno, G., Regnaut, S., and Goudet, J. (2005) Detecting the number of clusters of individuals using the software STRUCTURE: a simulation study. *Molecular Ecology* 14, 2611–2620.

Evans, L.M., Slavov, G.T., Rodgers-Melnick, E., Martin, J., Ranjan, P., *et al.* (2014) Population genomics of *Populus trichocarpa* identifies signatures of selection and adaptive trait associations. *Nature Genetics* 46, 1089–1096.

Excoffier, L., Hofer, T., and Foll, M. (2009) Detecting loci under selection in a hierarchically structured population. *Heredity* 103, 285–298.

Fahrenkrog, A.M., Neves, L.G., Resende, M.F.R., Dervinis, C., Davenport, R., *et al.* (2017) Population genomics of the eastern cottonwood (*Populus deltoides*). *Ecology and Evolution* 7, 9426–9440.

Frichot, E., Schoville, S.D., Bouchard, G., and François, O. (2013) Testing for associations between loci and environmental gradients using latent factor mixed models. *Molecular Biology and Evolution* 30, 1687–1699.

Gilbert, K.J. (2016) Identifying the number of population clusters with STRUCTURE : problems and solutions. *Molecular Ecology* 16, 601–603.

Goddard, M.E. and Hayes, B.J. (2009) Mapping genes for complex traits in domestic animals and their use in breeding programmes. *Nature Reviews Genetics* 10, 381–391.

Günther, T. and Coop, G. (2013) Robust identification of local adaptation from allele frequencies. *Genetics* 195, 205–220.

Hamrick, J.L. and Godt, M.J.W. (1996) Effects of life history traits on genetic diversity in plant species. *Philosophical Transactions of the Royal Society B: Biological Sciences* 351, 1291–1298.

Hardy, G.H. (1908) Mendelian proportions in a mixed population. *Science* 28, 49–50.

Hartl, D.L. and Clark, A.G. (2007) *Principles of Population Genetics*, 4th edn. Sinauer Associates, Sunderland, Massachusetts.

Ingvarsson, P.K. (2005) Nucleotide polymorphism and linkage disequilibrium within and among natural populations of European aspen (*Populus tremula* L., Salicaceae). *Genetics* 169, 945–953.

Jermstad, K.D., Bassoni, D.L., Jech, K.S., Ritchie, G.A., Wheeler, N.C., and Neale, D.B. (2003) Mapping of quantitative trait loci controlling adaptive traits in coastal Douglas fir. III. Quantitative trait loci-by-environment interactions. *Genetics* 165, 1489–1506.

Kado, T., Yoshimaru, H., Tsumura, Y., and Tachida, H. (2003) DNA variation in a conifer, *Cryptomeria japonica* (Cupressaceae sensu lato). *Genetics* 164, 1547–1559.

Kimura, M. (1968) Evolutionary rate at the molecular level. *Nature* 217, 624–626.

Krutovsky, K.V. and Neale, D.B. (2005) Nucleotide diversity and linkage disequilibrium in cold-hardiness- and wood quality-related candidate genes in Douglas fir. *Genetics* 171, 2029–2041.

Kurz, W.A., Dymond, C.C., Stinson, G., Rampley, G.J., Neilson, E.T., *et al.* (2008) Mountain pine beetle and forest carbon feedback to climate change. *Nature* 452, 987–990.

Lu, M., Krutovsky, K.V., Nelson, C.D., Koralewski, T.E., Byram, T.D., and Loopstra, C.A. (2016) Exome genotyping, linkage disequilibrium and population structure in loblolly pine (*Pinus taeda* L.). *BMC Genomics* 17, 730.

Lynch, M. (2010) Evolution of the mutation rate. *Trends in Genetics* 26, 345–352.

Mackay, T.F.C., Stone, E.A., and Ayroles, J.F. (2009) The genetics of quantitative traits: challenges and prospects. *Nature Reviews Genetics* 10, 565–577.

Meirmans, P.G. (2015) Seven common mistakes in population genetics and how to avoid them. *Molecular Ecology* 24, 3223–3231.

Neale, D.B. and Savolainen, O. (2004) Association genetics of complex traits in conifers. *Trends in Plant Science* 9, 325–330.

Nei, M. (1972) Genetic distance between populations. *American Naturalist* 106, 283–292.

Nei, M. (1973) Analysis of gene diversity in subdivided populations. *Proceedings of the National Academy of Sciences USA* 70, 3321–3323.

Nei, M. and Li, W.H. (1979) Mathematical model for studying genetic variation in terms of restriction endonucleases. *Proceedings of the National Academy of Sciences USA* 76, 5269–5273.

Neves, L.G., Davis, J.M., Barbazuk, W.B., and Kirst, M. (2013) Whole-exome targeted sequencing of the uncharacterized pine genome. *Plant Journal* 75, 146–156.

Nielsen, R. (2005) Molecular signatures of natural selection. *Annual Review of Genetics* 39, 197–218.

Nordborg, M. (2000) Linkage disequilibrium, gene trees and selfing: an ancestral recombination graph with partial self-fertilization. *Genetics* 154, 923–929.

Ossowski, S., Schneeberger, K., Lucas-Lledó, J.I., Warthmann, N., Clark, R.M., *et al.* (2010)

The rate and molecular spectrum of spontaneous mutations in *Arabidopsis thaliana*. *Science* 327, 92–94.

Patterson, N., Price, A.L., and Reich, D. (2006) Population structure and eigenanalysis. *PLoS Genetics* 2, e190.

Pelgas, B., Bousquet, J., Meirmans, P.G., Ritland, K. and Isabel, N. (2011) QTL mapping in white spruce: gene maps and genomic regions underlying adaptive traits across pedigrees, years and environments. *BMC Genomics* 12, 145.

Pritchard, J.K. and Rosenberg, N.A. (1999) Use of unlinked genetic markers to detect population stratification in association studies. *American Journal of Human Genetics* 65, 220–228.

Pritchard, J.K., Stephens, M., and Donnelly, P. (2000) Inference of population structure using multilocus genotype data. *Genetics* 155, 945–959.

Prunier, J., Gérardi, S., Laroche, J., Beaulieu, J., and Bousquet, J. (2012) Parallel and lineage-specific molecular adaptation to climate in boreal black spruce. *Molecular Ecology* 21, 4270–4286.

Prunier, J., Pelgas, B., Gagnon, F., Desponts, M., Isabel, N., et al. (2013) The genomic architecture and association genetics of adaptive characters using a candidate SNP approach in boreal black spruce. *BMC Genomics* 14, 368.

Rellstab, C., Gugerli, F., Eckert, A.J., Hancock, A.M., and Holderegger, R. (2015) A practical guide to environmental association analysis in landscape genomics. *Molecular Ecology* 24, 4348–4370.

Sax, K. (1923) The association of size differences with seed-coat pattern and pigmentation in *Phaseolus vulgaris*. *Genetics* 8, 552–560.

Silva-Junior, O.B. and Grattapaglia, D. (2015) Genome-wide patterns of recombination, linkage disequilibrium and nucleotide diversity from pooled resequencing and single nucleotide polymorphism genotyping unlock the evolutionary history of *Eucalyptus grandis*. *New Phytologist* 208, 830–845.

Siol, M., Wright, S.I., and Barrett, S.C.H. (2010) The population genomics of plant adaptation. *New Phytologist* 188, 313–332.

Slavov, G.T., DiFazio, S.P., Martin, J., Schackwitz, W., Muchero, W., et al. (2012) Genome resequencing reveals multiscale geographic structure and extensive linkage disequilibrium in the forest tree *Populus trichocarpa*. *New Phytologist* 196, 713–725.

Sork, V.L., Aitken, S.N., Dyer, R.J., Eckert, A.J., Legendre, P., and Neale, D.B. (2013) Putting the landscape into the genomics of trees: approaches for understanding local adaptation and population responses to changing climate. *Tree Genetics and Genomes* 9, 901–911.

Strauss, S.H., Lande, R., and Namkoong, G. (1992) Limitations of molecular-marker-aided selection in forest tree breeding. *Canadian Journal of Forest Research* 22, 1050–1061.

Subramanian, S. (2016) The effects of sample size on population genomic analyses – implications for the tests of neutrality. *BMC Genomics* 17, 123.

Tajima, F. (1989) Statistical method for testing the neutral mutation hypothesis by DNA polymorphism. *Genetics* 123, 585–595.

Valladares, F., Matesanz, S., Guilhaumon, F., Araújo, M.B., Balaguer, L., et al. (2014) The effects of phenotypic plasticity and local adaptation on forecasts of species range shifts under climate change. *Ecology Letters* 17, 1351–1364.

Verity, R. and Nichols, R.A. (2016) Estimating the number of subpopulations (K) in structured populations. *Genetics* 203, 1827–1839.

Waples, R.S. and Gaggiotti, O. (2006) What is a population? An empirical evaluation of some genetic methods for identifying the number of gene pools and their degree of connectivity. *Molecular Ecology* 15, 1419–1439.

Waples, R.S. (2006) A bias correction for estimates of effective population size based on linkage disequilibrium at unlinked gene loci. *Conservation Genetics* 7, 167–184.

Watterson, G.A. (1975) On the number of segregating sites in genetical models without recombination. *Theoretical Population Biology* 7, 256–276.

White, T.L., Adams, W.T., and Neale, D.B. (2007) *Forest Genetics*. CAB International, Wallingford, UK.

Wright, S. (1943) An analysis of local variability of flower color in *Linanthus parryae*. *Genetics* 28, 139–156.

Yan, J., Shah, T., Warburton, M.L., Buckler, E.S., McMullen, M.D., and Crouch, J. (2009) Genetic characterization and linkage disequilibrium estimation of a global maize collection using SNP markers. *PLoS One* 4, e8451.

Yeaman, S., Hodgins, K.A., Lotterhos, K.E., Suren, H., Nadeau, S., et al. (2016) Convergent local adaptation to climate in distantly related conifers. *Science* 353, 1431–1433.

Zhou, L., Bawa, R., and Holliday, J.A. (2014) Exome resequencing reveals signatures of demographic and adaptive processes across the genome and range of black cottonwood (*Populus trichocarpa*). *Molecular Ecology* 23, 2486–2499.

Zhou, Z., Jiang, Y., Wang, Z., Gou, Z., Lyu, J., et al. (2015) Resequencing 302 wild and cultivated accessions identifies genes related to domestication and improvement in soybean. *Nature Biotechnology* 33, 408–414.

5 Quantitative Genomics of Forest-tree Breeding

Introduction

Quantitative genetics is the study of the transmission and segregation of traits that follow a continuous distribution in the population. Such traits are also commonly referred to as "complex," and include the analysis of properties that are evaluated using binary or discrete values but are controlled by multiple loci. Sir Ronald Fisher (1890–1962) is considered the founder of the discipline for developing and applying mathematical and statistical concepts to the genetic analysis of traits that follow a more complex pattern of segregation than simple Mendelian characteristics. Fisher proposed that complex traits were controlled by an infinite number of unlinked and non-epistatic loci, each contributing a small effect, establishing the foundation for what became known as the infinitesimal model (Fisher, 1918). The infinitesimal model originated from the observation that the frequency distribution of most traits measured in populations can only be explained by the linear, additive contribution of large numbers of loci of small effect, rather than one or a few loci with a large effect(s) (Box 5.1). While deviations from the infinitesimal model have been detected for traits regulated by a few loci of moderate to large effect, most properties of immediate interest in agriculture and forestry (e.g. biomass yield, height, diameter at breast height, etc.) appear to follow the infinitesimal model.

Unknown to Fisher was the fact that the infinitesimal model exposed one of the most significant challenges to the identification of the loci that regulate complex traits, and their use in breeding. Under his model, the contribution of each individual locus to trait variance is small. Consequently, the statistical power to identify them is limited because it is difficult to separate the effect of an individual locus from that of many other independently segregating loci that contribute to the trait. Furthermore, the immediate and practical use of all loci that contribute to a complex trait in breeding programs is restricted because it requires tracking hundreds or thousands of them simultaneously. Until methods to genotype and follow the segregation of all genetic variation in genomes became available, the task was unfeasible. In fact, prior to the genomic era, it was impossible to directly validate the infinitesimal model other than using resemblance between relatives to estimate the collective contribution of all relevant loci in the genome to a quantitative trait (Falconer and Mackay, 1996). The collective analysis of large numbers of variable loci in individuals of breeding populations is now feasible for several species, and is bound to become an essential tool to advance genetic gains in future tree-breeding programs.

Forest-tree Breeding

Breeding of forest-tree species has a unique history when compared with the genetic improvement of various agricultural crops. The most primitive form of plant breeding started over 10,000 years ago with the development of agriculture and the habit of early farmers to select the surviving, most productive plants as seed sources for subsequent generations. By 2000 BC, today's most important food crops had been domesticated (Doebley *et al.*, 2006). Modern breeding of agricultural crops began with the discovery of Mendelian laws and the understanding that genetic components often

---

Box 5.1. The infinitesimal model

The number of possible genotypic classes increases exponentially (3^n) as an increasing number of loci are considered (one locus = AA, Aa, and aa; two loci = AABB, AABb, AAbb, AaBB, AaBb, Aabb, aaBB, aaBb, and aabb; and so on).

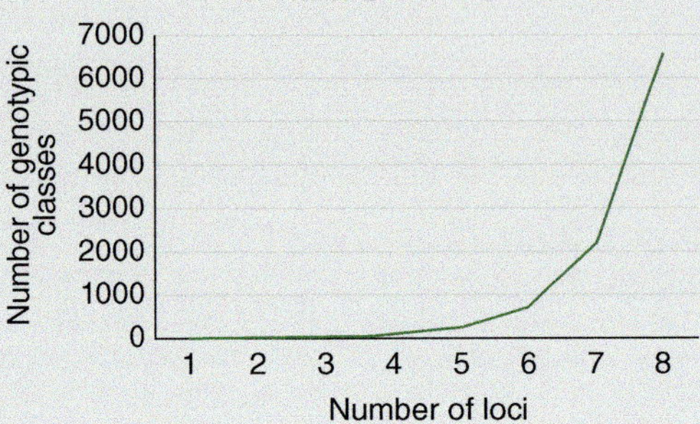

Moreover, assuming that individual alleles contribute additively and equally to trait variation, the distribution of the phenotypic data measured in a population rapidly takes a normal distribution. For two loci (A and B), the number of possible phenotypic classes is small ($n = 5$) and they are easily distinguishable (left panel below). If eight loci are considered, the number of phenotypic classes is much larger and less discrete (right panel below). Genetic analysis of well-studied quantitative traits, such as height in humans, suggest that hundreds to a few thousand genes regulate their expression (Marouli *et al.*, 2017). Under such genetic control, phenotypic classes become indistinguishable and form a continuous normal distribution.

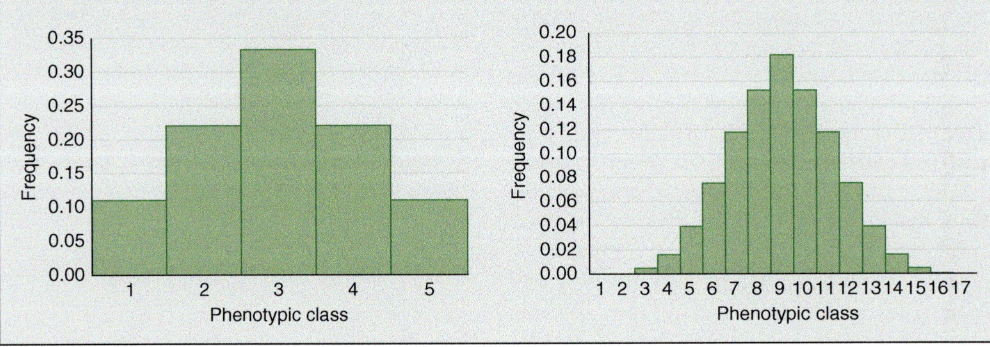

determine the phenotype. Today, due to their short generation time, the most intensively bred agricultural crops can undergo two or more cycles of breeding and selection per year. Ease of breeding and the small size of most agricultural crop plants also results in testing of large populations, permitting high-intensity selection. Consequently, all modern major food crops have limited similarity to their ancestors and require specific cultivation conditions that make them unlikely to persist in uncontrolled, natural environments.

In contrast to crop breeding, advanced forms of tree breeding that utilize genetic information only began in the mid-1900s (White *et al.*, 2007). The concept of applying genetics to improve forest productivity then rapidly expanded to several countries, particularly in North America, Asia, and Europe (Zobel and Talbert, 1984). The late start of tree breeding relative to agricultural food crops comes from the fact that during early human population growth, vast territories of natural forests were readily available for harvest. When wood became overexploited and scarce, it was readily replaced by lower-cost substitutes, such as coal, for energy production. However, the main hurdles to the adoption of tree breeding relate primarily to the fact that it is a complex and costly enterprise, compared with agricultural crops, for the following reasons:

1. *Long generation times.* Most tree species require an extended period of time to reach sexual maturity to produce flowers and seeds, resulting in lengthy breeding cycles. This limitation can be addressed to some extent by the use of strategies that induce early flowering in hardwoods (Yuceer *et al.*, 2003) and softwoods (Cecich *et al.*, 1994). However, these approaches have only been developed for a few species, can be genotype-specific, are labor intensive, and generally have not yet been widely used.

2. *Phenotyping.* Evaluating trees for growth properties, disease resistance, and other commercially useful traits often requires that they be at an advanced stage in their life cycle. For example, reliable growth properties are usually only obtained as the trees reach half of their rotation age (White *et al.*, 2007). Therefore, breeders have to wait a long time to identify and select the best individuals. Homogeneous field trials and care in the establishment of tests may lead to growth traits that better reflect the end-of-life-cycle properties, but this only addresses the issue to some extent. Properties such as wood-quality traits change over time (Zobel and Sprague, 1998), requiring that a certain stage of maturity be reached before selections can be made.

3. *Logistics of field trials.* Trees are large organisms that require significant amounts of space to grow and develop. Wide spacing among trees is often used in trials to minimize the effect of competition and let genotypes achieve their full genetic potential. Consequently, field trials are large and heterogeneous, resulting in a significant amount of environmental variance, requiring sophisticated field-plot designs and analysis methods to remove this source of error.

4. *Reproduction and mating system.* Many agricultural crops that have undergone a large number of cycles of breeding and selection can be selfed or crossed to close relatives (e.g. backcrossed). This approach allows the generation of immortalized inbred lines, which form the basis of breeding of the most productive agricultural crops worldwide. The same is difficult to achieve in forest-tree species. Hardwoods and conifers have a high genetic load in the form of deleterious alleles that are detrimental to growth and development when in a homozygous state (Griffin, 1983; Eriksson *et al.*, 2009). Therefore, breeding among relatives or selfing is unfeasible or plagued by a high incidence of arrested embryo development, seed mortality, or poor growth and development.

5. *Length of seed development.* For conifers, timing between fertilization and complete seed development can extend for 2 years or more. This phenomenon adds time and costs to breeding, and adds uncertainty in genetic improvement.

Because of these hurdles, tree genetic improvement programs are mostly only in their first few cycles of breeding and selection, and are only a few generations removed from their wild relatives. Thus, allele diversity and frequencies are still likely to largely reflect those observed in natural populations. The limited progress of tree-breeding programs, relative to agricultural crops, creates limitations, such as the lack of well-developed genetic resources. However, it also offers unique prospects to tree breeders, who have the opportunity to develop programs that accelerate genetic gain based on knowledge gained from agricultural crops, avoiding common mistakes, such as the undesirable loss of genetic diversity. In several agricultural crops, early loss of favorable alleles due to severe genetic bottlenecks has been shown to have major negative consequences for long-term genetic improvement and is now requiring the introgression of wild alleles in breeding populations (Tanksley, 1997). Through careful tracking of ancestral alleles, tree breeders have the opportunity to advance genetic improvement by selecting favorable alleles, while minimizing the loss of genetic diversity.

Despite the limitations of tree breeding, the first breeding cycles of most genetic improvement programs achieved exceptional gains (Fig. 5.1). Part of these gains was the result of advanced methods of silviculture (Rauscher and Johnsen, 2004). However, the improvement of genetic stocks also had a significant role, the importance of which will likely increase in future generations, as tree growers become more constrained by the costs of plantation establishment and maintenance. Since the 1950s, traditional methods have been applied for several cycles of tree breeding and selection. These methods and examples of their application are described in detail in other publications (e.g. White *et al.*, 2007). In this book, we extend the discussion to methods of genetically improving tree species by the use of advanced biotechnologies and genomics. Biotechnologies involve the use of molecular methods to introduce, modify, or delete functional genetic elements of the genome with the aim of producing a change in the phenotype. Biotechnological methods that involve altering existing or inserting novel genetic material are described in detail in Section II of this book. In parallel, methods of genomics-based breeding have also been gaining momentum as approaches to identify genetic elements that either contribute to phenotypes or are in linkage disequilibrium (LD) with them (see Chapter 4, this volume), and use this molecular information to enhance traditional breeding programs. These methods and examples of their application to tree breeding are described below.

Decomposition of Phenotypes

The phenotypic properties of an individual reflect the contribution of a combination of genetic and environmental factors, and their interactions. Thus, in principle, an individual's phenotype can be decomposed into these two elements: genetics (G) and the environment (E) (Eqn. 5.1). In the event that genes respond differently to the environment, then an interaction ($G \times E$) also occurs. For the sake of simplicity, the content that follows will focus exclusively on the genetic and environmental contributions to phenotypes:

$$P = G + E \tag{5.1}$$

For tree breeding, it is useful to separate the effect of genetics and environment. This is because breeders aim to select trees that have superior genetic value. While the environment can contribute to improved productivity, these changes are affected by silviculture practices.

When considering the genetic and environmental contributions to phenotypes, one often does not aim only to estimate these values for each individual. Instead, the variance of these components is equally important to estimate heritabilities (see below) and genetic gains. Eqn. 5.1 can be represented from the perspective of the variances associated with each element:

$$\sigma_P^2 = \sigma_G^2 + \sigma_E^2 \tag{5.2}$$

The contribution of σ_G^2 and σ_E^2 can, in principle, be estimated by eliminating all variation in either the genetic or environmental

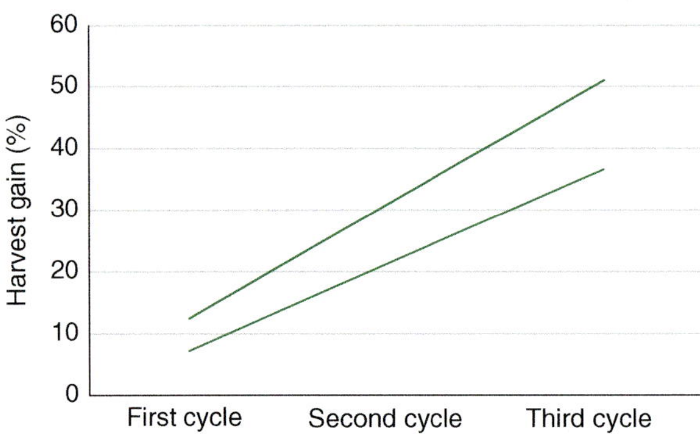

Fig. 5.1. Improvement in harvest yield of *Pinus elliottii* plantations grown in low (upper line) and high (lower line) rust hazard areas, derived from seeds collected in seed orchards established from selections made in the first, second, and third breeding cycles in the Cooperative Forest Genetics Research program at the University of Florida.

components. For instance, planting genetically identical copies of the same individual (commonly referred to as clones) at multiple sites results in σ_G^2 being equal to zero. Therefore, all variation that is observed in the phenotype of plants grown in different sites can only be the consequence of environmental variance. In this scenario, σ_P^2 is equal to σ_E^2. The opposite also applies. If genetically dissimilar individuals are planted on a single site that has no variation, then all observed variance must result from the variance measured among individuals, and σ_P^2 will be equal to σ_G^2. In reality, complete control of environmental variation is nearly impossible, and most studies aim to evaluate multiple individuals, families, or populations. Therefore, σ_G^2 is almost never equal to zero.

In addition to separating the contribution of genetic and environmental variance to phenotypes, it is also useful to dissect the genetic component into its three main elements: additive (A), dominance deviation (D), and epistatic interaction (I).

$$G = A + D + I \tag{5.3}$$

The additive effect and its variance correspond to the contribution of individual alleles to phenotypes. For breeding purposes, it is critical to distinguish the additive effect and its variance from other genetic components. This occurs because each parent contributes individual alleles to their progeny, not their genotypes. Therefore, to estimate a parental genetic contribution to the progeny phenotype, it is important to separate the additive value from the value derived from non-additive effects. For this reason, the additive value is also commonly referred to as the breeding value.

The total additive value of an individual can be estimated theoretically by simply summing the contribution of individual alleles at all loci that contribute to a trait. However, the two alleles at a locus can interact, so that their contribution is different from the sum of their individual effects. In this case, dominance deviation occurs. Thus, dominance corresponds to the difference between the effects of combining two alleles at a locus, relative to simply summing their

additive values. However, interactions may not occur solely between alleles at an individual locus. Instead, alleles at distinct loci may interact with each other, causing epistasis. Until methods to map quantitative trait loci (QTLs) were developed, it was impossible to identify the specific loci that regulated complex traits, and the contribution of each of their alleles to trait variation. To partially address this limitation, statistical methods to estimate the amount of genetic variation that exists for a given trait (i.e. its heritability) in a population were developed.

Heritability

In tree breeding, it is often necessary to estimate the proportion of the phenotypic variance that is caused by the genetic and environmental components. The proportion of the genetic variance that contributes to the total phenotypic variance is commonly referred to as the broad-sense heritability, or H^2:

$$H^2 = \frac{\sigma_G^2}{\sigma_P^2} \tag{5.4}$$

Because the total genotypic make-up of a progenitor is not passed on to its progeny, only individual alleles, the narrow-sense heritability (h^2) is more informative from a breeding perspective. The value of h^2 is estimated by the fraction of the total phenotypic variance contributed by the additive variance:

$$h^2 = \frac{\sigma_A^2}{\sigma_P^2} \tag{5.5}$$

Both H^2 and h^2 can be estimated based on analysis of the phenotypes of segregating families and the resemblance between relatives. For instance, progeny with the same two parents (full-sibs) are likely to resemble each other (or co-vary) more than progeny with only one parent in common (half-sibs). The degree of covariance between relatives is a well-studied subject in classical quantitative genetics and has been thoroughly described elsewhere (e.g. Falconer and Mackay, 1996).

Estimating heritability using genomic data

Traditionally, the covariance between relatives was calculated from pedigree information. However, pedigree information is often not available or is unreliable, particularly in natural populations that are not derived from a breeding program. This limitation has been overcome as abundant genomic data have become available for natural and breeding populations. The availability of genetic-marker data for individuals of a population can provide information about their relationship. This is because two individuals that share one or more ancestors are more likely to carry the same marker. As a consequence, the availability of methods to genotype genetic markers distributed genome-wide has led to the use of novel approaches to estimate the relationship among individuals (Ritland, 1996) and genetic variances in populations. Genetic variances obtained from genetic-marker data can also be partitioned into its different components (additive, dominant, and epistatic), regardless of the family structure of the population. Heritabilities can be estimated simply by regression of the phenotypic similarity between any two individuals on the portion of genome that is shared, based on genetic-marker data (Muñoz et al., 2014).

The use of genetic markers to estimate the genetic relationship among individuals and heritability has certain advantages that make it more suitable than estimates based on pedigree alone. For example, pedigree data do not account for errors in controlled pollinations, which can be detected by genetic markers. With pedigree information only, it is also not possible to calculate the precise percentage of shared alleles among related individuals. For instance, pairs of full-sibs share half of their alleles, on average. However, due to the random nature of genetic transmission (e.g. independent assortment) and linkage among alleles at different loci, individuals can share more of their genome with some of their siblings than others. Thus, the actual proportion shared between any two siblings will vary positively and negatively around this average.

Because marker information allows a better estimation of the number of alleles that are shared among individuals and their relationship, the approach is expected to result in better estimates of heritability. Heritability estimated for some of the most exhaustively studied traits in humans, such as height, is consistent with that obtained using traditional methods.

While the application of molecular markers for estimation of heritability has become widely used in the genetic analysis of complex traits in forestry, questions remain about the suitability of this approach (de los Campos et al., 2015). The estimation of heritability based on kinship implies that individuals that share common alleles also share similar phenotypes. This occurs because shared markers are expected to be in LD with alleles that control the phenotype. However, if the number of markers is insufficient to represent all genetic variants that contribute to a trait, or if LD between markers and QTL is incomplete, this pattern of allele sharing at markers and at causal loci may not be adequate to estimate heritability. This limitation can be particularly relevant if heritability is to be estimated in populations of largely unrelated individuals with low LD and insufficient marker density. Other limitations, including bias due to non-random distribution of causal variants with respect to LD, have also been identified (Yang et al., 2017).

Genomic Dissection of Complex Traits

Until the late 1980s, genetic analysis of complex traits in forest trees was focused primarily on estimating the proportion of the phenotypic variance controlled by heritable factors (heritability) and using that information to predict genetic gains from selection (White et al., 2007). Information on the amount of the phenotypic diversity, as determined by genetics, is sufficient to allow breeders to estimate how much improvement can be achieved in a generation. Quantitative genetic analysis of more complex pedigrees may also be used to estimate the contribution of additive and dominance effects to phenotypes and, consequently, estimate the values of using a given parent in multiple (general combining ability) or specific (specific combining ability) crosses.

Despite providing relevant information for breeding programs, the estimation of heritability does not offer a means to accelerate breeding. To use genetic information to accelerate breeding, loci that regulate trait variation need to be known to predict the genetic value of individuals. In this case, faster gain can be achieved by the early selection of individuals with higher genetic value, or by identifying the best parental combinations to create families with high mean genetic value. However, until the late 1980s few genetic resources were available to support the identification of loci that regulate trait variation. These loci are discovered by identifying genetic markers that co-segregate with traits of interest (see below). However, genetic markers that could be tested for co-segregation with phenotypes were very limited in number and genome coverage until the discovery of the polymerase chain reaction (PCR). Genetic markers developed using PCR established the foundation for the genetic dissection of traits and, eventually, the polymorphisms that regulate them.

Since the 1980s, increasingly sophisticated methods of quantitative genomic analysis have created the opportunity to detect loci that regulate complex traits and to use them to accelerate progress through breeding. Three strategies have become widely adopted: (i) linkage mapping, (ii) association mapping of QTLs; and (iii) genome-wide prediction methods. The three methods differ with respect to several key factors, including the type of population analyzed, the number of markers required, the QTL allelic diversity sampled, and the likelihood of uncovering specific polymorphisms responsible for complex traits (Table 5.1). A description of these methods, their application to forestry, and their advantages and disadvantages are described in the following section.

Linkage Mapping of QTLs

The first genetic approach proposed to uncover loci that regulate complex traits was linkage mapping of QTLs (Table 5.1). The general principle of QTL linkage mapping was first developed for the analysis of seed weight and pigmentation in common bean (*Phaseolus vulgaris*) (Sax, 1923). In the absence of molecular markers, Sax identified an association between "factors" that determine the color of the seed pigmentation and seed weight in the progeny of segregating crosses.

Linkage mapping of QTLs is typically performed in bi-parental segregating populations,

Table 5.1. Strategies for the quantitative genetic study of complex traits.

Type of study	Linkage mapping of QTLs	Association mapping of QTLs	Genomic selection
Aim	Identify loci that control a complex trait	Identify loci that control a complex trait	Predict a complex trait
Population	Bi-parental	Multi-parental	Multi-parental
Founders (*n*)	Known (2)	Unknown (2+)	Known (2+)
Generations since founders	1–2	Unknown	1+
Linkage disequilibrium	High	Low	Medium
Allelic diversity	Low	High	Medium
Genetic markers required	100s	100,000s to 1,000,000s	1,000s to 10,000s
Statistical power	High	Low	NA
Resolution	Low	High	NA

NA, not applicable.

although the principle is applicable to more complex pedigrees that include multiple parents and multi-generational families. In its simplest form, QTL analysis relies on tracking the segregation of alternative parental alleles in the progeny of a cross, and then comparing the phenotypes between the groups of individuals that inherited the different alleles (Fig. 5.2). If a statistically significant difference exists between the two groups, then there is evidence that one or more loci that regulate the trait are in LD with the marker. Because linkage mapping of QTLs is done in families founded by two parents and crossed for a single or a limited number of generations, the extent of LD in the population is large. Therefore, genotyping of relatively few genetic polymorphisms allows tracking of the haplotypes inherited from each parent. This advantage was particularly relevant in the early days of genomics, when the limited availability and high cost of genotyping made the analysis of large numbers of genetic markers prohibitive. With the dramatic decrease in cost and increase in throughput of genotyping methods, the advantage of this approach is no longer significant. The basic methodologies of linkage mapping of QTLs have been described extensively in the literature (e.g. Lynch and Walsh, 1998) and insightful reviews have been published (e.g. Mackay *et al.*, 2009).

Linkage-mapping QTL detection methods

The pioneering work from Karl Sax on linkage mapping of QTLs was based on the simple observation that seed weight was higher

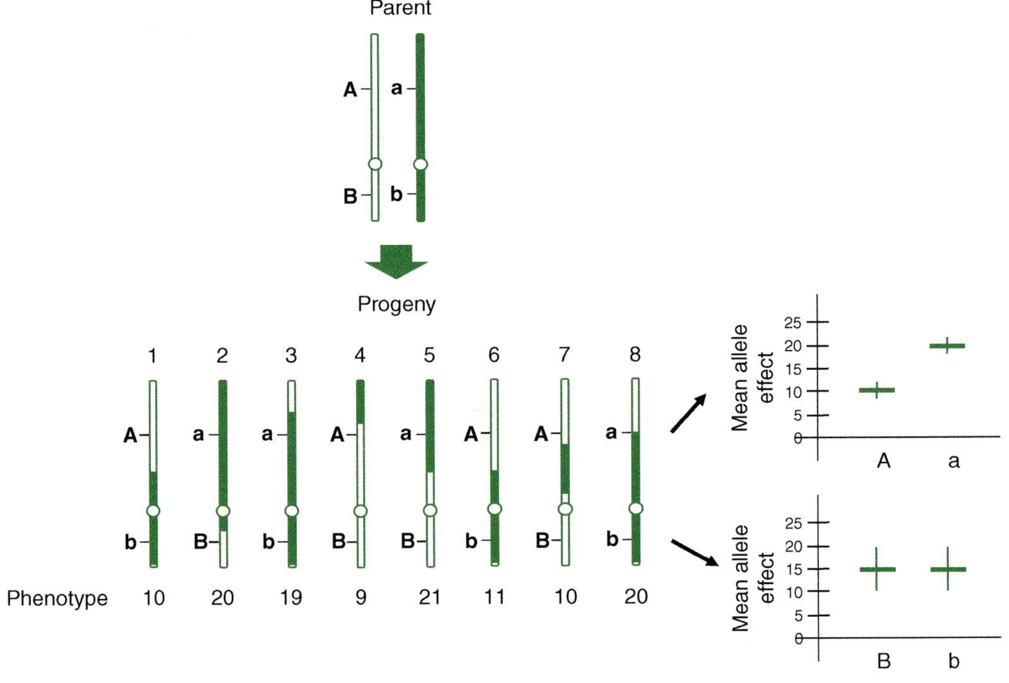

Fig. 5.2. Linkage mapping of QTLs. Alleles from two loci (A and B) segregate in the progeny of a heterozygous parent. For simplicity, only the segregation of alleles from one parent is shown. The mean phenotypic effect of each allele is estimated by the average of the phenotype of the individuals in the progeny that carry that allele. A QTL is detected when a significant difference between the mean effect of the two alleles at a locus is observed, as in locus A. In contrast, absence of a significant difference between phenotypes in individuals carrying the B and b alleles indicates that this locus does not contain a QTL.

in colored seeds than in white seeds (Sax, 1923). With the development of genotyping tools, analysis methods grew in sophistication to permit more precise inferences about the localization of QTLs (Lander and Botstein, 1989; Jansen, 1993). Over the last few decades, several methods of linkage mapping of QTLs were proposed that incrementally improved the ability to detect and precisely locate these loci (Doerge, 2002). The evolution of QTL linkage-mapping approaches and their general principles are described in the following sections.

Single-marker analysis

The simplest approach to mapping QTLs is based on the use of a t-test or analysis of variance (Soller *et al.*, 1976). In this case, the progeny are separated into groups that have inherited alternative parental alleles for a given locus and their phenotypic distributions are compared. If distributions between the genotypic groups are similar, the conclusion is that the locus is not in LD with a QTL. Different distributions suggest the presence of linkage between the locus and the QTL, which can be tested by calculating a t-statistic (in case of two genotypic classes) or an F-statistic (when three genotypic classes are present) to compare the averages of the marker genotype classes. More complex single-marker analysis methods can be applied using linear regression to account for other characteristics of a QTL to be considered, such as the separation of additive and dominance effects and $G \times E$ interactions.

The simplicity of the analysis and the fact that each marker is tested individually (it does not require the availability of a genetic map) favored the use of single-marker analyses, particularly until genome-wide marker coverage became possible. However, QTL location is inexact using this method because intervals between adjacent markers are not considered. The effect of individual markers is also likely to be underestimated when the QTL is not located in close genetic proximity to the marker. Depending on marker density, the power to detect QTLs is also affected if large genomic regions with no marker coverage exist.

Interval mapping

Interval mapping extends the single-marker analysis method of QTL detection by evaluating the space between consecutive genetic markers. This approach was first developed by Lander and Botstein (1989), but multiple statistical analysis alternatives have since been proposed (for a review of statistical methods of interval mapping, see Xu, 2013). The interval-mapping method proposed by Lander and Botstein uses maximum-likelihood estimation under a mixture model and remains an optimal approach to define the position of a QTL in the genetic space between two markers, based on the analysis of recombinants between them.

The principal advantage of interval mapping, compared with single-marker analysis, is that intermediate positions between markers are interrogated. When a genome sequence is available, it creates the opportunity to narrow the genomic interval and, consequently, the number of genes and polymorphisms that potentially underlie the QTL. Another advantage is that interval mapping does not require exclusion of individuals with a missing genotype at a marker. Furthermore, an improved quantification of the effect of a QTL can be obtained based on its estimated position, rather than relying on genotypes at the nearest genetic marker. Despite these improvements, interval mapping is still limited in that it considers a single-QTL model, which hinders the ability to detect and separate closely linked QTLs, as well as detecting interactions among QTLs.

Composite-interval mapping

While the introduction of interval mapping partially addressed the need to increase the accuracy of QTL locations by exploring the space between marker intervals, additional and incremental improvements were soon introduced that further improved resolution, including the use of composite-interval mapping and related approaches (Jansen, 1993; Zeng, 1994). Most linkage-mapping analysis studies aim to dissect the genetic loci that regulate complex traits, which are regulated by many loci in the genome.

Therefore, when an interval between markers is analyzed, many other genomic regions involved in trait regulation are likely to be segregating dependently (in the case of linkage to the genomic interval of interest) or independently in the population. Segregation of these alternative QTLs can mask the detection of the desired QTL. To address this potential limitation, additional markers can be included in the regression model as cofactors. This is achieved by fitting one QTL at a time in a given interval and simultaneously using other markers as cofactors to eliminate the variance attributed to other QTLs in trait variation. Simulations showed that this approach improved the power and resolution of QTL studies.

Other linkage-mapping QTL detection approaches

Numerous additional approaches to linkage mapping of QTLs have been proposed since interval and composite-interval mapping methods were first introduced. These include methods to account for the interaction between multiple loci that contribute to trait variation (Kao *et al.*, 1999) and the introduction of Bayesian methods for QTL detection (Ball, 2001; Yi and Shriner, 2008).

Statistical power and significance thresholds in linkage mapping of QTLs

The ability to detect a QTL in a linkage-mapping population depends on the difference between the effect of the alternative QTL alleles and the size of the population being phenotyped and genotyped (Fig. 5.3). Detection of a QTL is also dependent on the standard deviation within each genotypic class. For example, a small phenotypic difference between genotypic classes may be significant if there is no or a very small standard deviation within each class. In contrast, large differences between genotypic classes may not be detectable if the standard deviation within them is large. Thus, for phenotypes controlled by a single locus with high penetrance, the detection of the QTL is more likely compared with phenotypes controlled by many loci of small effect. This limitation can partially be overcome by increasing the size of the mapping population (Fig. 5.3). Having more individuals available within each genotypic class can allow the detection of significant (even if small) phenotypic differences.

Statistical thresholds are needed to define significant QTLs. However, in a linkage-mapping QTL study, multiple markers and marker intervals have to be tested for the

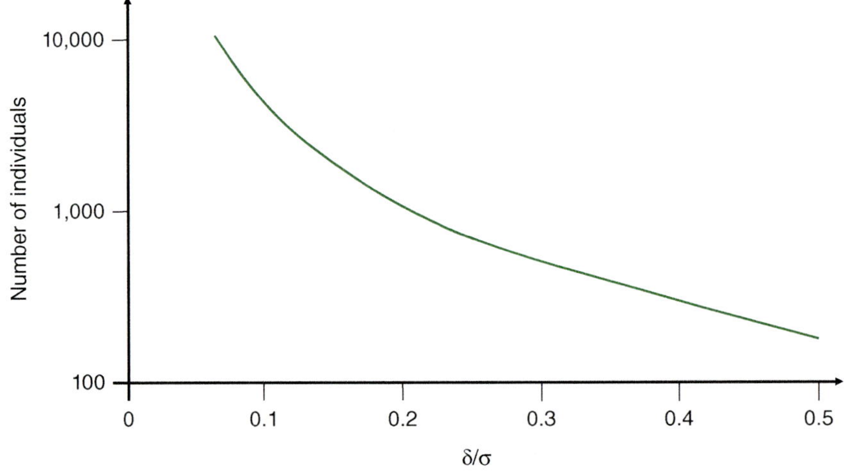

Fig. 5.3. Number of individuals required to detect a QTL by linkage mapping in a pseudo-backcross population. QTL effects are shown as the difference in phenotype between genotypic classes (δ), divided by the standard deviation of the trait within each marker class (σ). (Adapted from Mackay *et al.*, 2009.)

association between genotype and phenotype. Multiple testing becomes problematic, however, because commonly used thresholds applied to individual statistical tests need to be corrected. Alternatives to address the issue of multiple testing, such as using a Bonferroni correction, are not suitable in the context of linkage mapping of QTLs because statistical tests are not independent due to linkage between markers. Consequently, other approaches have been proposed, the most widely used of which is permutation randomization. Permutation resampling is applied by randomizing the phenotypic data among individuals in the linkage-mapping population (Churchill and Doerge, 1994).

Linkage-mapping studies in forestry

Despite the early development of methods to identify markers linked to QTLs, it was not until the mid-1990s that this approach was demonstrated in a forest species. In a pioneering study, QTLs for wood specific gravity were mapped in a full-sib cross between two loblolly pine (*Pinus taeda*) parents (Groover *et al.*, 1994). For QTL detection, a set of 146 RFLP markers was screened in a subset of the progeny that showed extreme phenotypic differences. Analysis of variance was used to identify genetic markers that were associated with the phenotype. These were then characterized in the entire population, resulting in the detection of five genomic regions associated with variation in wood specific gravity. This first demonstration was soon followed by the first QTL linkage-mapping studies that employed a PCR marker-based genetic map, in *Eucalyptus*, where loci that regulate vegetative propagation properties, growth, and wood quality were identified (Grattapaglia *et al.*, 1995, 1996). Since the publication of these pioneering studies, QTL detection using linkage mapping has become widely adopted in forestry. Despite the enthusiasm for this approach, direct applications to tree-breeding programs have been limited due to the difficulties in fine-mapping the QTLs to specific polymorphisms

(see section on Pitfalls and limitations of linkage mapping of QTLs).

QTL analysis for biotic and abiotic stresses

Long-lived forest-tree species are particularly vulnerable to disease because of their extended exposure to pathogens. Pest-management methods used with agricultural crops, including pesticide application, are also difficult to implement in extensive forest-tree plantations. Thus, development of families or clones that are tolerant or resistant to the primary pests is of great interest. Resistance to several plant diseases has been shown to be largely regulated by one or a few loci of large effect, simplifying their detection, isolation, and use in marker-assisted selection (Martin *et al.*, 1993). In forestry, the first QTL linkage-mapping studies for disease traits focused on identification of loci that impose resistance to two rust diseases, white pine blister rust caused by the fungus *Cronartium ribicola* in sugar pine (*P. lambertiana*) (Devey *et al.*, 1995), and fusiform rust caused by *Cronartium quercuum* in loblolly pine (Wilcox *et al.*, 1996). These tree diseases are among the most economically devastating in commercial forest-tree plantations in the USA, and affect several pine species. Both studies identified a single major genetic locus controlling resistance to specific rust strains, supporting the hypothesis that disease resistance is not necessarily polygenic in tree species. Most importantly, these early studies identified genetic markers in close linkage disequilibrium (LD) with the resistance locus in each species, a critical requirement for their use in selection of resistant individuals and for future cloning of the resistance gene. Nevertheless, multiple loci are likely to be involved in resistance to different strains of these rust species, implying that it may be necessary to stack various resistance alleles to achieve broad resistance (Kayihan *et al.*, 2005). For example, early studies of the genetic architecture of Monterey pine (*P. radiata*) resistance to *Dothistroma septospora* identified several QTLs, indicating that the trait is clearly quantitative (Devey *et al.*, 2004a). Several studies followed these first demonstrations of linkage mapping of disease-resistance loci in *Pinus*, including studies with *Eucalyptus* spp. (Freeman *et al.*, 2008) and *Populus* spp. (Jorge *et al.*, 2005).

In addition to disease pressure, trees also suffer from long-term exposure to abiotic sources of stress, including drought (Anderegg *et al.*, 2015) and, increasingly, elevated levels of ozone (Ollinger *et al.*, 1997). Drought tolerance-related traits have been the focus of studies that identified QTLs associated with carbon isotope composition and water-use efficiency. Carbon isotope discrimination measures the ratio of ^{13}C to ^{12}C and is indicative of drought tolerance in plants (Farquhar *et al.*, 1989). Water-use efficiency is defined as the amount of water needed by a plant to produce a given amount of biomass. QTLs have been mapped for both traits in *Pinus* (de Miguel *et al.*, 2014; Marguerit *et al.*, 2014); *Populus* (Monclus *et al.*, 2012); and *Eucalyptus* (Bartholomé *et al.*, 2015). QTLs for other major sources of abiotic stress, such as high levels of CO_2 (Ferris *et al.*, 2002; Rae *et al.*, 2006, 2007); ozone (Street *et al.*, 2011); and soil contaminants (Induri *et al.*, 2012) have also been reported.

QTL analysis for growth and wood properties

Together with disease resistance, most tree-breeding programs focus primarily on improving biomass yield and, to a lesser extent, wood quality. In contrast to disease resistance, tree biomass growth and wood properties are clearly controlled by a large number of genetic factors (Grattapaglia *et al.*, 2009), and have a very complex genetic architecture (Resende *et al.*, 2012a). Despite these challenges, biomass growth and properties have been the main traits of interest for linkage mapping of QTLs in forest-tree species. Following the pioneering linkage mapping analysis of QTLs for wood specific gravity in *Pinus* (Groover *et al.*, 1994) and for growth and wood-quality traits in *Eucalyptus* (Grattapaglia *et al.*, 1996), the scope was expanded to include several other commercial species. These included QTL studies in several species of *Pinus* (Sewell *et al.*, 2002; Weng *et al.*, 2002;

Brown *et al.*, 2003; Devey *et al.*, 2004b); *Eucalyptus* (Kirst *et al.*, 2004b; Thamarus *et al.*, 2004; Gion *et al.*, 2011); *Salix* (Tsarouhas *et al.*, 2002); and *Populus* (Novaes *et al.*, 2009).

QTL analysis for physiological and developmental traits

Tree growth is the consequence of a combination of diverse developmental and physiological factors that contribute to the phenotype (Grattapaglia *et al.*, 2009). For instance, leaf shape and area have been shown to impact tree growth in conifers and other plant species (Ridge *et al.*, 1986; Leverenz and Hinckley, 1990). Thus, several QTL studies have attempted to study the genetic regulation of these productivity-related traits by dissecting them into the physiological and developmental components believed to contribute to yield.

Tree phenology is characterized by the seasonal growth cycles, which are measured by the timing of bud set/flush or by the activation/cessation of cambial activity. These phenological traits have a significant impact on tree biomass yield because they define the length of the growing season. Most phenology QTL studies have focused on species within the genus *Populus* and related species (Frewen *et al.*, 2000; Rohde *et al.*, 2011; Fabbrini *et al.*, 2012). Phenological traits have been shown to have high heritability in tree species, and several studies in annual plants indicate that related properties, such as flowering time, are predominantly regulated by a few loci of large effect.

QTL analysis of molecular traits

Molecular phenotypes such as gene expression and protein accumulation can be treated statistically in the same way as whole-plant phenotypes, such as growth and wood quality. Thus, the regulatory loci of these molecular traits can also be identified using linkage mapping of QTLs. After being first proposed (Jansen and Nap, 2001), this approach was demonstrated in humans and model organisms (Schadt *et al.*, 2003), and shortly afterward was applied to *Eucalyptus*

(Kirst *et al.*, 2004, 2005). Other than identifying the regulatory loci that control the expression of individual genes, expression-QTL linkage mapping was also used in these early studies to uncover major regulatory hubs that control metabolic pathways and putative regulators of whole-plant phenotypes (Kirst *et al.*, 2004). While these initial reports were hampered by the lack of a genome sequence and a complete transcriptome, these limitations were overcome, resulting in more comprehensive studies that uncovered transcriptional networks and their genetic regulation in *Populus* and *Eucalyptus* (Drost *et al.*, 2010; Mizrachi *et al.*, 2017).

Pitfalls and limitations of linkage mapping of QTLs

The use of QTL linkage mapping offered the first opportunity for genetic dissection of quantitative phenotypes in forestry. Ultimately, the goal of these studies was to uncover the genes and polymorphisms that underlie the phenotypic variation, or to identify genetic markers to be used for breeding and selection of elite genotypes. However, linkage mapping of QTLs is severely constrained in many ways (Strauss *et al.*, 1992), some of which are described below:

1. *Limited genetic diversity sampled.* As an approach that characterizes the segregation of variable alleles between two parental genotypes, QTL analysis only captures a very limited portion of the total genetic variation that contributes to a trait in a species. Considering the extremely high levels of genetic diversity that are typically encountered in a tree population, such information can be of limited value in breeding populations.

2. *Linkage equilibrium between genetic markers and causative polymorphisms results in the limited application of discoveries across families and populations.* Markers detected as flanking QTLs are not usable as selection tools across populations because of the high linkage equilibrium in tree populations. Therefore, markers linked to a QTL in one family or population may be monomorphic, or not linked to it in another.

3. *Overestimation of QTL effects (Beavis effect).* Most QTL studies in tree populations have been done in relatively small populations, in the order of a few hundred individuals. William D. Beavis showed through simulations that the power to detect QTLs in such small populations was limited, and that when QTLs were detected, they were usually overestimated or false positives (Beavis, 1998).

4. *From QTLs to genes.* Linkage mapping of QTLs has typically resulted in the identification of broad genomic regions that contain one or more loci that regulate a trait. Unfortunately, these genome regions typically encompass hundreds of genes. An approach to uncover the causative locus that underlies a QTL is fine-mapping. This involves narrowing the QTL interval by sampling larger segregating populations and, consequently, recombination events between markers flanking the QTL. When a sufficiently narrow QTL interval exists, a reference genome sequence can be used to identify the specific genes and variants within that interval, and their specific function can then be verified using various biotechnological approaches. Attempts have been made to fine-map and clone a major QTL for *Melampsora* rust resistance in poplar. However, recombination suppression in the region of the locus has hampered the ability to narrow the interval sufficiently to determine the specific gene and polymorphism for the trait (Stirling *et al.*, 2001; Yin *et al.*, 2004).

Association Mapping of QTLs

Linkage mapping of QTLs in bi-parental populations was a first step toward detecting genomic regions that regulate complex traits. However, difficulties associated with identifying specific polymorphisms that control trait variation limited the value of this approach in tree-breeding programs. Other limitations further dampened the enthusiasm for QTL linkage mapping.

For the wide application of genomic approaches to tree-improvement programs, the ability to utilize genetic markers that are informative for breeding and selection across families is essential. Preferably, genetic markers should be useful across multiple breeding populations derived from the same founders. Due to the polygenic nature of most traits of relevance in forestry, genomic information should identify the loci that account for the largest fraction of the genetic variance for any given trait, instead of only those that segregate in a single cross. Achieving the goal of identifying all loci that contribute to traits and across breeding populations requires: (i) sampling large numbers of individuals or families that represent most of the genetic variance and (ii) characterizing genetic variation genome-wide, so that all of the polymorphisms that impact the trait can be accounted for. Association mapping was introduced to address some of the limitations of QTL linkage mapping (Table 5.1).

As with linkage mapping of QTLs, association mapping relies on the principle of detecting correlations between allele and trait variation. In other words, a significant difference between the phenotype measured in two groups of individuals, where each group carries a specific allele, defines a locus associated with the trait (Fig. 5.4). Association may be due to the locus being the regulator of the trait (the causative locus) or it being in LD with the causative locus. The main difference between a QTL linkage-mapping and an association-mapping approach is that the latter is typically carried out in populations of largely unrelated (i.e. genetically unstructured) individuals, or across large numbers of unrelated families. By sampling genetically unrelated individuals, association mapping captures a more extensive range of the genetic variation that impacts a trait, compared with QTL linkage mapping. By identifying polymorphisms associated with traits in populations of genetically unrelated individuals that last shared common ancestors many generations earlier, it is possible to identify polymorphisms that are closely linked or that cause genetic variation for the trait. In the absence of any major evolutionary force (e.g. genetic bottlenecks) that reduced effective population size, the large

number of historical recombination events that accumulated in these genomes greatly reduces the extent of LD. Therefore, polymorphisms correlated to a trait are likely to be in close genetic proximity, or to be the causative polymorphism that regulates part of the phenotypic variance.

The first demonstration of the use of association mapping in a plant species was in maize (*Zea mays*; Thornsberry *et al.*, 2001). In the analysis of a population of 92 maize inbred lines, the gene *Dwarf8*, previously implicated in flowering time by mutagenesis and QTL analysis, was shown to contain several polymorphisms associated with this trait. Studies in forest species soon followed, starting with the analysis of association between genetic polymorphisms in the *phyB2* gene and clinal variation, which affects the timing of bud set in Eurasian aspen (*Populus tremula*) (Ingvarsson *et al.*, 2006).

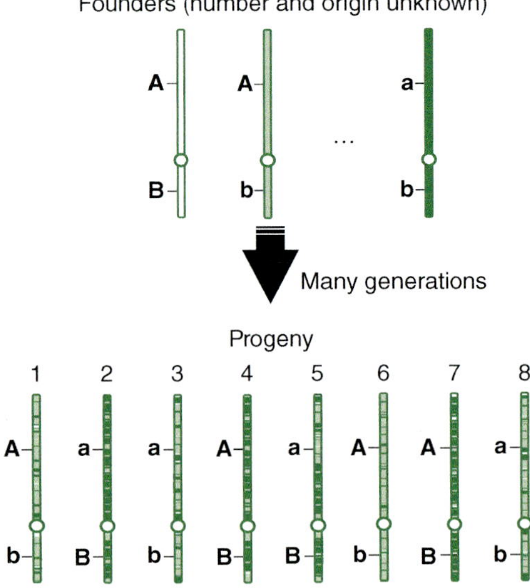

Fig. 5.4. Association mapping of QTL. The founders of an association population may be unknown in origin and number, and the progeny may be derived from multiple generations of recombination. As a consequence, the extent of LD blocks (represented by the chromosomal segments inherited by the progeny from each of the founders) is short. The effect of individual loci (A/a and B/b in the figure) can be tested for association with traits.

Association-mapping QTL detection methods

Single-locus association genetic analysis

Early association-mapping studies were applied predominantly to the identification of genetic loci that regulate clinal traits. Such traits are typically binary, separating subjects into case and control. In a scenario where only two outcomes are possible, and three genotypes are observed in a population (two homozygote classes and the heterozygote), an association between phenotype and genotype can be tested using Fisher's exact test.

While the simple approach of testing for associations using Fisher's exact test may be suitable under certain circumstances, early studies demonstrated the need to consider other factors that can impact the results of association analysis, such as the potential for detection of false, spurious associations due to unknown genetic structure in populations. Population genetic structure occurs when groups of individuals (a subset of the overall population, or subpopulation) have different allele frequencies relative to the general population (see section on Population Genomic Structure in Chapter 4, this volume). In this scenario, DNA polymorphisms will be associated with any trait that is significantly different between the subpopulation and the general population. One of the first reports of the negative effect of population genetic structure in association mapping resulted from a large-scale association-mapping study in humans (Knowler *et al.*, 1988). In this study, a significant correlation was detected between immunoglobulin G and type 2 diabetes, in a population of 4,920 Native Americans from the Pima and Papago tribes. However, the association was likely due to a Native American subpopulation that contained a high frequency of Caucasian alleles. Caucasians have an overall lower incidence of type 2 diabetes relative to Native Americans. Thus, any allele that is predominant in the Caucasian population, but less frequent among Native Americans, could be associated with the trait. Unknown population genetic effects have also been implicated in the dismissal of the first association-mapping results reported for a plant species (Thornsberry *et al.*, 2001). Further analysis of the maize *Dwarf8* locus and its association with flowering time in maize showed that the trait is strongly correlated with population structure (Larsson *et al.*, 2013).

An approach to address the negative consequences of genetic structure in association mapping is to use family-based tests of association. In this case, where a family relationship among individuals is known, methods such as the transmission/disequilibrium test (TDT) can be used (Spielman *et al.*, 1993). The TDT relies on the analysis of parents that are heterozygous for a locus tested for association with a disease, and evaluates the frequency that the alternative alleles are transmitted to affected progeny. While suitable for the analysis of populations with well-established family structure and binary traits, the method is not useful for application to most plant populations and traits. An alternative to TDT was proposed that applied novel methods to first uncover hidden genetic structure (Pritchard *et al.*, 2000a) and then followed up with classical statistical tests within subpopulations (Pritchard *et al.*, 2000b). However, these methods only focused on the analysis of binary traits.

In complex trait analysis of animal- and plant-breeding programs, quantitative characteristics have typically been analyzed statistically using linear models or single-marker regression. In the regression model, the effect of each individual marker is tested separately, and the marker is treated as a fixed effect. While the use of linear regression addresses the need for methods of association mapping that can accommodate continuous quantitative data, it does not account for the effect of population and family genetic structure. This issue was addressed by adopting the use of a mixed linear model, where a relationship matrix that describes the overall genetic relatedness among individuals at the family and subpopulation level is included as a random covariate (Yu *et al.*, 2006). The general assumptions about the approach remain the same as those with single-marker analysis, except that the covariance associated with relatedness among individuals is taken into account in the model. Since the introduction of mixed linear models to correct for family and population genetic structure in association genetic studies, other methods have been developed to address the computational challenge of testing millions of genetic polymorphisms in genome-wide association genetic studies (Segura *et al.*, 2012; Zhou and Stephens, 2012).

Multi-locus association-mapping analysis

Association-mapping studies commonly evaluate each individual polymorphism for its effect on a trait. However, functional units in the genome can extend beyond a single polymorphism to include several variants that occur within a coding or promoter region, if they all contribute (even if in small amounts) to the phenotype. Instead of analyzing individual polymorphisms separately, all variants within a functional unit (e.g. promoter or coding region) can be analyzed collectively for association with a trait. This approach offers several advantages over the analysis of individual polymorphisms. First, statistical power may be increased by considering the joint effect of multiple marginal associations within a genomic region. Second, the approach can significantly reduce the number of statistical tests to be performed and, consequently, reduce the need to accommodate the challenge of running multiple tests. The difficulty in the implementation of a multi-locus association-mapping approach is that not all the polymorphisms within a genomic region may contribute to the phenotype. For those polymorphisms that have an impact on traits, they may have different effects and in different directions (positive or negative) on the phenotype, relative to the reference allele.

Several approaches have been proposed to combine genetic polymorphisms across genes or genomic regions. These approaches can be separated into two broad categories: burden tests and variance-component tests. Burden tests collapse all rare variants into one genetic score and test it for association with the trait using a linear model (Lee *et al.*,

2014). Variance-component tests, such as the sequence or SNP kernel association test (SKAT), assign a higher weight to rare variants in the statistical model compared with the weight assigned with common variants (Ionita-Laza *et al.*, 2013). In general, burden tests are more powerful when most variants in a region are causal and the effects are in the same direction. The test was designed primarily to identify the contribution of rare mutations to disease phenotypes. These mutations are likely to be rare because there is selection pressure to maintain them at low frequencies, and they are expected to contribute negatively to the disease phenotype, relative to individuals carrying the reference allele. SKAT and other variance-component tests outperform burden tests when only a small proportion of the variants are causal and/or have different directional effects. Although such tests do not exclude common variants from their evaluation, they focus mainly on testing the effect of rare variants by upweighting rare-variant effects and downweighting common-variant effects and can, thus, lose substantial power when both rare and common genetic variants in a region influence a trait. Therefore, they are limited if there is a wide frequency spectrum of variants that impact a phenotype, ranging from rare to frequent.

A suite of additional tests that attempt to overcome the weakness of the first generation of multi-locus association-mapping tests has also been developed. For instance, the combined-sum tests SKAT-C and burden-C analyze common and rare variants separately and then combine the results into an overall test statistic in a way that both variant classes contribute to the test equally (Ionita-Laza *et al.*, 2013). The adaptive sum tests SKAT-A and burden-A differ from the combined-sum tests in that they test for association between markers and traits using different weight parameters and then report the result with the lower *P* value. SKAT-O is an omnibus test that combines SKAT and burden tests. Detailed descriptions of alternative tests have been given elsewhere (Lee *et al.*, 2014). Because the genetic architecture that regulates a complex trait is generally unknown, including the relevance of rare relative to common alleles, it is generally recommended that different burden and SKAT tests be utilized and compared, given that individual tests have distinct strengths and weaknesses.

Statistical power and significance in association mapping of QTLs

Similar to linkage mapping, the statistical power needed to detect QTLs by association mapping depends largely on the effect of a locus on the trait and the standard deviation within each genotypic class, as well as the size of the population being analyzed (Fig. 5.5). In contrast to linkage mapping, in association mapping of QTLs, the minor allele may occur at any frequency in a population (see a comparison of power when the minor allele is 10% (0.1) and 50% (0.5) in Fig. 5.5). As expected, the power to detect a QTL increases when the frequency of the two alleles approaches 50%. In addition, success in identifying a significant association between a marker and causative polymorphisms also depends on the LD between them.

As with linkage mapping, association analysis is also plagued by the fact that when multiple hypotheses are tested, the likelihood of making a type I error (observation of a false positive) increases. Association-mapping studies have largely adopted the use of Bonferroni correction to minimize false positives, by dividing the chosen false-positive rate (α) by the number of association mapping tests performed. Such an approach is conservative, as it ignores the fact that association-mapping tests are not always independent, particularly when considering loci that are in LD. Furthermore, as technology has evolved to permit the genotyping of most variants in the genome, the number of tests performed has increased from a few thousand to several million. Consequently, α is divided by a much larger denominator, resulting in very low thresholds to declare statistical significance of an association between the locus and the trait.

Several approaches have been used to limit the drawbacks of the overly conservative Bonferroni correction. Linkage disequilibrium pruning eliminates from analysis all loci that are strongly linked with one another, selecting only one locus in a block that exhibits LD. This approach reduces the number of statistical tests and addresses the main reason why Bonferroni correction is conservative, although the LD threshold selected to prune

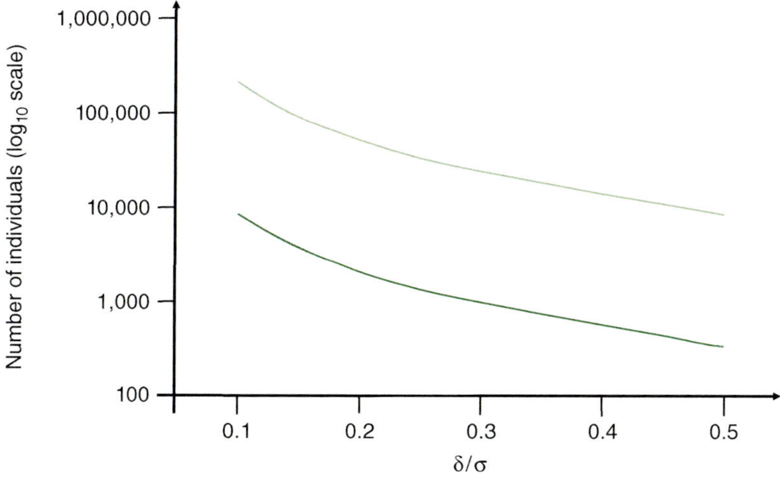

Fig. 5.5. Number of individuals required to detect QTLs in an association mapping population, in loci with a minor allele frequency of 0.5 (lower line) and 0.1 (upper line), and different QTL effects. The QTL effects are shown as the difference in phenotype between genotypic classes (δ), divided by the standard deviation of the trait within each marker class (σ). (Adapted from Mackay *et al.*, 2009.)

the data is arbitrary. Another approach involves pre-selecting for analysis only those loci with alternative alleles predicted to have a significant effect on the structure of proteins (e.g. causing nonsense mutations) or their expression. The drawback is that this assumes that the molecular effect of DNA sequence variants can be predicted. Finally, because the statistical power to detect a QTL by association mapping diminishes as the frequency of the alternative allele decreases, tests are often performed only on those loci where the minor allele frequency is higher than 5%. This strategy eliminates all loci with rare alleles in the population (the majority of loci in most tree populations), reducing the number of tests and the statistical threshold after a Bonferroni correction. This approach may exclude significant loci with rare alleles, but high phenotypic effect, from further analysis.

Association-mapping QTL studies in forestry

Candidate-gene association mapping

Most early studies that characterized genomic regions in breeding and natural-tree populations uncovered a pattern of low LD (Chapter 4, this volume). The low extent of LD implies that capturing all the genetic variation that may contribute to traits requires the genetic analysis of several million loci. For instance, if blocks of LD extend on average by 1,000 bp in a population of a species with a genome the size of *Populus trichocarpa* (480,000,000 bp or 0.48 Gbp), almost half a million loci would have to be genotyped to have at least one genetic marker representing each block. In comparison, assuming a similar extent of LD in a genome the size of *Pinus taeda* (22 Gbp), genotyping of 22 million loci would be necessary. This simplistic calculation does not consider that LD is highly variable throughout the genome of plants (Kim *et al.*, 2007; Gore *et al.*, 2009) and that more than two haplotypes may occur at given block of LD. Nonetheless, it demonstrates that a very large number of loci have to be genotyped to characterize all

the genetic variation that exists in a tree population. Considering the high cost and low throughput of genotyping platforms in the early days of genomics, association mapping studies in forestry opted to focus on specific candidate genes.

Candidate genes are selected based on the existence of one or more lines of evidence that suggest a role in a trait of interest. Candidate genes can be selected because they encode enzymes of pathways involved in the synthesis of a final product that is the trait of interest. For instance, genes in the phenylpropanoid pathway, involved in lignin monomer biosynthesis, have been adopted as candidate genes for regulation of lignin content (Wegrzyn *et al.*, 2010). Evidence from previous quantitative genetic studies can also be used to select candidate genes. In this case, genes localized within QTL regions detected in bi-parental populations have been considered potential candidates for trait regulation and evaluated using an association-genetics approach to fine-map the responsible locus (Muchero *et al.*, 2015). A third approach commonly adopted in the selection of candidate genes is based on their function in heterologous systems, particularly in the model plant *Arabidopsis thaliana* (Ingvarsson *et al.*, 2006; Cabezas *et al.*, 2015). Regardless of the selection approach, candidate genes are likely to represent a limited fraction of the loci that contribute to a phenotype. Nonetheless, these studies served as a proof of concept that association mapping of QTLs was possible in forestry, which was needed before genome-wide association studies could be conducted.

The use of an association-mapping approach using candidate genes was pioneered in forest genetics by the analysis of correlation between variation in the phytochrome gene *phyB2* and traits related to bud set and dormancy in *Populus tremula* (Ingvarsson *et al.*, 2006). Phytochrome genes had previously been implicated in tree dormancy (Howe *et al.*, 1998) and, consequently, represented an ideal candidate gene for evaluation. Furthermore, dormancy-related traits are highly heritable and appear to be largely controlled by a few loci of large effect (Bradshaw and Stettler, 1995). This evidence simplified the detection of significant associations between polymorphisms and the desired phenotype. In the initial association-mapping study, genotyping relied on PCR amplification, cloning, and sequencing of *phyB2* in a small population of 24 individuals that originated from larger populations distributed along the latitudinal range of *P. tremula* in Sweden. Genotypes were determined by aligning individual sequences to a reference and detection of genetic association with climate variables was performed by correlating genotype frequencies with clinal variation. The study demonstrated that loci located in highly variable regions were associated with photoperiod, showing for the first time that an association-mapping approach could be used to identify loci related to local adaptation in trees. Association-mapping studies greatly expanded the number of candidate genes in reports that followed, including a series of analyses of *Pinus taeda* genes and their association with wood properties and carbon discrimination traits (González-Martínez *et al.*, 2007, 2008). These studies explored an expanded set of known genes (20) for which 58 SNPs had previously been discovered (Brown *et al.*, 2004; González-Martínez *et al.*, 2006).

A major expansion in the number of loci characterized by association-genetics studies was made possible by the first generation of genotyping arrays, or chips (Chapter 1, this volume). This platform allowed an increase in the number of target loci from a few dozen (using SNPs) to several hundred. While still largely focused on polymorphisms discovered in selected candidate genes, researchers were no longer required to direct their genotyping efforts solely to a few polymorphisms with high potential of having a functional effect. Instead, numerous variants in candidate genes became the target of association-mapping studies. The use of a chip platform was introduced in the conifer *P. radiata*, where 149 loci selected from genes potentially involved in the control of wood quality and abiotic stress response were selected for association analysis in a population of 447 unrelated individuals (Dillon *et al.*, 2010). This first chip-based study uncovered a surprisingly high number of significant associations (10), where SNPs explained significant proportions (2–6%) of phenotypic variance. To confirm their effect on trait variation, these SNPs were then genotyped in an unrelated validation population, where a significant effect was verified for three of them. Other studies soon followed, including the analysis of several hundred SNPs in approximately 40 candidate genes related to wood chemical properties in *Populus trichocarpa* (Wegrzyn *et al.*, 2010) and *P. nigra* (Guerra *et al.*, 2013). Both studies confirmed the observation made in the conifer *Pinus radiata*, which indicated that, despite the high complexity of the traits of interest and the limited number of genes detected, significant associations could be identified for several of them. Finally, candidate-gene-based SNP chips were also developed for *P. pinaster*, containing genes related to biotic and abiotic stress responses, physical and chemical wood properties, phenology, and growth (Budde *et al.*, 2014; Cabezas *et al.*, 2015).

Genome-wide association mapping

The rapid advances in DNA sequencing throughput and a dramatic decrease in its cost (Chapter 1, this volume) led the possibility of expanding association mapping from candidate genes to the whole genome. Genome-wide association analysis applies the same principles of association genetic studies in candidate genes; however, it offers a range of advantages relative to the candidate-gene-based approach. Genome-wide analysis is not constrained by *a priori* knowledge about gene function. Instead, all existing polymorphisms are explored for association with phenotypic traits. Considering that the molecular understanding of the growth and development mechanisms of most relevant traits in forestry is in its infancy (Grattapaglia *et al.*, 2009), it is clear that the candidate-gene approach is severely limited. Furthermore, quantitative traits are likely to be regulated by many genes. Thus, Genome-wide analysis offers the opportunity to evaluate all possible genes and genetic variants that regulate a trait.

Two strategies can be used for genotyping all genetic variants in a genome: (i) resequencing the genome entirely or (ii) genotyping specific loci using platforms such as DNA chips (Chapter 1, this volume). Developing DNA chips for analysis of forest populations, while feasible, is still too costly, due to the large number of genetic variants that are commonly observed. While DNA chips that genotype tens of thousands of loci have been developed for *Eucalyptus* (Silva-Junior *et al.*, 2015) and *Populus* (Geraldes *et al.*, 2013), these only captured a fraction of variants in genomes and, thus, are not considered genome-wide studies. Thus, the most practical platform for genome-wide detection of genetic variants for association-mapping studies in trees has been to use a resequencing approach. The first study to resequence a large population of trees aimed to identify signatures of natural selection and adaptation in *Populus trichocarpa* (Evans *et al.*, 2014; discussed in detail in Chapter 4, this volume). Association mapping of QTLs was then carried out by correlating genotypes collected from this population with phenotypes for growth and wood quality, but focusing on genomic regions of QTLs detected by linkage mapping. While this study did not truly evaluate the association of polymorphisms in the entire genome with phenotypic traits, the availability of a database of genome-wide variants allowed the targeted analysis of association. Alternative strategies that do not rely on the resequencing of entire genomes, but instead a reduced representation, have also been used in poplar. For instance, a set of 18,153 genes was resequenced in a population of *P. deltoides* composed of 391 unrelated individuals, resulting in the detection of approximately 1 million genetic variants. While the population was small and the majority of loci had rare minor alleles (i.e. present in only a few individuals in the population), the analysis resulted in the detection of a potential causative SNP that regulates the ratio between syringyl and guaiacyl lignin monomers, in a well-known regulator of the phenotype (Fig. 5.6). Several SNPs were also detected in association with wood quality and growth traits, but further validation is necessary to confirm their role.

Pitfalls and limitations of association mapping of QTLs

The use of association mapping has similar objectives to linkage mapping of QTLs, namely the detection of DNA variants that underlie complex traits and their use in breeding. Association mapping addressed some of the most critical limitations of linkage-mapping studies by expanding the genetic diversity sampled and providing higher resolution. As a consequence, association mapping of QTLs is increasingly being adopted in forestry, particularly given that the dramatic reduction in the cost of sequencing by next-generation sequencing (NGS) methods has made it easier for polymorphisms to be identified and genotyped. However, several obstacles are likely to limit the value of association mapping for breeding applications, where knowledge of most loci that contribute to phenotypic variance is necessary. The main obstacles are:

1. *Highly complex phenotypes.* The genetic architecture of most traits of highest value in forestry (e.g. productivity or biomass growth) is likely to be highly complex and regulated by large numbers of loci of small effect. Detection of statistically significant genetic associations will require very large populations (Fig. 5.5). In contrast, the number of samples characterized in current association-mapping studies in forestry has been limited to a few hundred to a thousand individuals, due to the complexity and size of forestry field studies and the costs of genotyping and phenotyping such large numbers of individuals. Lessons learned from association mapping of complex traits, such as human height, suggest that several hundred thousand samples will be required to uncover loci that explain a few per cent of the genetic variation (Wood *et al.*, 2014; Marouli *et al.*, 2017).

2. *Abundance of rare alleles.* Natural and breeding populations of trees have generally been shown to contain an abundance of rare alleles. This occurs primarily because they have large effective population sizes (N_e), which limit the loss of new mutations by random genetic drift. While the extent of the contribution of rare variants to the

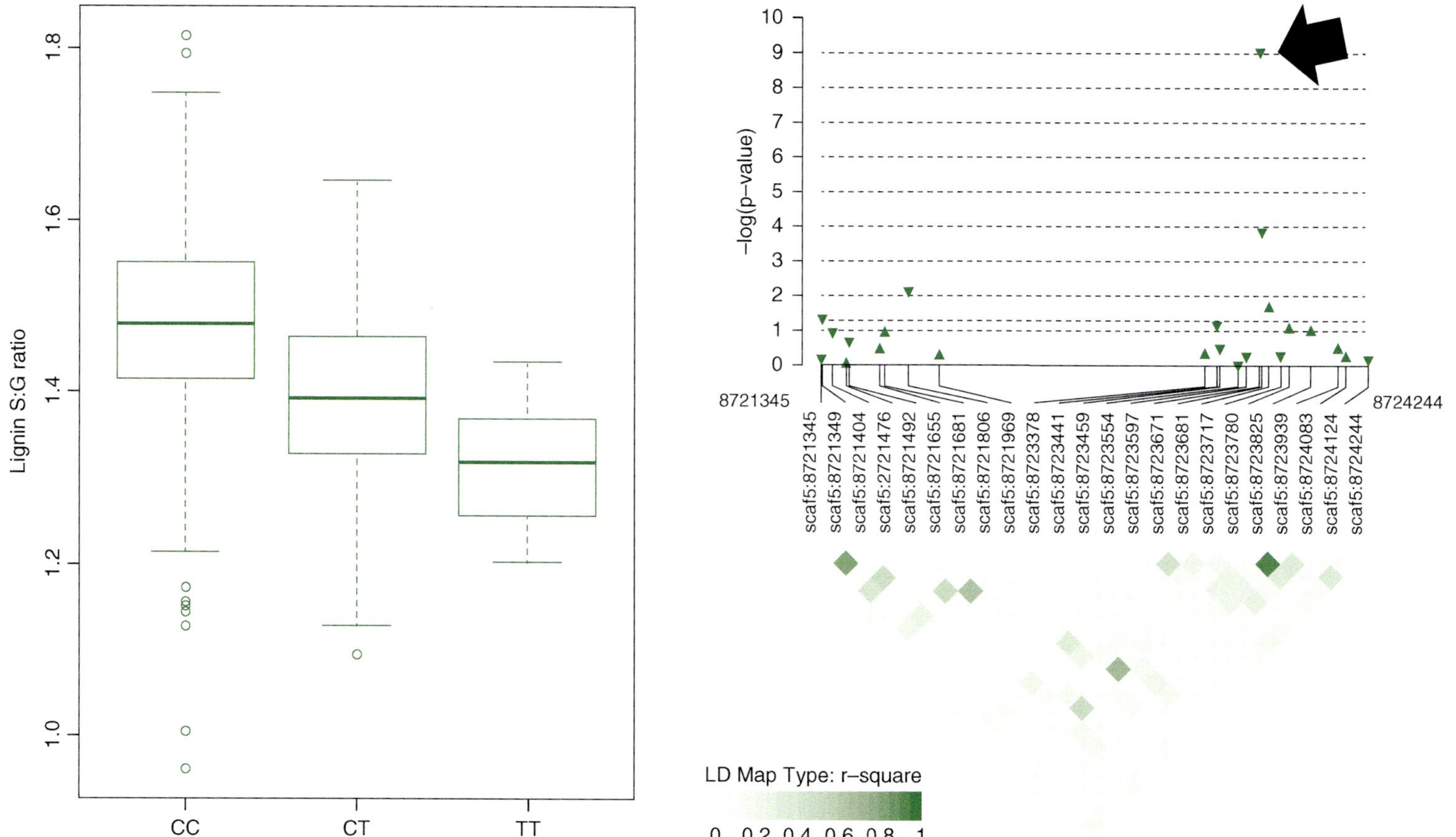

Fig. 5.6. Effect of a major SNP genetic marker (C/T) that regulates syringyl lignin:guaiacyl lignin (S:G) ratio in *Populus deltoides*, detected in the association population (left panel). LD in the region of the locus most highly associated with the S:G ratio showed rapid decay in adjacent loci (arrow, right panel). (Adapted from Fahrenkrog *et al.*, 2017.)

phenotypic variance is unclear, the detection of their effect on the phenotype is particularly difficult because few individuals carry the mutation. The abundance of rare variants also implies a much larger number of loci to genotype if a genome-wide association mapping study is desired.

3. *Limited LD.* The large effective size of tree populations has consequences for the extent of LD among loci, which has been shown generally to be very low. Therefore, unless putative causative polymorphisms are genotyped, there is a high likelihood that few other loci in LD will be significant.

Genomic Selection

While association mapping offered an alternative to linkage mapping of QTLs, many pitfalls remained. Both approaches can be invaluable for the identification of loci that regulate complex traits, creating an opportunity to understand the molecular mechanisms that control them. However, neither of these approaches has proven useful for tree breeding, largely because they only uncover a small proportion of the loci that contribute to variation in complex traits. To address the limitations of linkage and association mapping of QTLs, the use of genomic selection (GS, also known as genome-wide selection) was proposed (Meuwissen *et al.*, 2001). The general principle of GS relies on genotyping a dense set of genetic markers in a population, and phenotyping the same individuals for traits of interest. Based on this information, models that correlate an ensemble of markers to phenotypes are developed and used for breeding and selection of future generations (Fig. 5.7). Trait prediction using GS models is possible because they capture LD that exists between genetic markers and QTLs that contribute to a trait, as well as the genetic relationship between the training population and the subsequent generations derived from it. Contrary to association- and linkage-mapping approaches, the goal of GS is not to identify specific polymorphisms that regulate complex traits, based on a specific significance threshold. Instead, all variants are used in the prediction model, regardless of whether they are significantly associated with the trait.

Development of prediction models for GS

A large number of genetic markers is typically genotyped to develop GS prediction

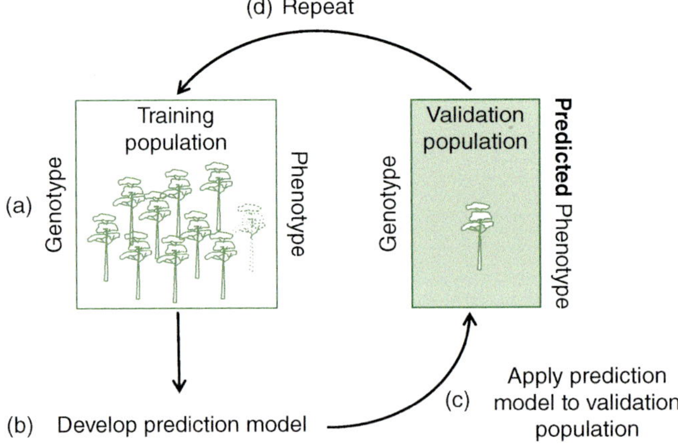

Fig. 5.7. Principles of GS. A training population (a; white box) is genotyped and phenotyped for traits of interest. Based on the collected data, a prediction model is developed (b). By using the genotype data from the validation population (green box) the phenotypes are predicted (c). The process is then repeated (d) with a different subset of individuals of the training population that, in the new iteration, become the validation population.

models in order to maximize the probability of having at least one marker in LD with each QTL, and to capture the relationship among individuals. Because the number of markers is usually far greater than the number of individuals, standard statistical approaches, such as ordinary least-squares, have poor predictive ability because marker effects are treated as fixed effects. This leads to multi-collinearity and overfitting among predictors. Instead, GS uses whole-genome regression, which combines the use of all the markers to predict phenotypes, but considers them to be a random effect. Development of improved methods to model large numbers of markers to predict phenotypes has been the most active area of research in GS and has resulted in variable performance under different genetic architectures of the complex trait (Daetwyler *et al.*, 2013; de los Campos *et al.*, 2013). Some of the most widely used methods to develop GS prediction models are described below.

Bayesian linear regression

The various Bayesian linear models that are applied to GS differ in their assumptions regarding the prior distribution of marker effects (de los Campos *et al.*, 2013; Gianola, 2013). For instance, in ridge regression, the Bayesian prior assumes that marker effects follow a normal distribution with a common variance. Models that fit loci with large effect on the trait have also been developed. These models rely on variable selection to remove markers of no effect and model variance heterogeneity. Extensive reviews and analysis of their performance to predict traits with different heritabilities and genetic architectures, as well as populations and species, have been published (e.g. Heslot *et al.*, 2012).

Reproducing kernel Hilbert space

Reproducing kernel Hilbert space (RKHS) combines features of kernel regression (non-parametric regression) with mixed-effect linear models (Gianola *et al.*, 2006). This method investigates patterns free of the structure imposed on linear models because RKHS is less parameterized, making it more suitable to accommodate effects of interactions. Therefore, RKHS is suitable for modeling non-additive effects and has been found to outperform parametric methods in cases where dominance and epistasis may be significant (Gonzalez-Recio *et al.*, 2008).

Genomic best linear unbiased predictor

Genomic best linear unbiased predictor (BLUP) is equivalent to traditional BLUP analysis of breeding traits (White *et al.*, 2007). The difference lies in the covariance matrix used to account for the relationship among the individuals in the population. Traditional BLUP uses the numerator relationship matrix (A) estimated from the known pedigree, usually with the recurrent method (Lynch and Walsh, 1998; Mrode, 2005). Genomic BLUP uses a relationship matrix estimated from genotypic information (G matrix) (Van-Raden, 2008; Powell *et al.*, 2010). In genomic BLUP, it is assumed that all markers with equal allele frequency have the same effect on additive and dominance variation, that is, all markers contribute similarly to the variation of the trait.

Model accuracy

Before GS prediction models can be used, they must be evaluated for accuracy. For this, the population is divided into a training group and a validation group. The individuals in the training group are used to develop the model. Individuals in the validation group have their genomic estimated breeding value (GEBV) calculated by multiplying the genotypes at all markers by their effect estimated from the model. The accuracy is determined by the correlation between the observed value and the GEBV in the validation group. The process is repeated many times in a cross-validation scheme, so that each individual is used in the training and validation groups (Fig. 5.7). For instance, in a tenfold cross-validation scheme, the complete population is divided so that 90% of the individuals are used as a training group to develop the model used to predict the trait in the remaining 10% (the validation group). The process is repeated by using a set composed of 10% of those previously used in the training group, but they are now employed for the validation. After repeating the process 10 times, all individuals will have been used once in the validation population. At each of the iterations, the correlation between the observed breeding value in the validation group and the GEBV of those same individuals is estimated. The average correlation and the deviation across multiple iterations is indicative of the accuracy of the model. Once a prediction model that reaches a minimum accuracy desired by the breeder is obtained, it can be applied to the next generation of individuals derived from the training population.

Factors that affect the accuracy of GS

For polygenic traits, it has been observed that the use of different prediction models and priors usually results in similar model accuracies (Heslot *et al.*, 2012; Pérez *et al.*, 2012; Resende *et al.*, 2012b). However, when diverse models were applied to traits that were thought to be oligogenic, such as pine rust resistance (Resende *et al.*, 2012b), accuracies were superior under priors that assumed variable selection, variance heterogeneity, or both.

The choice of approach used to model markers and phenotypes can be made *a posteriori* to the genotyping and phenotyping effort. However, several other factors related to the genetic properties of the breeding population and the trait being modeled can also have a significant impact on the accuracy of prediction models. Many of these factors can be determined *a priori* to the development of prediction models, and some are under the influence of the breeder who develops the populations. These factors include the number of molecular markers used to model phenotypes, the genetic architecture of the trait being predicted (number of QTLs and their effect), and the effective and census size of the training population. In addition to being discussed here, the impact of these factors on prediction accuracy in breeding and simulated populations has been reviewed elsewhere (Goddard, 2009; Grattapaglia and Resende, 2011; Heslot *et al.*, 2012).

Effective population size and LD between markers and QTLs

Linkage equilibrium is the random association of alleles at distinct loci. In contrast LD, is encountered when alleles observed at different loci are not independently associated (i.e. they tend to segregate together; Chapter 4, this volume). As described previously, GS relies on the ability to develop models that use markers distributed throughout the genome that are in LD with the actual QTLs. The longer the extent of LD, the fewer markers will be necessary to develop an accurate prediction model. Furthermore, with longer LD, it is more likely that the majority or all QTLs are in LD with at least one marker. Consequently, LD in a breeding population is an important indicator of the expected accuracy of prediction models. LD can be measured by the value of r^2 (Hartl and Clark, 2007). When considering the relationship between a marker and a QTL, r^2 is the proportion of variation caused by the alleles at a QTL that is explained by the marker. Simulations have shown an increase in the accuracy of over 10% in estimates of the GEBV when the average r^2 increased from 0.1 to 0.2 (Calus *et al.*, 2008).

Linkage disequilibrium is highly variable within the genome of plants (Kim *et al.*, 2007; Gore *et al.*, 2009) and is relatively difficult to measure on a genome-wide scale. Thus, an indirect measure of LD, the effective population size (N_e), has been used instead. N_e can be defined as the number of individuals of an idealized population that has the same amount of genetic drift as the current population (Hartl and Clark, 2007). The census population size of a population is usually larger than N_e. The effective population size can be estimated by various methods, including using genetic markers distributed in the genome (for methods of N_e estimation, see Chapter 4, this volume). The larger the value of N_e, the lower the extent of LD (Hill, 2009). Using a deterministic approach, it has been shown that N_e significantly impacts the accuracy of GS prediction models (Grattapaglia and Resende, 2011). The impact of N_e is particularly significant when marker density is low (≤ 2 markers per cM). This occurs because when a large N_e is used in a training population genotyped with only a few markers, many blocks of LD that may contain a QTL will not be "captured" by a genetic marker. In contrast, for training populations genotyped with a large number of loci (≥ 10 markers per cM), the difference in accuracy was small when compared with training populations with an N_e of 10 or 100.

Trait heritability and genetic architecture (number of QTLs)

Several GS studies in tree-breeding populations have found that models are more predictive for traits with higher heritability. For instance, predictive ability was lower for traits with narrow-sense heritability (h^2) of 0.2 compared with those with h^2 of 0.5 or higher (Resende *et al.*, 2012b). Similar outcomes were observed in simulation studies where an increase in h^2 from 0.2 to 0.6 resulted in an average increase of model accuracy of approximately 10% (Grattapaglia and Resende, 2011).

Number of markers

Just as the extent of LD in a breeding population has a significant impact on the accuracy of prediction models, the number of markers used to characterize blocks of LD is equally important. Markers that are widely distributed in the genome in sufficiently large numbers to be used to characterize each haplotype in a LD block will largely determine the accuracy of predictive models. If fewer markers than those necessary to characterize each haplotype are available, those regions that contain QTLs may not be sampled in the breeding population, resulting in lower accuracy. Deterministic approaches have suggested that accuracies reach their plateau at a density of 10–20 markers per cM (Grattapaglia and Resende, 2011). This suggests that for most tree species of commercial interest, such as *Pinus taeda* and *Eucalyptus grandis*, fewer than 20,000 widely distributed loci should suffice to generate suitable prediction models.

Size of the training population

Genomic selection prediction models are developed by regressing markers and phenotypes measured in a training population. The size of the training population has an immediate impact on the accuracy of prediction models because the more phenotypic observations there are, the better marker effects will be estimated.

Application of GS in forestry

Genomic selection was first evaluated in forestry by simulating the genetic properties of tree-breeding populations, and estimating the accuracy of prediction models obtained from these populations. A pioneering study evaluated various population and genotyping scenarios that can be controlled by the breeder, including the N_e, the number of individuals in the training population, and the number of genetic markers (Grattapaglia and Resende, 2011). Other factors evaluated included the genetic architecture of the predicted trait, such as its heritability, and the number of QTLs. Another study evaluated the implementation of GS under different breeding and seed-production scenarios, which mimicked those used for the genetic improvement of *Cryptomeria japonica*, and compared the genetic gains obtained through traditional phenotypic breeding (Iwata *et al.*, 2011). Overall, the results suggested that, for traits regulated by a large number of loci of small effect and moderate to low heritabilities, training populations of a few thousand individuals (2,000 or more), genotyped with 10,000–20,000 genetic markers would be sufficient to develop accurate models. These simulations also showed that the accuracy of GS models are similar to traditional breeding, but that the reduction in the length of the breeding cycle could result in a significant acceleration of genetic gains.

These early simulation studies were soon followed by demonstration in a real breeding population of *P. taeda*, which compared the performance of models developed for traits with different genetic architectures, as well as their performance across multiple environments and ages (Resende *et al.*, 2012a,b). The comparison of different prediction models showed little difference in accuracy for polygenic traits but a significant improvement for phenotypes controlled by a few loci of large effect in models that implemented variable selection. As expected, the heritability of traits also affected the accuracy of models, with higher heritabilities resulting in better accuracies. Finally, the studies also showed the limited value of applying models to predict traits across distinct environments, or to use growth data collected in early years to predict adult performance. The most relevant observation from these studies was that accuracies predicted from simulation studies were in good agreement with those from real populations, confirming that simulations could serve as a guide to determine the performance expectation of GS models in any given population. Since the demonstration of GS in a tree-breeding population of *Pinus* was first reported, numerous other studies have been published, largely confirming previous observations. These include other studies in *P. taeda* (Zapata-Valenzuela *et al.*, 2013) and *P. pinaster* (Bartholomé *et al.*, 2016; Isik *et al.*, 2016); other conifers, such as white spruce (*Picea glauca*; Beaulieu *et al.*, 2014a,b) and interior spruce (*Picea engelmannii × glauca*) (Gamal El-Dien *et al.*, 2015; Ratcliffe *et al.*, 2015); and the hardwood *Eucalyptus* (Müller *et al.*, 2017; Tan *et al.*, 2017).

Results from these studies also sparked an interest in evaluating other facets of genomics that are of particular interest to tree breeders, such as accounting for all genetic effects that contribute to phenotypic variance (additive, dominant, and epistatic) and their use in the development of GS prediction models that also incorporate these non-additive effects. Until genomic data became available, partitioning genetic variance into its additive and non-additive components had been difficult because the available family structure in most tree-breeding populations was not suitable for their estimation. The availability of extensive genetic-marker data made possible the development of additive- and dominance-realized genomic relationships for the estimation of variance components and the prediction of genetic values that include these effects. When applied to data from a breeding population of *P. taeda*, it was shown that non-additive components of the genetic variance can be significant and of similar magnitude to additive effects (Muñoz *et al.*, 2014). Furthermore, models that incorporate additive and dominance effects were developed to predict polygenic and oligogenic traits in the same population, but the inclusion of dominance only modestly improved predictions (de Almeida Filho *et al.*, 2016).

Pitfalls and limitations of genome-wide selection in forestry

The development of prediction models for use in GS has addressed many of the limitations of linkage and association mapping of QTLs. The most important distinction is that GS offers the opportunity for breeders to use genetic markers to account for most of the genetic variation that is associated with a complex trait. Consequently, GS allows the effective use of genomic data for tree breeding. Despite the potential of GS, there are still several unresolved issues:

1. *Going beyond the training population generation.* GS has proven effective in animal and crop breeding, where it is now widely implemented by using models for trait prediction in the next generations. However, studies in forestry have not yet assessed the true accuracy of prediction models after the generation in which the models were developed. Consequently, the proof of concept required for the definitive adoption of GS by many tree-breeding programs has not yet been achieved.

2. *Long-term accuracy of prediction models.* Preferably, prediction models should remain accurate over multiple generations to limit the need for phenotyping and genotyping populations beyond the training population. However, research in poultry and other species have shown that accuracies of predicted genomic values decrease substantially (~10%) in only one generation, although trait heritability, among other factors, determined the magnitude of that reduction (Weng *et al.*, 2016). It remains to be seen how often subsequent generations will have to be phenotyped and genotyped to retrain prediction models.

3. *Limited or no transferability of models across environments and populations.* Tree plantations are often established on highly heterogeneous sites, where they remain for several years until harvest, under variable pressure from biotic and abiotic sources of stress. This is in sharp contrast to agricultural crops and animal farming, where individuals are raised in highly controlled environments. Early work already suggests that differences in environment can negatively impact the transferability of models across sites in forestry (Resende *et al.*, 2012a). Furthermore, animal- and crop-breeding programs often have their origin in a relatively small population of founders. Therefore, there can be a high genetic relationship among individuals used in distinct breeding programs. This does not occur in forestry, where breeding programs typically utilize genetically distinct germplasm with limited shared ancestry.

Summary

Most traits of interest to tree breeders are quantitative and follow a continuous distribution driven by the contribution of many loci of small effect. Identifying such loci is difficult but of interest to support the selection of genetically superior individuals and to determine optimal mating designs. This is even more relevant in forest-tree breeding, where each cycle of improvement may take decades due to the long life cycle of trees, their long generation times, and the logistical complexities of field experiments. As a consequence of these challenges, most tree-improvement programs are in their first

cycles of breeding and selection, and are only a few generations removed from wild populations.

The genomics of forest trees has evolved rapidly in recent decades, with the development of NGS methods and the introduction of approaches that use genomic information to identify loci that contribute to complex traits. Genomic data have supported the separation of quantitative phenotypes into their genetic components and sources of variance, for estimation of heritabilities, without the need for specially designed mating populations. However, the most significant contribution of genomics has been in the identification of loci that contribute to complex traits, and in the use of genome-wide information to develop models that allow the prediction of phenotypes based on genomic data.

The use of genome-wide information to support breeding and selection started with the development of the first genetic maps in forestry, and their use in the identification of QTLs based on the co-segregation of genetic markers and traits. While of low resolution and limited in the diversity samples, QTL analysis became widely used in forestry studies for the detection of loci that contribute to biotic and abiotic stress resistance, growth and biomass yield and composition, and developmental properties. QTL analysis has also been applied to the study of molecular traits, such as gene expression, to understand how genes are controlled and interact with their regulators.

Association mapping was introduced in order to address some of the limitations of QTL studies, and because of the availability of new tools to characterize large parts of the genetic variability in genomes. Association mapping relies on similar principles to QTL analysis but is typically carried out in populations of largely unrelated individuals or families, capturing wide allele diversity at higher resolution because of the typical low extent of LD in these populations. However, association mapping also has a major limitation: very large populations are required to detect the contribution of alleles with small effects. As a consequence, most studies have identified only a few loci that contribute to the phenotypic variation. Nonetheless, association mapping has been used successfully to identify loci that are involved in variation for biomass yield and composition, and a variety of other traits.

Finally, an alternative to linkage and association mapping and the use of genomic data in breeding was created by the introduction of GS. In this case, dense sets of genetic markers are genotyped in populations that are also phenotyped for traits of interest. Models that correlate all genomic information to phenotypes can be developed, without attempting to identify the significance and effect of each of them individually, using Bayesian linear regression and other approaches. Early studies in forest-tree species showed that prediction models can provide accuracy and gains as high or higher than traditional breeding methods. Consequently, models have been developed for a wide variety of tree species and traits.

References

Anderegg, W.R.L., Schwalm, C., Biondi, F., Camarero, J.J., Koch, G., *et al.* (2015) Pervasive drought legacies in forest ecosystems and their implications for carbon cycle models. *Science* 349, 528–532.

Ball, R.D. (2001) Bayesian methods for quantitative trait loci mapping based on model selection: approximate analysis using the Bayesian information criterion. *Genetics* 159, 1351–1364.

Bartholomé, J., Mabiala, A., Savelli, B., Bert, D., Brendel, O., *et al.* (2015) Genetic architecture of carbon isotope composition and growth in *Eucalyptus* across multiple environments. *New Phytologist* 206, 1437–1449.

Bartholomé, J., van Heerwaarden, J., Isik, F., Boury, C., Vidal, M., *et al.* (2016) Performance of genomic prediction within and across generations in maritime pine. *BMC Genomics* 17, 604.

Beaulieu, J., Doerksen, T., Clément, S., MacKay, J., and Bousquet, J. (2014a) Accuracy of genomic selection models in a large population of open-pollinated families in white spruce. *Heredity* 113, 343–352.

Beaulieu, J., Doerksen, T.K., MacKay, J., Rainville, A., and Bousquet, J. (2014b) Genomic selection accuracies within and between environments and small breeding groups in white spruce. *BMC Genomics* 15, 1048.

Beavis, W. (1998) Limitations of QTL mapping. In: Paterson, A.H. (ed.) *Molecular Dissection of Complex Traits*. CRC Press, Boca Raton, Florida, pp. 145–162.

Bradshaw, H.D. and Stettler, R. (1995) Molecular genetics of growth and development in populus. IV. Mapping QTL with large effects on growth, form, and phenology traits in a forest tree. *Genetics* 139, 963–973.

Brown, G.R., Bassoni, D.L., Gill, G.P., Fontana, J.R., Wheeler, N.C., *et al.* (2003) Identification of quantitative trait loci influencing wood property traits in loblolly pine (*Pinus taeda* L.). III. QTL Verification and candidate gene mapping. *Genetics* 164, 1537–1546.

Brown, G.R., Gill, G.P., Kuntz, R.J., Langley, C.H., and Neale, D. (2004) Nucleotide diversity and linkage disequilibrium in loblolly pine. *Proceedings of the National Academy of Sciences USA* 101, 15255–15260.

Budde, K.B., Heuertz, M., Hernández-Serrano, A., Pausas, J.G., Vendramin, G.G., *et al.* (2014) *In situ* genetic association for serotiny, a fire-related trait, in Mediterranean maritime pine (*Pinus pinaster*). *New Phytologist* 201, 230–241.

Cabezas, J.A., González-Martínez, S.C., Collada, C., Guevara, M.A., Boury, C., *et al.* (2015) Nucleotide polymorphisms in a pine ortholog of the *Arabidopsis* degrading enzyme cellulase KORRIGAN are associated with early growth performance in *Pinus pinaster*. *Tree Physiology* 35, 1000–1006.

Calus, M.P.L., Meuwissen, T.H.E., de Roos, A.P.W., and Veerkamp, R. (2008) Accuracy of genomic selection using different methods to define haplotypes. *Genetics* 178, 553–561.

Cecich, R.A., Kang, H., and Chalupka, W. (1994) Regulation of early flowering in *Pinus banksiana*. *Tree Physiology* 14, 275–284.

Churchill, G.A. and Doerge, R. (1994) Empirical threshold values for quantitative trait mapping. *Genetics* 138, 963–971.

Daetwyler, H.D., Calus, M.P.L., Pong-Wong, R., de los Campos, G., and Hickey, J. (2013) Genomic prediction in animals and plants: simulation of data, validation, reporting, and benchmarking. *Genetics* 193, 347–365.

de Almeida Filho, J.E., Guimarães, J.F.R., e Silva, F.F., de Resende M.D.V., Muñoz, P., *et al.* (2016) The contribution of dominance to phenotype prediction in a pine breeding and simulated population. *Heredity* 117, 33–41.

de los Campos, G., Hickey, J.M., Pong-Wong, R., Daetwyler, H.D., and Calus, M.P.L (2013) Whole-genome regression and prediction methods applied to plant and animal breeding. *Genetics* 193, 327–345.

de los Campos, G., Sorensen, D., and Gianola, D. (2015) Genomic heritability: what is it? *PLoS Genetics* 11, e1005048.

de Miguel, M., Cabezas, J.-A., de María, N., Sánchez-Gómez, D., Guevara, M.-Á., *et al.* (2014) Genetic control of functional traits related to photosynthesis and water use efficiency in *Pinus pinaster* Ait. drought response: integration of genome annotation, allele association and QTL detection for candidate gene identification. *BMC Genomics* 15, 464.

Devey, M.E., Delfino-Mix, A., Kinloch, B.B., and Neale, D. (1995) Random amplified polymorphic DNA markers tightly linked to a gene for resistance to white pine blister rust in sugar pine. *Proceedings of the National Academy of Sciences USA* 92, 2066–2070.

Devey, M.E., Groom, K.A., Nolan, M.F., Bell, J.C., Dudzinski, M.J., *et al.* (2004a) Detection and verification of quantitative trait loci for resistance to *Dothistroma* needle blight in *Pinus radiata*. *Theoretical and Applied Genetics* 108, 1056–1063.

Devey, M.E., Carson, S.D., Nolan, M.F., Matheson, A.C., Te Riini, C., and Hohepa, J. (2004b) QTL associations for density and diameter in *Pinus radiata* and the potential for marker-aided selection. *Theoretical and Applied Genetics* 108, 516–524.

Dillon, S.K., Nolan, M., Li, W., Bell, C., Wu, H.X., and Southerton, S. (2010) Allelic variation in cell wall candidate genes affecting solid wood properties in natural populations and land races of *Pinus radiata*. *Genetics* 185, 1477–1487.

Doebley, J.F., Gaut, B.S., and Smith, B. (2006) The molecular genetics of crop domestication. *Cell* 127, 1309–1321.

Doerge, R.W. (2002) Mapping and analysis of quantitative trait loci in experimental populations. *Nature Reviews Genetics* 3, 43–52.

Drost, D.R., Benedict, C.I., Berg, A., Novaes, E., Novaes, C.R.D.B., *et al.* (2010) Diversification in the genetic architecture of gene expression and transcriptional networks in organ differentiation of *Populus*. *Proceedings of the National Academy of Sciences USA* 107, 8492–8497.

Eriksson, G., Schelander, B., and Åkebrand, V. (2009) Inbreeding depression in an old experimental plantation of *Picea abies*. *Hereditas* 73, 185–193.

Evans, L.M., Slavov, G.T., Rodgers-Melnick, E., Martin, J., Ranjan, P., *et al.* (2014) Population genomics of *Populus trichocarpa* identifies signatures of selection and adaptive trait associations. *Nature Genetics* 46, 1089–1096.

Fabbrini, F., Gaudet, M., Bastien, C., Zaina, G., Harfouche, A., *et al.* (2012) Phenotypic plasticity, QTL mapping and genomic characterization of bud set in black poplar. *BMC Plant Biology* 12, 47.

Fahrenkrog, A.M., Neves, L.G., Resende, M.F. Jr., Vazquez, A.I., de los Campos, G., *et al.* (2017) Genome-wide association study reveals putative regulators of bioenergy traits in *Populus deltoides*. *New Phytologist* 213, 799–811.

Falconer, D.S. and Mackay, T.F.C. (1996) *Introduction to Quantitative Genetics*, 4th edn. Longman, London.

Farquhar, G.D., Ehleringer, J.R., and Hubick, K. (1989) Carbon isotope discrimination and photosynthesis. *Annual Review of Plant Physiology and Plant Molecular Biology* 40, 503–537.

Ferris, R., Long, L., Bunn, S.M., Robinson, K.M., Bradshaw, H.D., *et al.* (2002) Leaf stomatal and epidermal cell development: identification of putative quantitative trait loci in relation to elevated carbon dioxide concentration in poplar. *Tree Physiology* 22, 633–640.

Fisher, R.A. (1918) The correlation between relatives on the supposition of Mendelian inheritance. *Transactions of the Royal Society of Edinburgh* 52, 399–433.

Freeman, J.S., Potts, B.M., and Vaillancourt, R. (2008) Few Mendelian genes underlie the quantitative response of a forest tree, *Eucalyptus globulus*, to a natural fungal epidemic. *Genetics* 178, 563–571.

Frewen, B.E., Chen, T.H., Howe, G.T., Davis, J., Rohde, A., *et al.* (2000) Quantitative trait loci and candidate gene mapping of bud set and bud flush in *Populus*. *Genetics* 154, 837–845.

Gamal El-Dien, O., Ratcliffe, B., Klápště, J., Chen, C., Porth, I., and El-Kassaby, Y.A. (2015) Prediction accuracies for growth and wood attributes of interior spruce in space using genotyping-by-sequencing. *BMC Genomics* 16, 370.

Geraldes, A., DiFazio, S.P., Slavov, G.T., Ranjan, P., Muchero, W., *et al.* (2013) A 34K SNP genotyping array for *Populus trichocarpa* : design, application to the study of natural populations and transferability to other *Populus* species. *Molecular Ecology Resources* 13, 306–323.

Gianola, D. (2013) Priors in whole-genome regression: the bayesian alphabet returns. *Genetics* 194, 573–596.

Gianola, D., Fernando, R.L., and Stella, A. (2006) Genomic-assisted prediction of genetic value with semiparametric procedures. *Genetics* 173, 1761–1776.

Gion, J.-M., Carouché, A., Deweer, S., Bedon, F., Pichavant, F., *et al.* (2011) Comprehensive genetic dissection of wood properties in a widely-grown tropical tree: *Eucalyptus*. *BMC Genomics* 12, 301.

Goddard, M. (2009) Genomic selection: prediction of accuracy and maximisation of long term response. *Genetica* 136, 245–257.

González-Martínez, S.C., Ersoz, E., Brown, G.R., Wheeler, N.C., and Neale, D. (2006) DNA sequence variation and selection of tag single-nucleotide polymorphisms at candidate genes for drought-stress response in *Pinus taeda* L. *Genetics* 172, 1915–1926.

González-Martínez, S.C., Wheeler, N.C., Ersoz, E., Nelson, C.D., and Neale, D. (2007) Association genetics in *Pinus taeda* L. I. Wood property traits. *Genetics* 175, 399–409.

González-Martínez, S.C., Huber, D., Ersoz, E., Davis, J.M., and Neale, D. (2008) Association genetics in *Pinus taeda* L. II. Carbon isotope discrimination. *Heredity* 101, 19–26.

González-Recio, O., Gianola, D., Rosa, G.J., Weigel, K.A., and Kranis, A. (2009) Genome-assisted prediction of a quantitative trait measured in parents and progeny: application to food conversion rate in chickens. *Genetics Selection Evolution* 41, 3.

Gore, M.A., Chia, J.-M., Elshire, R.J., Sun, Q., Ersoz, E.S., *et al.* (2009) A first-generation haplotype map of maize. *Science* 326, 1115–1117.

Grattapaglia, D., Bertolucci, F.L., Penchel, R., and Sederoff, R. (1996) Genetic mapping of quantitative trait loci controlling growth and wood quality traits in *Eucalyptus grandis* using a maternal half-sib family and RAPD markers. *Genetics* 144, 1205–1214.

Grattapaglia, D., Bertolucci, F.L., and Sederoff, R. (1995) Genetic mapping of QTL controlling vegetative propagation in *Eucalyptus grandis* and *E. urophylla* using a pseudo-testcross strategy and RAPD markers. *Theoretical and Applied Genetics* 90, 933–947.

Grattapaglia, D., Plomion, C., Kirst, M., and Sederoff, R. (2009) Genomics of growth traits in forest trees. *Current Opinion in Plant Biology* 12, 148–156.

Grattapaglia, D. and Resende, M.D.V. (2011) Genomic selection in forest tree breeding. *Tree Genetics & Genomes* 7, 241–255.

Griffin, A. (1983) Selfing effects in *Eucalyptus regnans*. *Silvae Genetica* 6, 216–221.

Groover, A., Devey, M., Fiddler, T., Lee, J., Megraw, R., *et al.* (1994) Identification of quantitative trait loci influencing wood specific gravity in an outbred pedigree of loblolly pine. *Genetics* 138, 1293–1300.

Guerra, F.P., Wegrzyn, J.L., Sykes, R., Davis, M.F., Stanton, B.J., and Neale, D. (2013) Association genetics of chemical wood properties in black poplar (*Populus nigra*) *New Phytologist* 197, 162–176.

Hartl, D.L. and Clark, D. (2007) *Principles of Population Genetics*. Sinauer Associates, Sunderland, Massachusetts.

Heslot, N., Yang, H.-P., Sorrells, M.E., and Jannink, J.-L. (2012) Genomic selection in plant breeding: a comparison of models. *Crop Science* 52, 146–160.

Hill, W.G. (2009) Estimation of effective population size from data on linkage disequilibrium. *Genetics Research* 38, 209–216.

Howe, G.T., Bucciaglia, P.A., Hackett, W.P., Furnier, G.R., Cordonnier-Pratt, M.M., and Gardner, G. (1998) Evidence that the phytochrome gene family in black cottonwood has one PHYA locus and two PHYB loci but lacks members of the PHYC/F and PHYE subfamilies. *Molecular Biology and Evolution* 15, 160–175.

Induri, B.R., Ellis, D.R., Slavov, G.T., Yin, T., Zhang, X., *et al.* (2012) Identification of quantitative trait loci and candidate genes for cadmium tolerance in *Populus*. *Tree Physiology* 32, 626–638.

Ingvarsson, P.K., García, M.V., Hall, D., Luquez, V., and Jansson, S. (2006) Clinal variation in *phyB2*, a candidate gene for day-length-induced growth cessation and bud set, across a latitudinal gradient in European aspen (*Populus tremula*). *Genetics* 172, 1845–1853.

Ionita-Laza, I., Lee, S., Makarov, V., Buxbaum, J.D., and Lin, X. (2013) Sequence kernel association tests for the combined effect of rare and common variants. *American Journal of Human Genetics* 92, 841–853.

Isik, F., Bartholomé, J., Farjat, A., Chancerel, E., Raffin, A., *et al.* (2016) Genomic selection in maritime pine. *Plant Science* 242, 108–119.

Iwata, H., Hayashi, T., and Tsumura, Y. (2011) Prospects for genomic selection in conifer breeding: a simulation study of *Cryptomeria japonica*. *Tree Genetics & Genomes* 7, 747–758.

Jansen, R.C. (1993) Interval mapping of multiple quantitative trait loci. *Genetics* 135, 205–211.

Jansen, R.C. and Nap, J. (2001) Genetical genomics: the added value from segregation. *Trends in Genetics* 17, 388–391.

Jorge, V., Dowkiw, A., Faivre-Rampant, P., and Bastien, C. (2005) Genetic architecture of qualitative and quantitative *Melampsora larici-populina* leaf rust resistance in hybrid poplar: genetic mapping and QTL detection. *New Phytologist* 167, 113–127.

Kao, C.H., Zeng, Z.B., and Teasdale, R. (1999) Multiple interval mapping for quantitative trait loci. *Genetics* 152, 1203–1216.

Kayihan, G.C., Huber, D.A., Morse, A.M., White, T.L., and Davis, J. (2005) Genetic dissection of fusiform rust and pitch canker disease traits in loblolly pine. *Theoretical and Applied Genetics* 110, 948–958.

Kim, S., Plagnol, V., Hu, T.T., Toomajian, C., Clark, R.M., *et al.* (2007) Recombination and linkage disequilibrium in *Arabidopsis thaliana*. *Nature Genetics* 39, 1151–1155.

Kirst, M., Myburg, A.A., de León, J.P.G., Kirst, M.E., Scott, J., and Sederoff, R. (2004) Coordinated genetic regulation of growth and lignin revealed by quantitative trait locus analysis of cDNA microarray data in an interspecific backcross of eucalyptus. *Plant Physiology* 135, 2368–2378.

Kirst, M., Basten, C.J., Myburg, A.A., Zeng, Z.-B., and Sederoff, R. (2005) Genetic architecture of transcript-level variation in differentiating xylem of a *Eucalyptus* hybrid. *Genetics* 169, 2295–2303.

Knowler, W.C., Williams, R.C., Pettitt, D.J., and Steinberg, A. (1988) Gm3;5,13,14 and type 2 diabetes mellitus: an association in American Indians with genetic admixture. *American Journal of Human Genetics* 43, 520–526.

Lander, E.S. and Botstein, D. (1989) Mapping Mendelian factors underlying quantitative traits using RFLP linkage maps. *Genetics* 121, 185–199.

Larsson, S.J., Lipka, A.E., and Buckler, E. (2013) Lessons from *Dwarf8* on the strengths and weaknesses of structured association mapping. *PLoS Genetics* 9, e1003246.

Lee, S., Abecasis, G.R., Boehnke, M., and Lin, X. (2014) Rare-variant association analysis: study designs and statistical tests. *American Journal of Human Genetics* 95, 5–23.

Leverenz, J.W. and Hinckley, T. (1990) Shoot structure, leaf area index and productivity of evergreen conifer stands. *Tree Physiology* 6, 135–149.

Lynch, M. and Walsh, B. (1998) *Genetics and Analysis of Quantitative Traits*. Sinauer Associates, Sunderland, Massachusetts.

Mackay, T.F.C., Stone, E.A., and Ayroles, J. (2009) The genetics of quantitative traits: challenges and prospects. *Nature Reviews Genetics* 10, 565–577.

Marguerit, E., Bouffier, L., Chancerel, E., Costa, P., Lagane, F., *et al.* (2014) The genetics of water-use efficiency and its relation to growth in maritime pine. *Journal of Experimental Botany* 65, 4757–4768.

Marouli, E., Graff, M., Medina-Gomez, C., Lo, K.S., Wood, A.R., *et al.* (2017) Rare and low-frequency coding variants alter human adult height. *Nature* 542, 186–190.

Martin, G., Brommonschenkel, S., Chuwongse, J., Frary, A., Ganal, M., *et al.* (1993) Map-based cloning of a protein kinase gene conferring disease resistance in tomato. *Science* 26, 1432–1436.

Meuwissen, T.H.E., Hayes, B.J., and Goddard, M. (2001) Prediction of total genetic value using genome-wide dense marker maps. *Genetics* 157, 1819–1829.

Mizrachi, E., Verbeke, L., Christie, N., Fierro, A.C., Mansfield, S.D., *et al.* (2017) Network-based integration of systems genetics data reveals pathways associated with lignocellulosic biomass accumulation and processing. *Proceedings of the National Academy of Sciences USA* 114, 1195–1200.

Monclus, R., Leplé, J.-C., Bastien, C., Bert, P.-F., Villar, M., *et al.* (2012) Integrating genome annotation and QTL position to identify candidate genes for productivity, architecture and water-use efficiency in *Populus* spp. *BMC Plant Biology* 12, 173.

Mrode, R.A. (2005) *Linear Models for the Prediction of Animal Breeding Values*, 2nd edn. CAB International, Wallingford, UK.

Muchero, W., Guo, J., DiFazio, S.P., Chen, J.-G., Ranjan, P., *et al.* (2015) High-resolution genetic mapping of allelic variants associated with cell wall chemistry in *Populus*. *BMC Genomics* 16, 24.

Müller, B.S.F., Neves, L.G., de Almeida Filho, J.E., Resende, M.F.R., Muñoz, P.R., *et al.* (2017) Genomic prediction in contrast to a genome-wide association study in explaining heritable variation of complex growth traits in breeding populations of *Eucalyptus*. *BMC Genomics* 18, 524.

Muñoz, P.R., Resende, M.F.R., Gezan, S.A., Resende, M.D.V., de los Campos, G., *et al.* (2014) unraveling additive from non-additive effects using genomic relationship matrices. *Genetics* 198, 1759–1768.

Novaes, E., Osorio, L., Drost, D.R., Miles, B.L., Boaventura-Novaes, C.R.D., *et al.* (2009) Quantitative genetic analysis of biomass and wood chemistry of *Populus* under different nitrogen levels. *New Phytologist* 182, 878–890.

Ollinger, S.V., Aber, J.D., and Reich, P. (1997) Simulating ozone effects on forest productivity: interactions among leaf-, canopy-, and stand-level processes. *Ecological Applications* 7, 1237–1251.

Pérez, P., Gianola, D., González-Camacho, J.M., Crossa, J., Manès, Y., and Dreisigacker, S. (2012) Comparison between linear and non-parametric regression models for genome-enabled prediction in wheat. *G3: Genes, Genomes, Genetics* 2, 1595–1605.

Powell, J.E., Visscher, P.M., and Goddard, M.E. (2010) Reconciling the analysis of IBD and IBS in complex trait studies. *Nature Reviews Genetics* 11, 800–805.

Pritchard, J.K., Stephens, M., and Donnelly, P. (2000a) Inference of population structure using multilocus genotype data. *Genetics* 155, 945–959.

Pritchard, J.K., Stephens, M., Rosenberg, N.A., and Donnelly, P. (2000b) Association mapping in structured populations. *American Journal of Human Genetics* 67, 170–181.

Rae, A.M., Ferris, R., Tallis, M.J., and Taylor, G. (2006) Elucidating genomic regions determining enhanced leaf growth and delayed senescence in elevated CO_2. *Plant, Cell & Environment* 29, 1730–1741.

Rae, A.M., Tricker, P.J., Bunn, S.M., and Taylor, G. (2007) Adaptation of tree growth to elevated CO_2, quantitative trait loci for biomass in *Populus*. *New Phytologist* 175, 59–69.

Ratcliffe, B., El-Dien, O.G., Klápště, J., Porth, I., Chen, C., *et al.* (2015) A comparison of genomic selection models across time in interior spruce (*Picea engelmannii × glauca*) using unordered SNP imputation methods. *Heredity* 115, 547–555.

Rauscher, H.M. and Johnsen, K. (eds) (2004) *Southern Forest Science: Past, Present, and Future*. Southern Research Station, Asheville, North Carolina. Available at: www.srs.fs.usda.gov/pubs/gtr/gtr_srs075.pdf (accessed July 15, 2019).

Resende, M.F.R., Muñoz, P., Acosta, J.J., Peter, G.F., Davis, J.M., *et al.* (2012a) Accelerating the domestication of trees using genomic selection: accuracy of prediction models across ages and environments. *New Phytologist* 193, 617–624.

Resende, M.F.R., Muñoz, P., Resende, M.D.V., Garrick, D.J., Fernando, R.L., *et al.* (2012b) Accuracy of genomic selection methods in a standard data set of loblolly pine (*Pinus taeda* L.) *Genetics* 190, 1503–1510.

Ridge, C.R., Hinckley, T.M., Stettler, R.F., and van Volkenburgh, E. (1986) Leaf growth characteristics of fast-growing poplar hybrids *Populus trichocarpa × P. deltoides*. *Tree Physiology* 1, 209–216.

Ritland, K. (1996) Marker-based method for inferences about quantitative inheritance in natural populations. *Evolution* 50, 1062–1073.

Rohde, A., Storme, V., Jorge, V., Gaudet, M., Vitacolonna, N., *et al.* (2011) Bud set in poplar—genetic dissection of a complex trait in natural and hybrid populations. *New Phytologist* 189, 106–121.

Sax, K. (1923) The association of size differences with seed-coat pattern and pigmentation in *Phaseolus vulgaris*. *Genetics* 8, 552–560.

Schadt, E.E., Monks, S.A., Drake, T.A., Lusis, A.J., Che, N., *et al.* (2003) Genetics of gene expression surveyed in maize, mouse and man. *Nature* 422, 297–302.

Segura, V., Vilhjálmsson, B.J., Platt, A., Korte, A., Seren, Ü., *et al.* (2012) An efficient multi-locus mixed-model approach for genome-wide association studies in structured populations. *Nature Genetics* 44, 825–830.

Sewell, M.M., Davis, M.F., Tuskan, G.A., Wheeler, N.C., Elam, C.C., *et al.* (2002) Identification of QTL influencing wood property traits in loblolly pine (*Pinus taeda* L.). II. Chemical wood properties. *Theoretical and Applied Genetics* 104, 214–222.

Silva-Junior, O.B., Faria, D.A., and Grattapaglia, D. (2015) A flexible multi-species genome-wide 60K SNP chip developed from pooled resequencing of 240 Eucalyptus tree genomes across 12 species. *New Phytologist* 206, 1527–1540.

Soller, M., Brody, T., and Genizi, A. (1976) On the power of experimental designs for the detection of linkage between marker loci and quantitative loci in crosses between inbred lines. *Theoretical and Applied Genetics* 47, 35–39.

Spielman, R.S., McGinnis, R.E., and Ewens, W. (1993) Transmission test for linkage disequilibrium: the insulin gene region and insulin-dependent diabetes mellitus (IDDM). *American Journal of Human Genetics* 52, 506–516.

Stirling, B., Newcombe, G., Vrebalov, J., Bosdet, I., and Bradshaw, H. (2001) Suppressed recombination around the *MXC3* locus, a major gene for resistance to poplar leaf rust. *Theoretical and Applied Genetics* 103, 1129–1137.

Strauss, S.H., Lande, R., and Namkoong, G. (1992) Limitations of molecular-marker-aided selection in forest tree breeding. *Canadian Journal of Forest Research* 22, 1050–1061.

Street, N.R., James, T.M., James, T., Mikael, B., Jaakko, K., *et al.* (2011) The physiological, transcriptional and genetic responses of an ozone-sensitive and an ozone tolerant poplar and selected extremes of their F_2 progeny. *Environmental Pollution* 159, 45–54.

Tan, B., Grattapaglia, D., Martins, G.S., Ferreira, K.Z., Sundberg, B., and Ingvarsson, P. (2017) Evaluating the accuracy of genomic prediction of growth and wood traits in two *Eucalyptus* species and their F_1 hybrids. *BMC Plant Biology* 17, 110.

Tanksley, S.D. (1997) Seed banks and molecular maps: unlocking genetic potential from the wild. *Science* 277, 1063–1066.

Thamarus, K., Groom, K., Bradley, A., Raymond, C.A., Schimleck, L.R., *et al.* (2004) Identification of quantitative trait loci for wood and fibre properties in two full-sib pedigrees of *Eucalyptus globulus*. *Theoretical and Applied Genetics* 109, 856–864.

Thornsberry, J.M., Goodman, M.M., Doebley, J., Kresovich, S., Nielsen, D., and Buckler, E.

(2001) *Dwarf8* polymorphisms associate with variation in flowering time. *Nature Genetics* 28, 286–289.

Tsarouhas, V., Gullberg, U., and Lagercrantz, U. (2002) An AFLP and RFLP linkage map and quantitative trait locus (QTL) analysis of growth traits in Salix. *Theoretical and Applied Genetics* 105, 277–288.

Wegrzyn, J.L., Eckert, A.J., Choi, M., Lee, J.M., Stanton, B.J., *et al.* (2010) Association genetics of traits controlling lignin and cellulose biosynthesis in black cottonwood (*Populus trichocarpa*, Salicaceae) secondary xylem. *New Phytologist* 188, 515–532.

Weng, C., Kubisiak, T., Nelson, C., and Stine, M. (2002) Mapping quantitative trait loci controlling early growth in a (longleaf pine × slash pine) × slash pine BC 1 family. *Theoretical and Applied Genetics* 104, 852–859.

Weng, Z., Wolc, A., Shen, X., Fernando, R.L., Dekkers, J.C.M., *et al.* (2016) Effects of number of training generations on genomic prediction for various traits in a layer chicken population. *Genetics Selection Evolution* 48, 22.

White, T.L., Adams, W.T., and Neale, D. (2007) *Forest Genetics*. CAB Internation, Wallingford, UK.

Wilcox, P.L., Amerson, H.V., Kuhlman, E.G., Liu, B.H., O'Malley, D.M., and Sederoff, R. (1996) Detection of a major gene for resistance to fusiform rust disease in loblolly pine by genomic mapping. *Proceedings of the National Academy of Sciences USA* 93, 3859–3864.

Wood, A.R., Esko, T., Yang, J., Vedantam, S., Pers, T.H., *et al.* (2014) Defining the role of common variation in the genomic and biological architecture of adult human height. *Nature Genetics* 46, 1173–1186.

Xu, S. (2013) Interval mapping. In: *Principles of Statistical Genomics*. Springer, New York, pp. 109–129.

Yang, J., Zeng, J., Goddard, M.E., Wray, N.R., and Visscher, P. (2017) Concepts, estimation and interpretation of SNP-based heritability. *Nature Genetics* 49, 1304–1310.

Yi, N. and Shriner, D. (2008) Advances in Bayesian multiple quantitative trait loci mapping in experimental crosses. *Heredity* 100, 240–252.

Yin, T.-M., Difazio, S.P., Gunter, L.E., Jawdy, S.S., Boerjan, W., and Tuskan, G. (2004) Genetic and physical mapping of *Melampsora* rust resistance genes in *Populus* and characterization of linkage disequilibrium and flanking genomic sequence. *New Phytologist* 164, 95–105.

Yu, J., Pressoir, G., Briggs, W.H., Vroh Bi, I., Yamasaki, M., *et al.* (2006) A unified mixed-model method for association mapping that accounts for multiple levels of relatedness. *Nature Genetics* 38, 203–208.

Yuceer, C., Kubiske, M.E., Harkess, R.L., and Land, S. (2003) Effects of induction treatments on flowering in *Populus deltoides*. *Tree Physiology* 23, 489–495.

Zapata-Valenzuela, J., Whetten, R.W., Neale, D., McKeand, S., and Isik, F. (2013) Genomic estimated breeding values using genomic relationship matrices in a cloned population of loblolly pine. *G3: Genes Genomes Genetics* 3, 909–916.

Zeng, Z.B. (1994) Precision mapping of quantitative trait loci. *Genetics* 136, 1457–1468.

Zhou, X. and Stephens, M. (2012) Genome-wide efficient mixed-model analysis for association studies. *Nature Genetics* 44, 821–824.

Zobel, B. and Talbert, J. (1984) *Applied Forest Tree Improvement*. Wiley, New York.

Zobel, B.J. and Sprague, J. (1998) *Juvenile Wood in Forest Trees*. Springer, Berlin/Heidelberg.

6

Principles of Forest Biotechnology

Introduction

Biotechnology is the exploitation of a biological process for the production of something that is desired by humans. There are innumerable examples of biotechnology, including the use of yeast to leaven bread or for the fermentation of sugars to produce ethyl alcohol, which humans have been doing for centuries. Although forest biotechnology encompasses more than just genetic engineering, that will be the focus of Section II. We will begin in Chapter 6 by describing how it is done and then provide several examples of the ways in which it has been used, for both basic and applied research, in Chapters 7–10.

Historically, tree improvement was done via conventional breeding. Generally the goal was to introgress the genetic information that controlled a desired trait into an individual that already possessed many other desirable characteristics. Because of recombination and segregation, it was necessary to make repeated crosses to obtain an individual with all of the traits that were wanted; the greater the number of traits being sought, the greater the number of crosses that were needed. With their long juvenile periods, it has frequently not been possible to do advance-generation breeding with forest-tree species and, as a result, they have not been domesticated to the same extent as annual, agronomic crops. Genetic engineering has allowed us to overcome this obstacle, by inserting a small number of genes that control the traits of interest. Thus, what we hoped to accomplish with conventional breeding is often no different than what we do using recombinant DNA technology, but we can expedite the process through genetic engineering.

Although genomic selection will allow for more rapid selection and domestication of forest-tree species, it doesn't preclude the need for genetic engineering. For example, we may want to introduce a trait that is controlled by a gene that does not exist in the genome of the species of interest, or a sexually compatible one. Moreover, genomic editing (presently) needs to be done using genetic engineering.

Producing Transgenic Plants

Agrobacterium-mediated transformation

In order to genetically engineer a plant, one must be able to insert a gene into the genome of an individual plant cell and then induce that modified cell to divide and differentiate into cells that comprise a whole plant. The former process is referred to as transformation and the latter, regeneration.

The most common method of transforming plant cells exploits the naturally occurring gene-insertion mechanism of *Agrobacterium tumefaciens*, the causative agent of crown gall disease (reviewed in detail by Lee and Gelvin, 2008). This bacterium contains a closed-circular piece of double-stranded DNA called the tumor-inducing (Ti) plasmid. During infection, the bacterium inserts a segment of its Ti plasmid, called transfer DNA (T-DNA), into the plant's nuclear genome (Fig. 6.1). This T-DNA consists of genes encoding enzymes that catalyze the synthesis of two plant growth regulators, cytokinin and auxin, which together control, among other things, cell proliferation. Expression of these bacterial genes in the plant cell results in the formation of a tumor, within which the bacterium resides. The T-DNA also contains genes encoding enzymes that catalyze the synthesis of unique

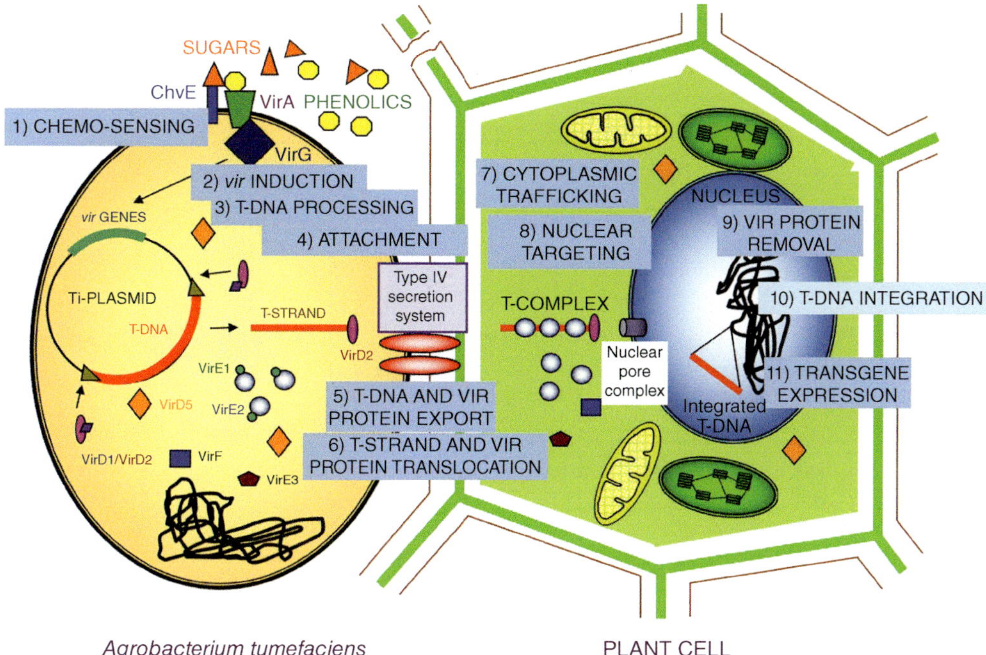

Fig. 6.1. The mechanism by which *A. tumefaciens* inserts T-DNA into the genome of plant cells. The virulence (*vir*) genes, which reside on the tumor-inducing (Ti) plasmid, facilitate transfer of the T-DNA that is present between the left- and right-hand border sequences. Among other functions (labelled 1–11), the *Vir* genes encode proteins that line a channel through which the T-DNA moves, protect the T-DNA from nuclease activity in the cytosol of the plant cell and contain a nuclear-targeting sequence. (Figure provided by Stan Gelvin, Department of Biological Sciences, Purdue University, Indiana.)

amino acids, called opines, that the plant cannot utilize, but serve as a nutritional source for the bacterium.

Agrobacterium tumefaciens does not select specific genes to be shuttled into the plant. The T-DNA is defined by specific 25-bp border sequences; any genes located between the left- and right-hand borders get inserted into the plant genome. The functions involved in T-DNA transfer are performed by the expression products of genes that reside on the Ti plasmid backbone but are outside the T-DNA borders. These genes are collectively known as the virulence (*Vir*) genes, and they can act *in trans*. Disarmed strains of *A. tumefaciens*, which are incapable of causing disease, are used to transform plant cells. These strains are produced by removing the T-DNA from the wild-type Ti plasmid but keeping the *Vir* genes intact, resulting in what is called a "helper plasmid" (Fig. 6.2A).

Genetic engineers assemble another plasmid (a wide-host-range vector) that contains, between the T-DNA border sequences, the genetic material that is to be inserted into plant cells. Other essential elements are also added to the backbone of this synthetic plasmid, outside the borders, such as an origin of replication and a selectable-marker gene. The former affects the number of copies of the plasmid that accumulate in the bacterial host; the latter imparts resistance to a selection agent, such as an antibiotic, to which non-transformed cells are susceptible. This synthetic plasmid, which is known as a binary vector, is transformed into a disarmed strain of *A. tumefaciens* (Fig. 6.2B).

Generally, a minimum of two genes is included between the T-DNA borders of the binary vector: the gene of interest and a second selectable-marker gene. Genes that are inserted in the plant genome are called

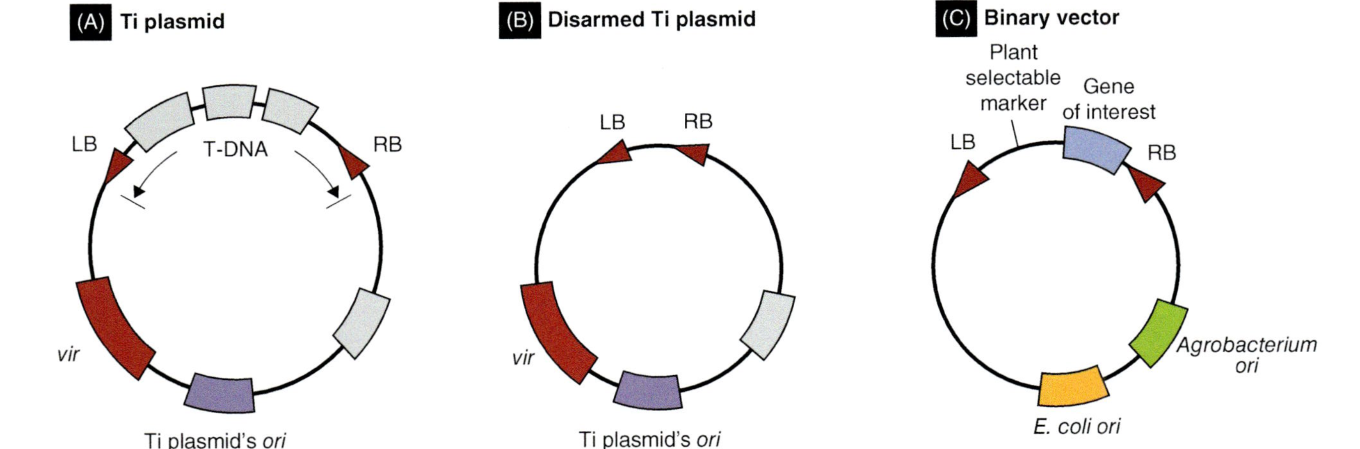

Fig. 6.2. The binary-vector system for transforming plant cells. (A, B) Removing the T-DNA from the Ti plasmid of *A. tumefaciens* (A) results in the formation of a disarmed Ti plasmid (B), also known as a "helper plasmid." (C) A binary vector is constructed by inserting the genes to be inserted into the plant cell's genome (usually the gene of interest and a selectable-marker gene) between the left- and right-hand border sequences (LB and RB, respectively), which are derived from *A. tumefaciens*. Origins of replication (*ori*), which regulate the number of copies of the plasmid produced by the host, are on the vector backbone. The binary vector is transformed into a strain of *A. tumefaciens* harboring the helper plasmid. The *vir* genes located on the helper plasmid are able to act *in trans* to facilitate transfer of the DNA between the border sequences on the binary vector into the nucleus of a plant cell. (Figure reprinted from www.thermofisher.com; © 2018 Thermo Fisher Scientific, Inc. Used under permission.)

transgenes. The gene of interest imparts the desired trait, such as herbicide tolerance, insect resistance, precocious flowering, and altered wood properties. The selectable-marker gene located on the binary vector backbone (outside the T-DNA borders) is usually under the control of a prokaryotic promoter and is used to ensure that the plasmid is maintained by the bacterial host. A promoter is a segment of DNA that controls the timing, location, and extent of a gene's expression. The binary vector is not essential for survival of the bacterium; therefore, if the strain containing it is not grown in the presence of the selection agent to which the selectable-marker gene imparts resistance, the binary vector can be lost when the bacterium undergoes mitosis. The selectable-marker gene that lies within the T-DNA must be under the control of a promoter that will function within a eukaryotic cell, because it will be used to select plant cells that have been transformed (see section on plant regeneration below). If a promoter that functions in both prokaryotes and eukaryotes is used to drive expression of the selectable marker within the T-DNA, this gene can be used to maintain the binary vector in the bacterial host as well as selecting transformed cells in the explant (i.e. a piece of the plant that has been excised). The nopaline synthase gene promoter from *A. tumefaciens* is often used in this way.

In some cases, the gene of interest also serves as a selectable marker, such as an herbicide-tolerance gene. If so, the active ingredient for the herbicide is incorporated into the medium on which the co-cultivated explants are cultured. When a glyphosate (active ingredient in the herbicide Roundup®)-tolerance gene was used in this way, the transformation efficiency (i.e. the percentage of co-cultivated explants that give rise to transgenic plants) was greater than what could be achieved when using an antibiotic-resistance gene (R. Meilan, unpublished data).

An explant is co-cultivated in a suspension culture of *A. tumefaciens* cells that harbor both the binary vector and helper plasmid, which facilitates transfer of the T-DNA of the binary vector into plant cells. However, not all of the cells that comprise the explant will be transformed by the bacterium. The selectable-marker gene provides a way of eliminating those cells that are not transformed. The most commonly used selectable marker is the neomycin phosphotransferase gene (*NPTII*), which imparts resistance to kanamycin. Untransformed plant cells are completely susceptible to this antibiotic and die when exposed to it. When co-cultivated explants are cultured on a solid medium containing kanamycin, only cells containing *NPTII* survive. Because the selectable-marker gene is directly linked to the gene of interest, it too should be present in transformed plant cells. It is important to have the selectable-marker gene located nearest the left-hand T-DNA border. The right-hand border is anchored to the plant chromosome first, but the T-DNA insertion process may be aborted prematurely. If this occurs when the selectable-marker gene is situated near the right-hand border, selected transformed cells may not contain all, or even part, of the gene of interest, leading to the regeneration of "false-positive" plants (i.e. transgenic plants that do not contain the gene of interest).

Insertion of T-DNA is a random process; it is extremely unlikely that two transformed cells will have the T-DNA inserted at the same locus. A plant regenerated from a cell containing the T-DNA insert at a specific location is referred to as an "independent event" or a "line." Genomic DNA surrounding the site of insertion affects the efficiency with which the transgene is expressed; this is referred to as a "position effect." Because of this differential expression, many lines are often produced for each plant genotype–binary vector combination. Lines produced in this way, exhibiting a range of transgene expression levels, are functionally equivalent to an allelic series. *Agrobacterium tumefaciens* often inserts more than one copy of the T-DNA, at a given locus and/or scattered throughout the genome, and there is evidence that an increase in copy number can lead to post-transcriptional transgene silencing (Tang *et al.*, 2007), often referred to as co-suppression. At a given locus, multiple inserts can be oriented in tandem (head to tail) or as inverted repeats (head to head).

If only a single transgenic line is examined, one cannot exclude the possibility that the observed phenotype is caused by the insertion of the transgene into another, unrelated gene, the inactivation of which—via insertional mutagenesis—leads to the observed phenotype. Because such random inactivation events would be extraordinarily unlikely to lead to a consistent phenotype in more than one independent transgenic line, one can confidently conclude that an observed phenotype is caused by the transgene.

Matrix attachment regions (MARs) are DNA sequences that bind to protein matrices within a cell's nucleus to form DNA loop structures (Spiker and Thompson, 1996). It is thought that when transgenes are flanked by MARs, they form an independent chromatin domain and hence are protected from the effects of DNA adjacent to the insertion site (Hall *et al.*, 1991). In earlier literature, MARs were referred to as scaffold-attachment regions; however, because the abbreviation SAR was already being used in the scientific literature (for "systemic acquired resistance"), MAR was subsequently used instead to avoid confusion. Li *et al.* (2008) showed that when T-DNA was flanked with MAR elements, transgene expression was enhanced and stabilized.

The cottonwood leaf beetle (*Chrysomela scripta*) is a major insect pest of *Populus* spp. Genes encoding protein toxins produced by *Bacillus thuringiensis* (Bt) have often been used to engineer insect-resistant plants (Chapter 10, this volume). *Chrysomela scripta* is susceptible to the protein encoded by a Bt gene known as *Cry3A*. In addition to adult *C. scripta* being much more tolerant of Bt than the larval stage, they also do not consume much leaf material. Most defoliation is done by the larvae; adults rely heavily on stored metabolic reserves. As a result, adult *C. scripta* is usually not killed by endogenous Bt toxins. Meilan *et al.* (2000) engineered various hybrid poplar (genus *Populus*) genotypes to contain T-DNA that included a *Cry3A* gene bordered by MARs. Transgene expression levels were so high in these trees that dead adult *C. scripta* were found on field-grown transgenic trees (R. Meilan, unpublished data). This result was unexpected and unprecedented, and is thought to be the direct result of MAR-enhanced expression.

Biolistics

Another DNA delivery system, biolistics, involves coating microscopic beads (e.g. gold or tungsten) with DNA. This approach is often used for species that are not susceptible to *A. tumefaciens* infection. The beads are propelled at an explant, usually with a burst of compressed air as the driving force. The first-generation biolistic delivery system used a 0.22-caliber nail-gun cartridge to accelerate the microparticles; hence, it was called a "gene gun." Once inside the cell, DNA that detaches from the bead can recombine with a plant chromosome. The biolistic approach is often less efficient than *Agrobacterium*-mediated transformation because of: (i) cellular damage from the impact of the beads, some of which aggregate; (ii) the need for the DNA to dissociate from the projectile; (iii) digestion of the transgene by cytosolic enzymes (i.e. nucleases); and (iv) the need for recombination to occur. The process for selecting cells transformed biolistically is the same as that described above.

Plant regeneration

Transformed cells that survive selection are used to regenerate a whole plant. The two main routes for *in vitro* regeneration are organogenesis and embryogenesis. In the former, cells are coaxed to divide and differentiate into specific organ types (e.g. shoots and roots). Both types of regeneration are done by successive transfers of co-cultivated explants to media containing the proper type and concentration of plant growth regulators, mainly cytokinins and auxins, along with other additives (e.g. vitamins, sugars, various salts). Varying the cytokinin:auxin ratio or substituting other plant growth regulators, such as gibberellins, can result in callus formation, shoot organogenesis, shoot elongation, and root organogenesis. A very efficient transformation and regeneration protocol has been developed for *Populus* spp. (Fig. 6.3) (Meilan and Ma, 2006). Organogenesis is the preferred route for plant regeneration, but it has only be accomplished with a limited number of tree species (Wang, 2006).

Fig. 6.3. Steps involved in the production of transgenic poplar. (A) Non-transgenic plantlets growing *in vitro*. (B) Removing an explant with a scalpel. (C) Cutting leaf discs with a paper hole punch. (D) Floating leaf discs on sterile, distilled water. (E) Suspension culture of *A. tumefaciens* in which the leaf discs are swirled (i.e. co-cultivated). (F) Leaf discs that had been co-cultivated arrayed on callus-induction medium. (G) Callus formation on leaf discs. (H) Shoot formation on callus masses that had been transferred to a shoot-induction medium. (I) Shoots excised from callus masses grown on shoot-elongation medium. (J) Elongated shoots grown on a root-induction medium. (K) Plantlets being propagated on basal medium. (L) Fully regenerated transgenic plants. The whole process generally takes about 8 months to complete.

Embryogenesis is the process by which cells differentiate directly into an embryo, similar to what is contained within a plant seed. This approach can be problematic because immature embryos are often used as the explant that gets co-cultivated with *A. tumefaciens*. These immature embryos are not highly differentiated, so the cells comprising them will more readily give rise to mature embryos after they are transformed. Because this embryonic tissue, which is the result of the fusion of two gametes, is harvested prematurely, nothing is known about the phenotype of the tree that would result if the embryo had been allowed to finish developing, germinate, and grow into a mature plant. In addition, it is often difficult to get embryos generated *in vitro* to germinate.

A variation on this theme is somatic embryogenesis. In this approach, somatic cells are induced to differentiate into embryos. Because explants are excised from tissues of mature trees, many of which are deemed to be genetically superior, their performance is known in advance. This technology allows the generation of clonal planting stock, which may not be genetically engineered.

The production of transgenic poplar plants is a lengthy process. A genotype designated INRA 717-1B4 (*Populus tremula* × *P. alba*) is widely used by researchers because it is more amenable to transformation by *A. tumefaciens* and *in vitro* regeneration than virtually any other poplar genotype that has been tested. Even for this genotype, it can take up to 8 months to go from co-cultivation of leaf discs to the regeneration of rooted plants. Shortening the time needed to complete this process would greatly enhance the value of this model plant. Payyavula *et al.* (2009) developed a cell-culture system for *Populus*. Under their standard conditions, the density of 717-1B4 cells in their suspension cultures increases by >400% every 2 weeks. Another advantage of the 717-1B4 cell-culture system is that the 717-1B4 cells can be reprogrammed, as needed, to undergo organogenesis and whole-plant regeneration (C.-J. Tsai, Georgia, 2017, personal communication).

Controlling Gene Expression

Upregulation

Not all of a plant's genes are transcriptionally active at all times. Some are transiently expressed at a certain time of the year, at a given stage of development, and/or in a specific tissue or cell type. The expression of a gene's coding sequence (i.e. transcription, processing of mRNA, and translation of mRNA into protein) is carefully regulated by various control sequences. As stated above, promoters are elements of DNA that direct the timing, location, and extent of a gene's expression. Constitutive promoters allow high levels of expression, in (nearly) all tissues, all of the time. The constitutive promoter most commonly used for plants is derived from the cauliflower mosaic virus (CaMV) *35S* gene. Other promoters may allow tissue- and/or temporal-predominant expression, or can be activated through treatment with a specific inductive agent. For example, the promoter from the *Arabidopsis thaliana RESPONSIVE TO DESICCATION 29A* (*rd29A*) gene is elicited by salt, temperature, and osmotic stress, and results in prolonged stress-inducible expression in nearly all plant tissues, beginning as early as 20 min and continuing for 24 h after imposition of the stress (Yamaguchi-Shinozaki and Shinozaki, 1993). However, work by Caldwell (2012) revealed that when used to drive gene expression in *Populus*, the *rd29A* promoter does not respond to stress in the way that has been reported for *A. thaliana*.

Mohamed *et al.* (2001) used hybrid cotton-wood (*Populus trichocarpa* × *P. deltoides*) to study a copper-inducible gene expression system that was first demonstrated in tobacco. The system consists of a highly expressed, copper-activated transcription factor (*ACE1*), along with a synthetic promoter that combines an *ACE1*-binding site (the metallothionein response element) and a minimal CaMV *35S* promoter, to control expression of the *uidA* gene. This gene, which is also known as *GUS*, is the most commonly used reporter gene in plant biology. It encodes the enzyme β-glucuronidase, which catalyzes the conversion of a colorless substrate to an insoluble, blue-colored product that precipitates in cells expressing it (Jefferson, 1987).

Unexpectedly, Mohamed *et al.* (2001) observed high levels of *GUS* expression in their transgenic lines in the absence of the *ACE1* gene. When the *ACE1* gene was present, *GUS* expression occurred even in the absence of copper, and showed a parabolic dependence on copper at concentrations up to 50 μM, above which it inhibited expression. The authors speculated that this was due to an endogenous factor that interacts with *ACE1* at levels of transcription afforded by the minimal *35S* promoter. Given these results, this copper-inducible promoter is not useful in poplar. This and the previous example reveal that promoters do not always function consistently across heterologous species, and highlights the importance of testing them in the species of choice.

Filichkin *et al.* (2006) demonstrated the utility of an alcohol-inducible promoter for controlling the expression of the *GUS* reporter gene in *Populus*. GUS activity could be detected fluorometrically after 5 days of treatment with ethanol at concentrations as low as 0.5%. They also showed that prolonged induction by ethanol vapors significantly increased the GUS activity in leaves from both *in vitro*- and greenhouse-grown plants.

Enhancers are the portion of a promoter that can elevate a gene's expression level, and can act *in trans*. An enhancer element from one promoter can be fused to another to boost its expression level, and some transcriptional enhancers can impart tissue specificity (Ben-fey *et al.*, 1990). There are also translational enhancers, such as the Ω element from tobacco mosaic virus (Gallie and Kado, 1989). Finally, there are downstream control elements, terminators, which signal the cessation of transcription.

One way to ascertain a gene's function is through its overexpression. This is done by isolating the gene's coding sequence, assembling a binary vector in which this sequence is inserted between a constitutive promoter and a terminator, and producing a transgenic plant containing this transcriptional unit. The phenotype exhibited by the regenerated plant may provide an indication of the role played by the gene. There is frequently sufficient sequence similarity between poplar and *A. thaliana* that a poplar ortholog can restore function to an *A. thaliana* line that has a mutation in the corresponding gene. There is a collection containing knock-outs (produced via insertional mutagenesis) for nearly every one of the approximately 27,000 genes in *A. thaliana*. Seeds for mutant lines maintained in this collection are available from the Arabidopsis Biological Resource Center (www.arabidopsis.org/portals/mutants/stockcenters.jsp, accessed July 15, 2019). *Arabidopsis thaliana* can be transformed very easily using what is called the "floral dip" method (Clough and Bent, 1998). Because of the ease with which *A. thaliana* can be transformed, candidate genes are often evaluated in this plant before expending the effort needed to produce transgenic poplar plants.

Another way to demonstrate functionality is to downregulate or block the expression of a native gene. This approach is discussed below.

Downregulation

RNA interference

One form of post-transcriptional gene silencing, RNA interference (RNAi), is a powerful tool to help elucidate gene function. An RNAi construct is made by fusing a portion of the gene that is being targeted for suppression with its reverse complement. A short piece of DNA (often an intron) is inserted between these two sequences, to serve as a hinge. When this genetic construct is expressed in a transgenic plant, the reverse-complementary sequence folds back to bind to its homologous partner, forming a hairpin structure that is used to produce segments of double-stranded RNA. This duplex is cleaved into what is called small interfering RNA (siRNA). The siRNA is unwound and one of the strands, the guide strand, binds to its complementary sequence in the target mRNA, leading to its degradation. The RNA-induced silencing complex (RISC) is a multi-protein complex that uses the siRNA as a template to find the complementary sequence on mRNA, leading to its cleavage (Fig. 6.4) (Ahlquist, 2002).

RNAi rarely blocks all expression of the target gene, and the extent to which the target gene is downregulated depends on the promoter used to drive expression of the RNAi vector, and position effects. This may be advantageous under certain circumstances, but there are situations under which a complete knockout of the target gene is desired.

MicroRNAs

MicroRNAs (miRNAs) are genomically encoded RNAs that are not translated into protein, but they help regulate native-gene expression by targeting specific mRNAs for degradation. One example is the *Corngrass1* (*Cg1*) gene, which was identified while screening a retrotransposon-mutated population of maize (*Zea mays*) and encodes a miR156-class miRNA (Chuck *et al.*, 2007). It is thought to regulate the expression of genes that are involved in the initiation of meristems and lateral organs. The action of this miRNA is mediated by modulating the expression of genes containing a Squamosa-binding protein box, which is complementary to the mature *Cg1* miRNA (Xing *et al.*, 2010). *Cg1* appears to be a master regulator that controls the expression of several other genes, thus affecting many phenotypes. In an effort to boost its value as a bioenergy crop, the *Cg1* gene has been overexpressed in numerous lines of poplar (Rubinelli *et al.*, 2013).

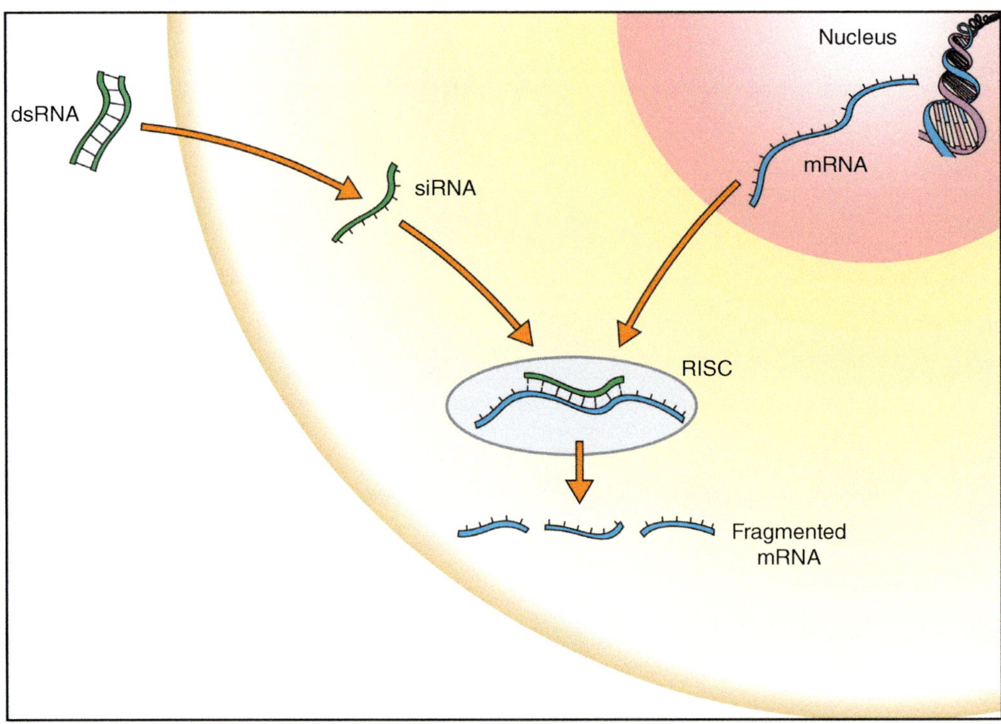

Fig. 6.4. The mechanism by which RNAi downregulates gene expression. A short piece of DNA (often an intron) is inserted between two complementary sequences of DNA (from the gene that is being targeted), to serve as a hinge. When this genetic construct is expressed in a plant cell, the reverse-complementary sequence folds back to bind to its homologous partner, forming a hairpin structure, which ultimately is used to produce segments of double-stranded RNA. This duplex is cleaved into siRNA. The siRNA is unwound and one of the strands, the guide strand, binds to its complementary sequence in the target mRNA, leading to its degradation. (After Robinson, 2004.)

Virus-induced gene silencing

Virus-induced gene silencing (VIGS) is a technique that utilizes an RNA-based virus defense system to target degradation of mRNA encoded by native genes in the host. This genetic engineering tool provides an alternative to transformation-based RNAi, and enables functional analysis and silencing of gene families (Becker and Lange, 2010). So far, VIGS has only been tested in herbaceous model and crop plants, but the findings could be extended to trees. In addition, it may be advantageous in forest-tree species that are recalcitrant to transformation. Effective VIGS-based vectors are unavailable for forest trees, but a good candidate to develop a system for trees is poplar mosaic virus, which naturally infects genotypes within the genus *Populus* (Busov *et al.*, 2009).

Practical applications of overexpression and downregulation

Overexpression and downregulation can provide opportunities to develop improved varieties of trees with environmentally beneficial traits, in addition to the typical commercially important traits such as herbicide tolerance and insect resistance. For example, cytochrome P450 2E1 is a mammalian enzyme with broad specificity for a variety of halogenated compounds, many of which are environmental pollutants and known carcinogens (e.g. chloroform, trichloroethylene, carbon tetrachloride, benzene, and vinyl chloride). Overexpression of the gene that encodes this enzyme in poplar led to vastly increased rates of uptake and metabolism of these pollutants (Doty *et al.*, 2007).

The use of plants to decontaminate a site—termed phytoremediation—is less expensive and less destructive than the available alternatives, and is preferable because the contaminant is metabolized rather than accumulated in the plant.

In addition, when Zawaski *et al.* (2011) used RNAi to suppress *SHORT INTER-NODES* (*SHI*) and the related *STYLISH1* (*STY1*) gene in *Populus*, their transgenic plants exhibited improved root and shoot growth, longer fibers, and an increased proportion of xylem tissue. Moreover, Li *et al.* (2011) reported on their genome-wide search for drought stress-induced miRNAs in *Populus euphratica*, many of which could be useful for protecting yield in areas where water is limiting. Other examples of the power of this technology can be found in Chapters 7–10 (this volume).

Genome Engineering

Minichromosomes

Many commercially important plant species have been transformed via *A. tumefaciens* or biolistics, but these methods have their drawbacks. For example, a limited amount of DNA—a few genes at most—can be inserted with each transformation event, the DNA is inserted at random positions within the genome, transgene insertion can disrupt the integrity of the host genome, and position effects can result in unpredictable transgene expression levels (Houben and Schubert, 2007). As a result, it has been exceedingly difficult to differentially regulate the expression of several genes simultaneously in plants, especially given the need to transform cells and regenerate plants repeatedly.

Chromosome-based vector systems have been used to simultaneously transfer several genes into an individual cell. Engineered microchromosomes have been developed for mammalian systems, but not as much progress has been made with plants. The two general strategies involve modifying a native chromosome (top-down approach) or constructing an artificial chromosome using cloned chromosomal elements (bottom-up approach). For the latter, centromeric sequences and a selectable-marker gene are transformed into a cell, where assembly is mediated by endogenous processes. *De novo* assembly inside cells is difficult to control and has only been accomplished in a restricted number of mammalian cell lines (Houben and Schubert, 2007). Limited progress has been made with plants. For example, meiotic transmission of an *in vitro*-assembled autonomous minichromosome has been reported for maize (Carlson *et al.*, 2007).

Homologous recombination

Being able to direct recombination at a specific location in the genome would be very useful for engineering trees. Using this technology, a system has been developed to maintain a heterozygous condition in mouse, using what is called a "balancer chromosome" (Yu and Bradley, 2001). In some plants, heterosis (i.e. hybrid vigor) can be achieved with a few loci or even a single heterozygous gene (Krieger *et al.*, 2010); therefore, developing a balancer system for plants has implications for plant breeding.

Directed translocations in plants have other important applications. It has been shown how chromosomal rearrangement led to a change in the usual chromosome number in Brassicaceae of eight to five in *A. thaliana*. Karyotypic alterations such as these ordinarily result in sexual incompatibility between species (Chan, 2010). By directing chromosomal translocations, it may be possible to prevent genetically engineered trees from crossing with wild relatives, which is important for transgene confinement (Chapter 7, this volume).

In many cases, it is desirable to insert DNA at a specific location in the plant genome. Two site-specific recombination systems that have been used in the past are Cre–Lox and FLP/FRT; the former has been shown to work well in plants (Gilbertson, 2003). Repetitive use of this system could permit the insertion of several transgenes at a single locus. Homologous recombination frequency has been enhanced in maize by cutting the target site with a sequence-specific zinc-finger nuclease (see below) (Shukla *et al.*, 2009; Li *et al.*, 2010). By incorporating a Lox or FRT site at a specific location in the genome, it is possible to achieve precise chromosomal rearrangements with single-nucleotide accuracy (Chan, 2010).

Chromosome engineering provides many opportunities for accelerating the genetic improvement of trees. For example, mini-chromosome engineering techniques could facilitate the stacking of several genes and/or network regulators, so that trees can be tailored for specific purposes. This could include, for example, improving biomass productivity, pest resistance, and water-use efficiency in trees to enhance their utility as dedicated bioenergy crops. One of the biggest challenges will be the development of efficient transformation systems for introducing these large synthetic genetic constructs into plant genomes. In addition, a better understanding of gene and protein function and how gene networks are regulated is needed. Responding to these challenges requires a combination of synthetic and systems biology tools for analyzing large datasets, and computer modeling.

Zinc-finger nucleases

Zinc-finger nucleases (ZFNs) provide one approach to creating mutant lines (Durai *et al.*, 2005). They involve the use of enzymes that are engineered to create a double-stranded break at a specific target locus, leading to directed changes in the DNA sequence. These modifications can include point mutations or targeted gene insertions. This approach overcomes the randomness of traditional mutagenesis and transgene insertion. Employing this approach, Townsend *et al.* (2009) demonstrated high-frequency ZFN-stimulated targeting of the acetolactate synthase genes in tobacco. Specific mutations in these genes are known to confer resistance to imidazolinone and sulphonylurea herbicides. Similarly, Shukla *et al.* (2009) disrupted the *ZmIPK1* gene in maize, leading not only to herbicide tolerance but also to a reduction in phytate, a component of maize feed that is particularly

difficult for animals to digest. Developing ZFN systems for forest-tree species could enhance genetic-improvement capabilities. It could also help to overcome problems of epigenetic transgene silencing and to mimic intermediate levels of downregulation, leading to the creation of partially defective alleles (e.g. splice variants).

Another example of a programmable genome-editing tool is transcription activator-like effector nucleases (TALENs) (Joung and Sander, 2013). As with ZFNs, this technique relies on DNA-binding domains that are fused to the catalytic domain of a restriction endonuclease. Molecular tools that target specific genomic loci through protein–DNA interactions require the assembly of a new protein for each site. As a result, ZFNs and TALENs cannot be used for high-throughput applications.

Clustered regularly interspaced short palindromic repeats

Genome editing has been revolutionized by the discovery and development of a system known as clustered regularly interspaced short palindromic repeats (CRISPR). This system has been co-opted from bacteria, which employ it as protection against viral attack. CRISPR can be used to insert a new sequence at a specific location in the genome or make precise changes in the sequence of a particular gene. Type II CRISPR involves the use of Cas9, a nuclease that cleaves both strands of genomic DNA. Its action triggers the cell's double-strand break-repair mechanism, which utilizes either the non-homologous end-joining or the homology-directed repair pathway (Overballe-Petersen *et al.*, 2013). The former results in insertions and/or deletions (indels), which disrupt the targeted gene; the latter can be used to insert a novel sequence (Fig. 6.5). CRISPR has been demonstrated to work in a wide variety of prokaryotic and eukaryotic organisms, including bacteria, fungi, plants, and animals (Wang *et al.*, 2016).

Zhou *et al.* (2015) were the first to modify a tree genome with this technology. Using a single construct, they were able to simultaneously knock out two genes within the *4CL* family (in the phenylpropanoid pathway), and

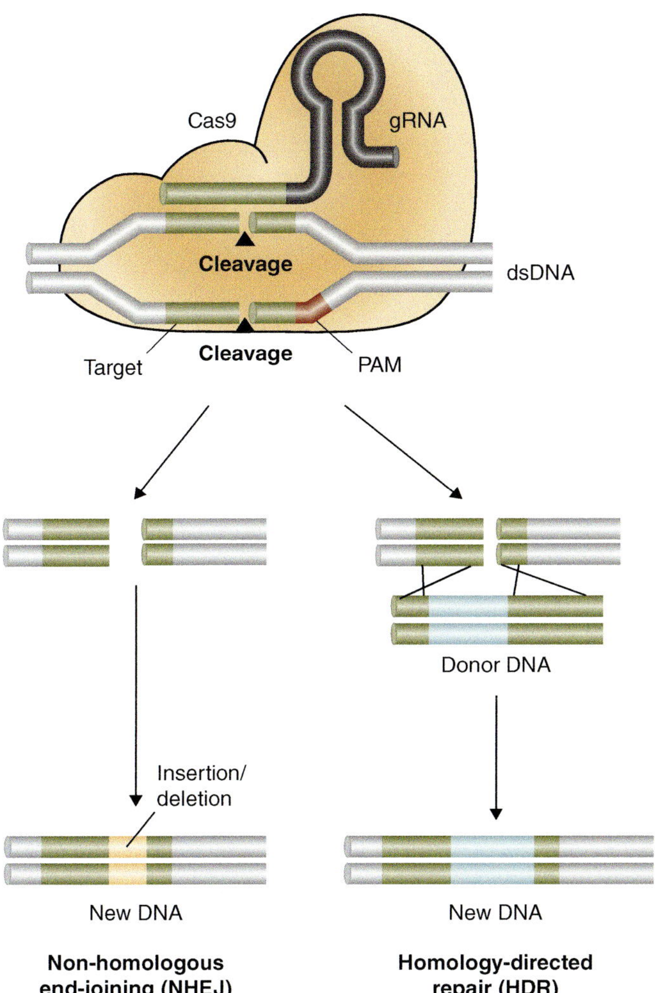

Fig. 6.5. Gene editing using the CRISPR/Cas9 system. Guide RNA (gRNA) binds to the homologous region within the target gene. The Cas9 nuclease cleaves both strands of genomic DNA. Because it has ends that are homologous to sequences on either side of the break, donor DNA will be inserted via a homology-directed repair mechanism. If donor DNA is absent, the cell's double-strand break-repair mechanism will utilize the non-homologous end-joining pathway, resulting in insertions and/or deletions, which disrupt the targeted gene (Overballe-Petersen *et al.*, 2013). (Figure reprinted from www.neb.com, 2019, with permission from New England Biolabs.)

every transformant that was analyzed exhibited bi-allelic modification. Tsai and Xue (2015) caution that the frequent occurrence of sequence polymorphisms in out-crossing species, such as those within the genus *Populus*, as well as other tree species, can cause CRISPR/Cas9 to be ineffective. Thus, they emphasize the importance of avoiding SNPs when designing guide RNA (gRNA). Various tools available for genome editing in *Populus* have been described by Xue and Tsai (2015) and Xue *et al.* (2015), including a facile tool that is useful for designing gRNAs.

Multiplex genome editing using the CRISPR/Cpf1-crRNA array system has recently been demonstrated in mammals. Using the Cpf1-crRNA architecture, it was possible to concatenate strings of gRNAs into an array under the control of a single polymerase III promoter (U6 or U3) to edit several genes concurrently (Zetsche *et al.*, 2017). The Cpf1 systems from an *Acidaminococcus* and a *Lachnospiraceae* bacterium have recently been used successfully in herbaceous plants (Tang *et al.*, 2017; Xu *et al.*, 2017).

Gene Discovery

Targeting induced local lesions in genomes

Targeting induced local lesions in genomes (TILLING) is a reverse-genetics approach that allows the detection of a mutation in a particular gene. This technique requires both an efficient mutagenesis method, such as the chemical mutagen ethyl methanesulfonate, and a robust DNA screening method that is capable of detecting single-base substitutions (i.e. point mutations) in the affected gene. Successful TILLING depends on the formation of heteroduplexes that develop when more than one allele is amplified via the polymerase chain reaction (PCR) (Chapter 1, this volume). When the amplicons are heated and then cooled slowly, a protrusion ("bubble") forms where there is a mismatch between the two strands of DNA that hybridize. The bulging strand can be cleaved using endonucleases. The cleavage products are then separated based on size using various techniques, such as gel electrophoresis (Chapter 1, this volume) (Kurowska *et al.*, 2011). Eco-TILLING is a method that uses TILLING techniques to look for natural mutations that occur among individuals in wild populations (Comai *et al.*, 2004). Gilchrist *et al.* (2006), who were the first to use Eco-TILLING in poplar, demonstrated that it could be used as an effective SNP discovery tool.

Activation tagging

Cisgenesis and intragenesis are techniques that involve inserting into the plant genome a DNA fragment from the species that is being genetically engineered or from cross-compatible species, respectively. Activation tagging is a forward-genetics approach that can be used to produce mutant lines to help identify native genes that control key traits (Busov *et al.*, 2003; Harrison *et al.*, 2007; Busov *et al.*, 2011). Activation tagging relies on a binary vector that contains strong enhancer elements located near the T-DNA border. Insertion of the vector in close proximity to a native gene tends to cause its upregulation and a gain-of-function, dominant mutation (Ye *et al.*, 2011). Several genes have been discovered through activation tagging in poplar. These include: *GA 2-oxidase*, a gibberellin catabolism gene that appears to be involved in regulating tree stature (Busov

et al., 2003); *LATERAL ORGAN BOUNDARIES DOMAIN*, a gene that is a positive regulator of phloem formation during secondary growth in poplar (Yordanov *et al.*, 2010); and *BIG LEAF*, which regulates leaf size by promoting cell proliferation (Yordanov *et al.*, 2017). About 3,000 activation-tagged mutant poplar lines have been generated through pilot studies conducted in various laboratories. These tagged genes are potentially useful for producing transgenic and cisgenic poplar using a reverse-genetics approach that can provide insight into poorly understood processes in trees, and may reveal targets for their genetic improvement.

Gene and enhancer trapping

Gene and enhancer trapping are alternative DNA insertion strategies that are based on detecting gene expression, instead of observing mutant phenotypes (Springer, 2000). Gene-trap constructs include a reporter gene but lack an operational promoter, whereas enhancer-trap constructs contain a reporter gene under the control of a minimal promoter. In contrast to a selectable marker, a reporter gene allows visualization of a gene's expression. When the reporter gene is inserted near the promoter of an endogene, the expression pattern of the reporter will be similar to that of the native gene. The genomic region flanking the insertion site can easily be isolated (because it has been tagged with a novel sequence) and characterized for its function (e.g. by up- or downregulating its expression). Groover *et al.* (2004) demonstrated the potential of trapping as a method for discovering novel genes in trees.

High-throughput Analyses of Gene Expression

Microarrays have been used to study the simultaneous expression patterns of thousands of genes. They are produced by anchoring unique sequences of DNA (either oligonucleotides or cDNA) to a glass slide in neatly arrayed microscopic spots (the probe). Spotting is done using very fine needles and robotics. mRNA (the target) is isolated from both control and treated plants; each RNA sample is labeled with a different-colored fluorescent dye. The labeled RNA is allowed to hybridize to its complementary, single-stranded DNA probe before the microarray is scanned. A laser then excites each fluorescent dye at a specific wavelength, and emissions are captured digitally. The fluorescent images for the control and treated samples are superimposed and the resulting color of each spot reveals whether a given gene is expressed differentially in the two samples.

With dramatic improvements in throughput and reductions in the cost of sequencing, microarrays are being replaced by RNA sequencing (RNA-seq.). Using next-generation sequencing (NGS), it is now possible to inexpensively reveal all of the mRNA present in a biological sample at the time it was collected. This complete assemblage of transcripts is known as the transcriptome. With this technology, it is now possible to determine which genes are being expressed and to what extent at a given moment. This could include a certain stage of development or in response to a specific treatment. Challenges associated with this approach include: deciding what tissue or cell type to sample, and when, and determining which of the thousands of expression products is directly responsible for an observed phenotype. For more information on high-throughput analyses, see Chapter 1 (this volume).

Summary

To genetically engineer a tree, a single, transformed cell must be regenerated into a whole plant. *Agrobacterium tumefaciens* is commonly used as a vehicle to transform (introduce genetic material) into the chromosome of a plant cell, but other approaches, such as biolistics, are available. The two general approaches to regenerating a plant are organogenesis and embryogenesis. The latter is the preferred route but has only been accomplished with a limited number of tree species, and is a labor-intensive and time-consuming process. Shortening the time needed to regenerate plants *in vitro* can improve the utility of trees for various end uses, expedite research, and deliver useful products to the consumer more rapidly.

Promoters are elements of DNA that regulate the timing, location, and extent of a gene's expression. Overexpression of an introduced gene (transgene) is one way to elucidate gene function. This is often done by fusing a constitutive promoter to the gene of interest, which is introduced into the plant; alternatively, inducible promoters may be employed. Frequently, promoters perform differentially across heterologous species. Gene function can also be elucidated by downregulating a target gene through a process known as RNAi. VIGS provides an alternative to transformation-based RNAi.

Insertion of a transgene by *A. tumefaciens* is a random process, and it is extremely unlikely for two transformed cells to have the gene inserted at the same locus. A plant regenerated from a cell containing the inserted DNA at a specific location is referred to as a "line." Genomic DNA surrounding the insertion site affects the extent to which the transgene is expressed; this is referred to as a "position effect." Because of this differential expression, several lines are often produced using a given genetic construct. Lines produced in this way, exhibiting a range of transgene expression levels, are functionally equivalent to an allelic series.

Many techniques have been developed to alter genomic sequence at a specific location. These include site-specific recombination systems, ZFNs and TALENs. However, none of these approaches is suitable for high-throughput applications.

In some cases, it may be desirable to knock out expression of an endogenous gene entirely. Discovery of the mechanism known as CRISPR has made this possible. This system can be used to insert a new sequence at a specific location in the genome or make precise changes in the sequence at a particular locus. Using an alternative version of this technology, it is possible to edit several genes simultaneously.

TILLING, activation tagging, and gene and enhancer trapping are all techniques that have been used for gene discovery, by modifying gene expression. Transcriptome analysis using microarrays has been used to study the simultaneous expression of thousands of genes. This technique has been used to evaluate gene-expression patterns and to help elucidate functionality, but it can also be useful for gene discovery. With dramatic improvements in throughput and reductions in the cost of sequencing, microarrays have largely been replaced by RNA-seq. Challenges associated with RNA-seq include deciding which tissue or cell type to sample, and when, and determining which of the thousands of expression products is directly responsible for an observed phenotype.

References

Ahlquist, P. (2002) RNA-dependent RNA polymerase, viruses, and RNA silencing. *Science* 296, 1270–1273.

Becker, A. and Lange, M. (2010) VIGS-genomics goes functional. *Trends in Plant Science* 15, 1–4.

Benfey, P.N., Ren, L., and Chua, N.H. (1990) Tissue-specific expression from CaMV 35S enhancer subdomains in early stages of plant development. *EMBO Journal* 9, 1677–1684.

Busov, V.B., Meilan, R., Pearce, D.W., Ma, C., Rood, S.B., and Strauss, S.H. (2003) Activation tagging of a dominant gibberellin catabolism gene (*GA 2-oxidase*) from poplar that regulates tree stature. *Plant Physiology* 132, 1283–1291.

Busov, V.B., Strauss, S.H., and Pilate, G. (2009) Transformation as a tool for genetic analysis in *Populus*. In: Jansson, S., Bhalerao, R.P., and Groover, A.T. (eds) *Genetics and Genomics of Populus*. Springer, New York, pp. 113–133.

Busov, V., Yordanov, Y., Gou, J., Meilan, R., Ma, C., *et al.* (2011) Activation tagging is an effective gene tagging system in *Populus*. *Tree Genetics and Genomes* 7, 91–101.

Caldwell, M.T. (2012) Altering abscisic acid biosynthesis to improve water-use efficiency in *Populus*. MSc thesis, Purdue University, West Lafayette, Indiana.

Carlson, S.R., Rudgers, G.W., Zieler, H., Mach, J.M., Luo, S., et al. (2007) Meiotic transmission of an *in vitro* assembled autonomous maize minichromosome. *PLoS Genetics* 3, 1965–1974.

Chan, S.W.L. (2010) Chromosome engineering: power tools for plant genetics. *Trends in Biotechnology* 28, 605–610.

Chuck, G., Cigan, A.M., Saeteurn, K., and Hake, S. (2007) The heterochronic maize mutant *Corngrass1* results from overexpression of a tandem microRNA. *Nature Genetics* 39, 544–549.

Clough, S.J. and Bent, A.F. (1998) Floral dip: a simplified method for *Agrobacterium*-mediated transformation of *Arabidopsis thaliana*. *Plant Journal* 16, 735–743.

Comai, L., Young, K., Till, B.J., Reynolds, S.H., Greene, E.A., *et al.* (2004) Efficient discovery of DNA polymorphisms in natural populations by Ecotilling. *Plant Journal* 37, 778–786.

Doty, S.L., James, C.A., Moore, A.L., Vajzovic, A., Singleton, G.L., *et al.* (2007) Phytoremediation of volatile environmental pollutants with transgenic trees. *Proceedings of the National Academy of Science USA* 104, 16816–16821.

Durai, S., Mani, M., Kandavelou, K., Wu, J., Porteus, M.H., and Chandrasegaran, S. (2005) Zinc finger nucleases: custom-designed molecular scissors for genome engineering of plant and mammalian cells. *Nucleic Acids Research* 33, 5978–5990.

Filichkin, S.A., Meilan, R., Busov, V.B., Ma, C., Brunner, A.M., and Strauss, S.H. (2006) Alcohol-inducible gene expression in transgenic *Populus*. *Plant Cell Reports* 25, 660–667.

Gallie, D.R. and Kado, C.I. (1989) A translational enhancer derived from tobacco mosaic virus is functionally equivalent to a Shine–Dalgarno sequence. *Proceedings of the National Academy of Science USA* 86, 129–132.

Gilbertson, L. (2003) Cre–lox recombination: Cre-ative tools for plant biotechnology. *Trends in Biotechnology* 21, 550–555.

Gilchrist, E.J., Haughn, G.W., Ying, C.C., Otto, S.P., Zhuang, J., *et al.* (2006) Use of ecotilling as an efficient SNP discovery tool to survey genetic variation in wild populations of *Populus trichocarpa*. *Molecular Ecology* 15, 1367–1378.

Groover, A., Fontana, J., Dupper, G., Ma, C., Martienssen, R., *et al.* (2004) Gene and enhancer trap tagging of vascular-expressed genes in poplar trees. *Plant Physiology* 134, 1742–1751.

Hall, G. Jr, Allen, G.C., Loer, D.S., Thompson, W.F., and Spiker, S. (1991) Nuclear scaffolds and scaffold-attachment regions in higher plants. *Proceedings of the National Academy of Science USA* 88, 9320–9324.

Harrison, E.J., Bush, M., Plett, J.M., McPhee, D.P., Vitez, R., *et al.* (2007) Diverse developmental mutants revealed in an activation-tagged population of poplar. *Canadian Journal of Botany* 85, 1071–1081.

Houben, A. and Schubert, I. (2007) Engineered plant minichromosomes: a resurrection of B chromosomes? Plant Cell 19, 2323–2327.

Jefferson, R.A. (1987) Assaying chimeric genes in plants: the GUS gene fusion system. *Plant Molecular Biology Reporter* 5, 387–405.

Joung, J.K. and Sander, J.D. (2013) TALENs: a widely applicable technology for targeted genome editing. *Nature Reviews Molecular Cell Biology* 14, 49–55.

Krieger, U., Lippman, Z.B., and Zamir, D. (2010) The flowering gene *SINGLE FLOWER TRUSS* drives heterosis for yield in tomato. *Nature Genetics* 42, 459–463.

Kurowska, M., Daszkowska-Golec, A., Gruszka, D., Marzec, M., Szurman, M., Szarejko, I., and Maluszynski, M. (2011) TILLING - a shortcut in functional genomics. *Journal of Applied Genetics* 52, 371–390.

Lee, L.-Y. and Gelvin, S.B. (2008) T-DNA binary vectors and systems. *Plant Physiology* 146, 325–332.

Li, B., Qin, Y., Duan, H., Yin, W., and Xia, X. (2011) Genome-wide characterization of new and drought stress responsive microRNAs in *Populus euphratica*. *Journal of Experimental Botany* 62, 3765–3779.

Li, J.Y., Brunner, A.M., Meilan, R., and Strauss, S.H. (2008) Matrix attachment region elements have small and variable effects on transgene expression and stability in field-grown *Populus*. *Plant Biotechnology Journal* 6, 887–896.

Li, Z., Moon, B.P., Xing, A., Liu, Z.B., McCardell, R.P., *et al.* (2010) Stacking multiple transgenes at a selected genomic site via repeated recombinase-mediated DNA cassette exchanges. *Plant Physiology* 154, 622–631.

Meilan, R. and Ma, C. (2006) Poplar (*Populus* spp.). In: Wang, K. (ed.) *Methods in Molecular Biology*, Vol. 344, Agrobacterium *Protocols*, 2nd edn. Humana Press, Totowa, New Jersey, pp. 143–151.

Meilan, R., Ma, C., Cheng, S., Eaton, J.A., Miller, L.K., *et al.* (2000) High levels of Roundup® and leaf-beetle resistance in genetically engineered hybrid cottonwoods. In: Blatner, K.A., Johnson, J.D., and Baumgartner, D.M. (eds) *Hybrid Poplars in the Pacific Northwest: Culture, Commerce and Capability*. Washington State University Cooperative Extension Bulletin MISC0272, Pullman, Washington, pp. 29–38.

Mohamed, R., Meilan, R., and Strauss, S.H. (2001) Complex behavior of a copper-inducible gene expression system in transgenic poplar. *Forest Genetics* 8, 69–72.

Overballe-Petersen, S., Harms, K., Orlando, L.A.A., Mayar, J.V.M., Rasmussen, S., *et al.* (2013) Bacterial natural transformation by highly fragmented and damaged DNA. *Proceedings of the National Academy of Science USA* 110, 19860–19865.

Payyavula, R., Babst, B., Nelsen, M., Harding, S., and Tsai, C.-J. (2009) Glycosylation-mediated phenylpropanoid partitioning in *Populus tremuloides* cell cultures. *BMC Plant Biology* 9, 151.

Robinson, R. (2004) RNAi therapeutics: how likely, how soon? *PLoS Biology* 2, e28.

Rubinelli, P.M., Chuck, G., Li, X., and Meilan, R. (2013) Constitutive expression of microRNA *Corngrass1* in poplar affects axillary meristem outgrowth, internode length, and lignin quantity and composition. *Biomass and Bioenergy* 54, 312–321.

Shukla, V.K., Doyon, Y., Miller, J.C., DeKelver, R.C., Moehle, E.A., *et al.* (2009) Precise genome modification in the crop species *Zea mays* using zinc-finger nucleases. *Nature* 459, 437–441.

Spiker, S. and Thompson, W.F. (1996) Nuclear matrix attachment regions and transgene expression in plants. *Plant Physiology* 110, 15–21.

Springer, P.S. (2000) Gene traps: tools for plant development and genomics. *Plant Cell* 12, 1007–1020.

Tang, W., Newton, R.J., and Weidner, D.A. (2007) Genetic transformation and gene silencing mediated by multiple copies of a transgene in eastern white pine. *Journal of Experimental Botany* 58, 545–554.

Tang, X., Lowder, L.G., Zhang, T., Malzahn, A.A., Zheng, X., *et al.* (2017) A CRISPR–Cpf1 system for efficient genome editing and transcriptional repression in plants. *Nature Plants* 3, 17018.

Townsend, J.A., Wright, D.A., Winfrey, R.J., Fu, F.L., Maeder, M.L., *et al.* (2009) High-frequency modification of plant genes using engineered zinc-finger nucleases. *Nature* 459, 442–445.

Tsai, C.-J. and Xue, L.-J. (2015) CRISPRing into the woods. *GM Crops and Food* 6, 206–215.

Wang, H., La Russa, M., and Qi, L.S. (2016) CRISPR/Cas9 in genome editing and beyond. *Annual Review of Biochemistry* 85, 227–264.

Wang, K. (ed.) (2006) Methods in Molecular Biology, vol. 344: *Agrobacterium* Protocols, vol. II, 2nd edn. Humana Press, Totowa, New Jersey.

Xing, S., Salinas, M., Höhmann, S., Berndtgen, R., and Huijser, P. (2010) miR156-targeted and non-targeted SBP-box transcription factors act in concert to secure male fertility in *Arabidopsis*. *Plant Cell* 22, 3935–3950.

Xu, R., Qin, R., Li, H., Li, D., Li, L., *et al.* (2017) Generation of targeted mutant rice using a CRISPR–Cpf1 system. *Plant Biotechnology Journal* 15, 713–717.

Xue, L.-J. and Tsai, C.-J. (2015) AGEseq: analysis of genome editing by sequencing. *Molecular Plant* 8, 1428–1430.

Xue, L.-J., Alabady, M.S., Mohebbi, M., and Tsai, C.-J. (2015) Exploiting genome variation to improve next-generation sequencing data analysis and genome editing efficiency in *Populus tremula × alba* 717-1B4. *Tree Genetics and Genomes* 11, 82.

Yamaguchi-Shinozaki, K. and Shinozaki, K. (1993) Characterization of the expression of a desiccation-responsive *rd29* gene of *Arabidopsis thaliana* and analysis of its promoter in transgenic plants. *Molecular and General Genetics* 236, 331–340.

Ye, X., Busov, V., Zhao, N., Meilan, R., McDonnell, L.M., *et al.* (2011) Transgenic *Populus* trees for forest products, bioenergy, and functional genomics. *Critical Reviews in Plant Sciences* 30, 415–434.

Yordanov, Y., Ma, C., Yordanova, E., Meilan, R., Strauss, S.H., and Busov, V.B. (2017) *BIG LEAF* is a regulator of organ size and adventitious root formation in poplar. *PLoS One* 12, e0180527.

Yordanov, Y.S., Regan, S., and Busov, V. (2010) Members of the *LATERAL ORGAN BOUNDARIES DOMAIN* transcription factor family are involved in the regulation of secondary growth in *Populus*. *Plant Cell* 22, 3662–3677.

Yu, Y. and Bradley, A. (2001) Engineering chromosomal rearrangements in mice. *Nature Reviews Genetics* 2, 780–790.

Zawaski, C., Kadmiel, M., Ma, C., Ying, G., Xiangning, J., *et al.* (2011) *SHORT INTERNODES*-like genes regulate shoot growth and xylem proliferation in *Populus*. *New Phytologist* 191, 678–691.

Zetsche, B., Heidenreich, M., Mohanraju, P., Fedorova, I., Kneppers, J., *et al.* (2017) Multiplex gene editing by CRISPR–Cpf1 using a single crRNA array. *Nature Biotechnology* 35, 31–34.

Zhou, X., Jacobs, T.B., Xue, L.-J., Harding, S.A., and Tsai, C.-J. (2015) Exploiting SNPs for biallelic CRISPR mutations in the outcrossing woody perennial *Populus* reveals 4-coumarate:-CoA ligase specificity and redundancy. *New Phytologist* 208, 298–301.

7 Approaches to Genetically Engineering Flowering Control in Trees

Introduction

Plant domestication is the conversion of a plant adapted, via natural selection, for survival, growth, and reproduction in the wild into one that has been genetically altered, through conventional breeding and/or genetic engineering, to yield products useful to humans, and/or to grow in locations other than those to which it has become adapted. When cultivated using modern agricultural systems, domesticated plants are often significantly more productive than their wild progenitors. Because of their long juvenile phases, trees have not been domesticated to the same extent as annual herbaceous plants. Due the extent of their domestication, many agronomic crops have few or no wild relatives with which they are interfertile. This is not true for trees, many of which are wind pollinated and obligate out-crossers.

All cells produced by a genetically engineered plant contain the transgene, including its gametes. After a genetically engineered tree undergoes maturation, release of its sexual propagules can result in the establishment of progeny containing the transgene outside the plantation where the transgenic parent was grown, in the wild. In order for this to occur, however, the genetically engineered tree needs to be sexually compatible with a wild tree and, together, they need to produce viable offspring. Thus, the transgenic and wild trees would have to be sufficiently close, physically, for their wind-born pollen to remain viable long enough under the prevailing environmental conditions for a successful pollination event to occur. There is concern that if an introduced gene imparts a selective advantage to trees harboring it, they could become weedy or invasive. Therefore, a durable method for mitigating the risk of transgene spread and persistence in the environment is likely to be required before genetically engineered trees can be grown for commercial purposes (Meilan, 2006).

Papaya (*Carica papaya*) is a notable exception to this restriction. Papaya was introduced into Hawaii and is the basis of a thriving industry there. Much of the crop is grown on small, family-owned farms. Papaya orchards were being devastated by the papaya ringspot virus, which is transmitted by numerous species of aphid, such as *Myzus persicae*. Papaya trees infected with this virus are stunted, produce disfigured fruit and have lower yields (Tennant *et al.*, 1994). Transgenesis was used to engineer virus-resistant papaya using an approach that was first suggested by Beachy (1993). It involves utilization of a phenomenon known as coat-protein-mediated resistance. Constitutive expression of a gene that encodes a viral coat protein interferes with disassembly of viral particles in transgenic cells, thus imparting resistance. Federal regulators allowed fully fertile, transgenic, virus-resistant papaya to be grown commercially in Hawaii, primarily for two reasons. First, there are no interfertile wild relatives of papaya in Hawaii. Second, there is a large, physical barrier—the Pacific Ocean—that prevents the transgene from spreading to wild trees outside plantations where the transgenic papaya are grown.

Transgene Confinement

To achieve transgene confinement, researchers are attempting to genetically engineer trees that do not produce functional flowers (Meilan *et al.*, 2001; Lemmetyinen and Sopanen, 2004; Brunner *et al.*, 2007; Fritsche *et al.*, 2018). The various approaches being used to engineer sterility will be discussed below. Much of this experimental work has been done with poplar (genus *Populus*), which is now widely accepted as the model species for studying forest-tree biology, owing to its small genome, ever-increasing genomic sequence resources, an extensive metabolomics database, rapid growth, and the relative ease with which it can be clonally propagated *ex vitro* and transformed and regenerated *in vitro* (Bradshaw *et al.*, 2000; Wullschleger *et al.*, 2002; Jansson and Douglas, 2007).

Besides helping to alleviate regulatory and public concerns about transgenic trees, reproductive sterility can prevent "genetic pollution" from non-transgenic trees. Some growers are planting thousands of acres with a single genotype of poplar within its native range. Although recombination and segregation occur during gamete formation, there are at most two alleles for each locus that are transmitted during fertilization. At anthesis, the pollen cloud downwind of a large clonal population will be greatly enriched for pollen from the plantation-grown trees. Seedlings that are sired by these clones will have reduced genetic diversity relative to those resulting from fertilization by pollen derived from a diverse mixture of wild trees.

Many tree species are grown as crops in regions of the world where they are not native. These introduced species sometimes become invasive in their new environments (Hughes, 1994; Richardson, 1998), where they can drastically alter the ecosystems that they invade. Engineered sterility would help prevent the spread of progeny from the plantation-grown trees and stem gene flow to wild individuals with which they are sexually compatible (Skinner *et al.*, 2000). This is not as problematic for species that are dioecious, such as poplar, but occasionally they will produce a small number of bisexual catkins (Dickmann, 2001).

Leaves are the primary photosynthetic organ in trees; because they have a finite leaf area, trees have a limited capacity to produce photosynthate. Reproductive growth is a very energetically demanding process. During years that trees are investing heavily in reproductive effort, there is a measurable decrease in their vegetative growth (Eis *et al.*,

1965; Tappeiner, 1969; Teich, 1975). Because trees that are engineered for reproductive sterility will not be using some of their limited pool of carbohydrates for producing reproductive structures (e.g. pollen, seeds, catkins), it is assumed that sterile trees will grow faster than trees of the same genotype and age that are allowed to flower (Lännenpää *et al.*, 2005). Moreover, sterility will eliminate the production of nuisance reproductive structures, such as pollen, which may be useful for hayfever sufferers, and cones from trees, such as sweet gum (*Liquidambar styraciflua*), which are an annoyance to landscapers and homeowners. Thus, having trees that are genetically engineered only for flowering control may have value that extends beyond genetic confinement.

DiFazio *et al.* (2012) showed that absolute sterility may not be necessary for effective transgene containment. Based on model simulations, they concluded that, for a wide range of transgenes, extensive pollen flow to large populations of interfertile wild relatives will not necessarily result in the establishment, in the wild, of a substantial number of offspring containing the transgene. They predicted that the amount of transgene spread will depend greatly on ecological context and cultural practices, and can be greatly restricted or prevented through management strategies or by using genes that mitigate gene flow, such as those that impart reproductive sterility.

Various degrees of fertility may also be achieved by altering ploidy level (e.g. triploids or aneuploids), or by altering the expression of genes that specifically regulate floral development or control the transition to maturation. However, any approach that does not result in absolute, durable sterility would likely not satisfy regulatory concerns.

In some cases, delaying the onset of initial flowering may be sufficient for preventing transgene escape. Various tree species have different ages at which they acquire the competence to produce flowers. For example, the juvenile period for poplar is 5–7 years (Braatne *et al.*, 1996), whereas Douglas-fir (*Pseudotsuga menziesii*) does not flower until it is about 20 years old (Stein and Owston, 2008). When poplars are coppiced, they revert to a juvenile state (i.e. their internal clocks are reset). If transgenic poplars are grown in dedicated energy plantations that are regularly coppiced prior to the age at which they undergo maturation, there would be little risk of transgene flow.

In addition to delaying flowering, it may also be possible to hasten its onset through genetic engineering, leading to briefer juvenile periods and thus shorter breeding cycles. It may also be desirable to engineer sterility that is reversible, with an appropriate stimulus, so that a tree can temporarily be used for sexual reproduction.

Despite indications that one or more flowering-control strategies can be useful for engineering transgene confinement, no single approach has satisfied the essential requirements for deregulation, which is needed before transgenic trees can be deployed commercially. Investigators continue to search for ways to engineer complete sterility that can be confirmed in juvenile trees, which have not yet acquired the competence to flower, so the effectiveness of the strategy can be evaluated sooner. Finally, in order to engineer durable sterility, it must be shown that transgene expression is stable following multiple rounds of propagation, after various developmental events, under a variety of environmental conditions, and over several growing seasons, without negatively impacting growth.

Engineering Flowering Control

Sterility

Several approaches to engineering reproductive sterility have been tested in trees (Fritsche *et al.*, 2018). These are summarized below. Tests are under way to determine which of these will cause sterility that is sufficiently specific, absolute, and lasting. In the end, a combination of approaches may be required to achieve durable sterility.

Cell ablation

Cell ablation involves tissue-specific expression of a gene whose translation product would lead to the elimination of specific cell types

(Meilan *et al.*, 2001). In early attempts at utilizing cell ablation to engineer sterility in poplar, researchers relied on heterologous promoters, which were thought to cause floral-specific expression in tobacco (*Nicotiana tabacum*) and *Arabidopsis thaliana* (Koltunow *et al.*, 1990; Hackett *et al.*, 1992; Wang *et al.*, 1993), to control the expression of cytotoxin genes, such as *DTA* (Greenfield *et al.*, 1983) and *barnase* (Hartley, 1988). The former encodes the A subunit of the diphtheria toxin. Without the B subunit, the toxin cannot be taken up by cells; thus, the A subunit is only lethal to cells within which the *DTA* gene product accumulates. The DTA protein interferes with ribosome translocation along the mRNA; therefore, it blocks protein synthesis, ultimately leading to cell death. The *barnase* gene encodes a ribonuclease, which digests RNA, also resulting in cell death. Generally, the version of the *barnase* gene that is used to engineer sterility lacks a signal peptide sequence, which confines the protein to the cytoplasm (Lännenpää *et al.*, 2005). Another protein known as barstar specifically binds to the barnase protein in a way that blocks its enzymatic activity.

Skinner *et al.* (2003) experimented with the promoter from the *Populus trichocarpa* ortholog of the *DEFICIENS* gene, *PTD*, which was thought to be expressed only in cells destined to become flowers (Sheppard *et al.*, 2000). Skinner and colleagues used the promoter from *PTD* to drive expression of the *GUS* reporter gene and *DTA* in *A. thaliana*, tobacco, and poplar. They showed that the *PTD* promoter directs expression of *GUS* early in the developing floral organs in *A. thaliana* and poplar (Fig. 7.1A,B). These lines were co-transformed with the *A. thaliana LEAFY* (*LFY*) gene, which was under the control of a constitutive promoter from the cauliflower mosaic virus (CaMV) *35S* gene. Overexpression of *LFY* had previously been shown to induce precocious flowering in poplar (Weigel and Nilsson, 1995). Expression of the PTD::*DTA* (promoter::coding sequence) fusion in *A. thaliana* resulted in either ablation or altered development of petals and stamens (Fig. 7.1C), and the absence of petals, stamens, and carpels in tobacco (Fig. 7.2A). The PTD::*DTA* construct also appeared to prevent flowers from developing on poplar co-transformed with 35S::*LFY* (Fig. 7.2B). Expression of PTD::*DTA* had no significant effects on growth in tobacco (J.S. Skinner, unpublished data); however, when these constructs were transformed into poplar, the resulting transgenic plants exhibited reduced vegetative growth (Fig. 7.3), suggesting low-level ("leaky") expression of the *DTA* transgene in non-target (vegetative) tissues.

Lännenpää *et al.* (2005) used a similar approach to prevent flower development in silver birch (*Betula pendula*). The promoter from *BpFRUITFULL-LIKE1* (*BpFULL1*), a gene involved in the early stages of flower development in birch, was used to drive the expression of *barnase*. In the wild, the juvenile period for native birches is between 5 and 10 years (Longman and Wareing, 1959).

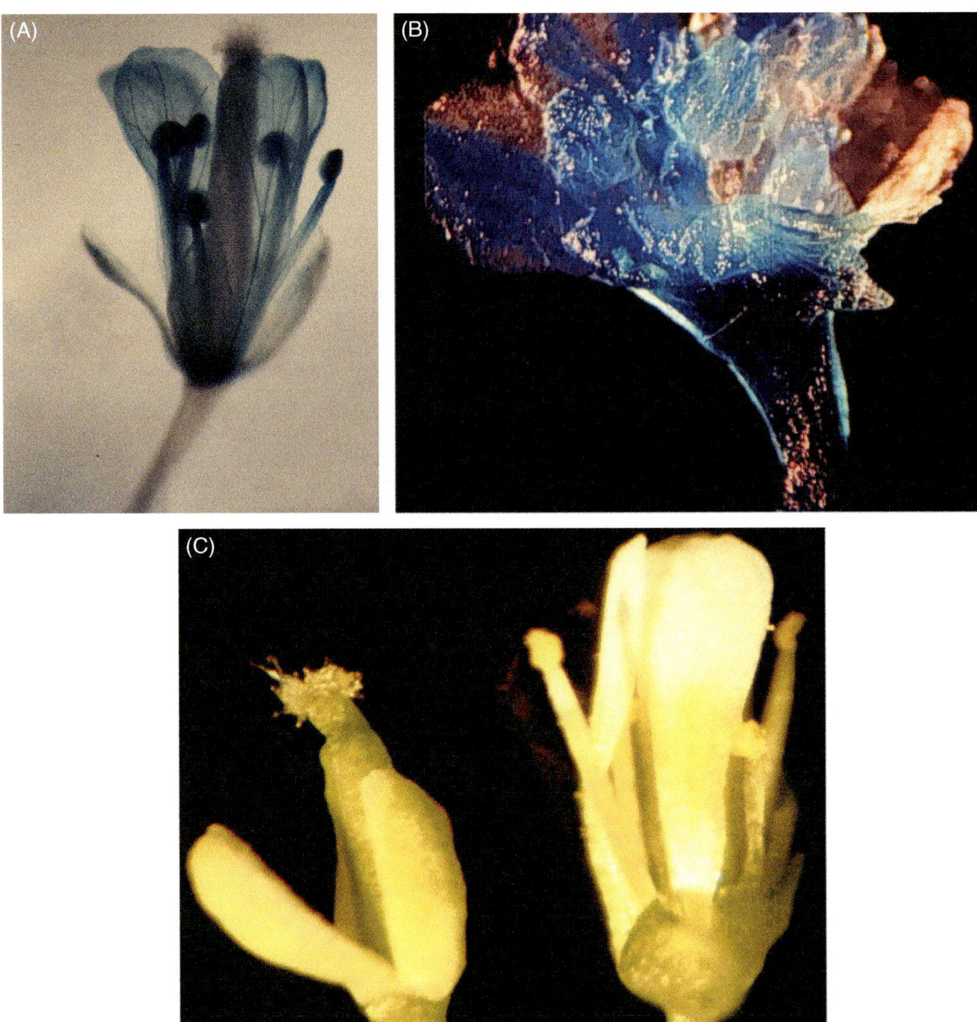

Fig. 7.1. Specificity of the promoter from the *PTD* gene. (A) PTD::*GUS* expression in *Arabidopsis thaliana* floral parts. (B) PTD::*GUS* expression in floral parts of an early-flowering (co-transformed with 35S::*LFY*) aspen. Zones stained blue in (A) and (B) indicate *GUS* reporter-gene expression. (C) The effects of PTD:: *DTA* on flower development in *A. thaliana*. A transgenic flower is shown on the left, a non-transgenic control on the right. Expression of the PTD::*DTA* fusion resulted in either ablation or altered development of petals and stamens. (After Ellis *et al.*, 2001 (A, B) and Skinner *et al.*, 2003 (C).)

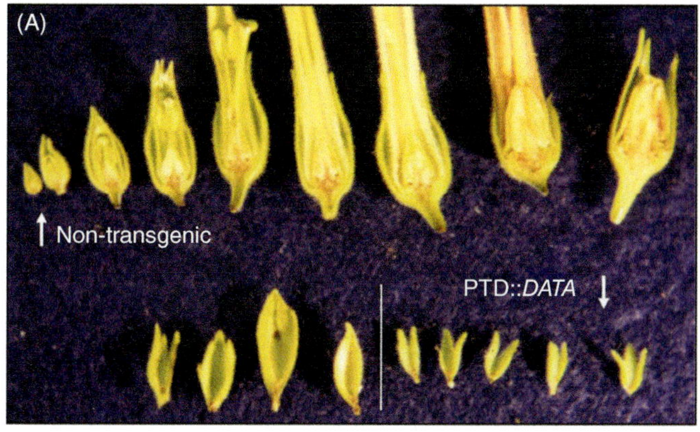

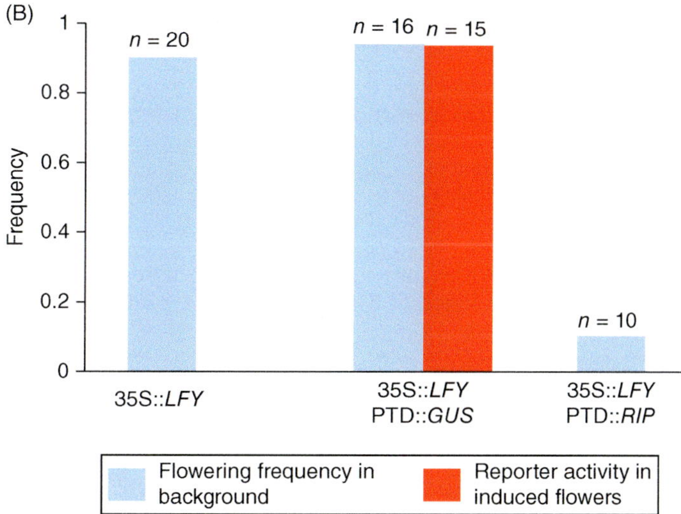

Fig. 7.2. Effects of PTD::*DTA* on flower development in (A) tobacco and (B) aspen. (A) The upper panel shows normal flower development. Florets from transgenic tobacco plants consisted of only sepals (lower panel). (B) The 35S::*LFY* construct was used to induce precocious flowering in aspen. (After Skinner *et al.*, 2003.)

When grown in a greenhouse, birches begin flowering when they are about 3 years old (Viherä-Aarnio and Ryynänen, 1994). To obtain results sooner, Lännenpää and colleagues used an early-flowering genotype of birch, BPM5, which begins to produce inflorescences 3–5 months after rooting, when grown under continuous light in a greenhouse. The BpFULL1::*barnase* fusion was transformed into *A. thaliana*, tobacco, and BPM5. In all three species, inflorescence development was blocked. Many of their transgenic birch lines had a bush-like appearance and their growth rates were much slower than those of wild-type plants. Four of their non-flowering lines exhibited a normal phenotype. It is thought that the bushy phenotype and dichotomous branching were caused by early expression of *barnase* in the shoot apex, leading to its death before differentiating into an inflorescence meristem. Lateral cells may have then given rise to two separate meristems.

Wei *et al.* (2007) tested an attenuation system that was designed to prevent the harmful effects of floral promoter::cytotoxin fusions on vegetative growth of transgenic plants. Poplar was transformed with the

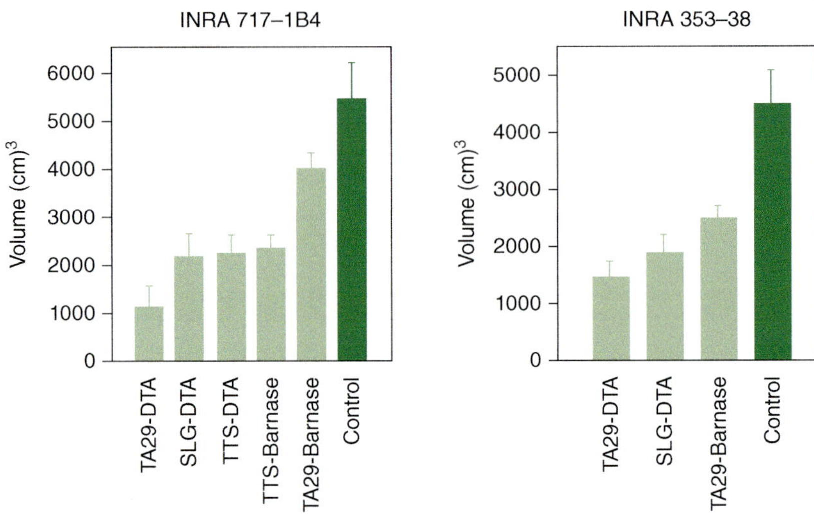

Fig. 7.3. Volumetric growth of hybrid poplar trees that were genetically engineered for reproductive sterility. The tobacco (*Nicotiana tabacum*) tapetum-specific TA-29 promoter, a tobacco transmitting tract-specific (TTS) promoter, or a *Brassica* S-locus glycoprotein (SLG) promoter fused to either the gene encoding Barnase (Barn) or the A subunit of the diphtheria toxin (DTA) were introduced into female (INR 717-1B4, *Populus tremula* x *P. alba*) or male (INRA 353-38, *P. tremula* x *P. tremuloides*) clones of hybrid aspen. (After Skinner *et al.*, 2000.)

barnase gene under the control of the promoter from *PTFL*, the *P. trichocarpa* ortholog of *LFY*. This promoter is highly expressed in developing male and female inflorescences; however, the *PTLF* promoter also causes measurable expression in leaf primordia, young leaves, and other vegetative tissues (Rottmann *et al.*, 2000). The *barstar* gene was included on the same T-DNA as the *barnase* gene but was driven by either a basal *35S* promoter (35SBP; 35SBP was fused to the tobacco mosaic virus Ω element, a translation enhancer) or the nopaline synthase (NOS) promoter. It was hoped that limited *barstar* expression would attenuate the effects of *barnase* in vegetative tissues.

It was not possible to regenerate plants from cells transformed with only the PTLF:: *barnase* fusion (i.e. unattenuated), leading to speculation that there was sufficient *barnase* expression in vegetative tissues to be lethal. Including the barstar-attenuation constructs allowed recovery of transgenic plants, but the transformation rate was reduced by nearly one-third of that normally seen in the poplar genotype that was used for this study. Seven percent of the attenuated transgenic lines had highly abnormal morphology, which was obvious in the early stages of regeneration; these lines had high barnase:barstar expression ratios. In a preliminary greenhouse study, the attenuated plants looked normal and

grew at the same rate as wild-type and transgenic plants missing the *barnase* gene. Unexpectedly, after one or two growing seasons in the field, the growth rate of nearly all of the attenuated lines was greatly reduced, and the extent of stunting was related to their barnase:barstar expression ratio. The authors noted that these results exemplify the importance of conducting field trials during early stages of research to identify the potentially negative effects of transgene expression in trees.

The tapetum is a layer of cells lining the inner surface of the anther and nourishes the developing pollen. Brunner *et al.* (2007) assembled a binary vector that contained a putative tapetal-specific promoter from the tobacco *TA29* gene to drive the expression of *barnase*. Field-grown poplars that were transformed with this construct exhibited a high degree of male sterility (Fig. 7.4), and the transgene had no obvious effects on vegetative growth.

Dominant-negative mutations

Another way to interfere with expression of a gene in a plant is to transform it with a mutated endogene sequence that, when expressed, will negate the normal function of that gene or its product. This is referred to as a

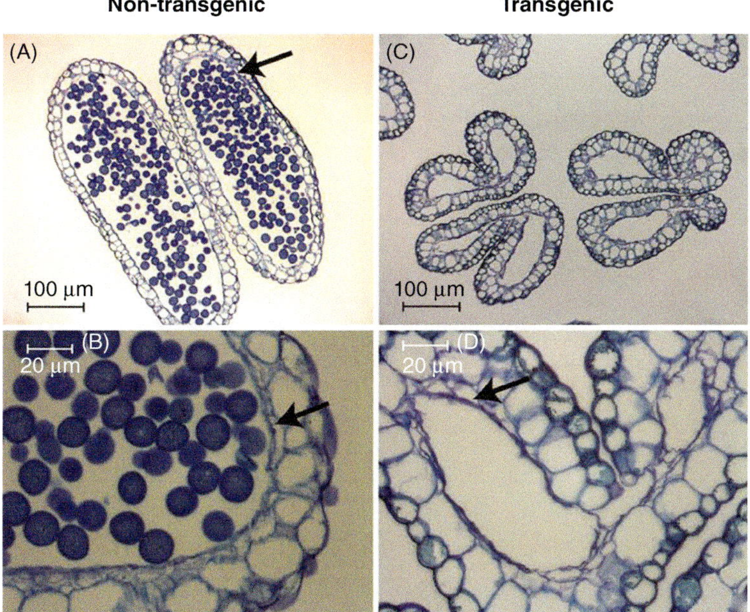

Fig. 7.4. Transverse sections of nearly mature anthers from a transgenic, putatively male-sterile, field-grown poplar and a non-transgenic control poplar of the same age. Samples were fixed, dehydrated, embedded in glycol methacrylate plastic, sectioned, and mounted on slides. Sections were stained in 0.5% toluidine blue O in citrate buffer. Arrows indicate the tapetal layer, which is absent or disorganized in transgenic plants. Magnification: ×100 (A, C); ×400 (B, D). (From Brunner *et al.*, 2007.)

dominant-negative mutation. This approach has been used for genes that encode transcription factors, such as MADS-box genes (see below), that are required for normal floral development (Jeon *et al.*, 2000; Ferrario *et al.*, 2004). It is effective because many transcription factors are part of multi-protein complexes. If done correctly, the introduced mutation will lead to a protein that can interact with other proteins that make up the complex but prevent the latter from binding to its target DNA sequence (Brunner *et al.*, 2007).

Based on the results of work with rice (*Oryza sativa*) and mammalian MADS-box genes, Brunner *et al.* (2007) used site-specific mutagenesis to substitute the amino acids predicted to be needed for dimerization and/or DNA binding in the *AGAMOUS* (*AG*) and *APETALA* genes. Constitutive expression of these mutated genes in *A. thaliana* led to strong loss-of-function phenotypes at a frequency of about 30%. The same constructs were tested in poplar and sweet gum (Brunner *et al.*, 2007).

Downregulation

Reducing or eliminating the expression of one or more genes involved in floral development is another potential way to control flowering. Floral development is controlled by a complicated network of transcription factors (both activators and repressors), many of which contain a highly conserved sequence motif known as the MADS box. Lemmetyinen *et al.* (2004) reported on two MADS-box genes from silver birch, *BpMADS1* and *BpMADS6*, which are orthologs of the *A. thaliana SEPALLATA3* and *AG* genes, respectively. *In situ* hybridization revealed that *BpMADS1* is only expressed early in inflorescence meristem development. They detected transcripts of *BpMADS6* in both male and female inflorescences, but not in roots, leaves, or shoot tips. When transformed into an early-flowering birch genotype, a 35S::*BpMADS1* antisense construct caused the conversion of some inflorescences into vegetative shoots. A 35S::*BpMADS6* antisense construct resulted in a complete absence of stamens in some male inflorescences. Their results suggest that *BpMADS6* is involved in flower formation, whereas *BpMADS1* participates in both inflorescence and flower development.

Genes involved in floral development can also be silenced using an approach known as RNA interference (RNAi) (Chapter 6, this volume). The level of target-gene expression seems to affect the efficacy of RNAi suppression. In addition, it has been shown that the silencing effect can spread from the target to a nearby sequence. This "transitive suppression" seems to occur only when the target is a transgene, not an endogenous gene (endogene). Due to regulatory redundancy and gene duplications, it may be necessary to simultaneously silence more than one gene in order to develop an effective sterility system for trees (Brunner *et al.*, 2007). More than one gene can be silenced with a single RNAi vector by either using a sequence that is highly conserved among all the target genes or by concatenating unique fragments of multiple genes (reviewed by Watson *et al.*, 2005). In addition, it is now possible to edit several genes simultaneously using a single CRISPR construct (Chapter 6, this volume).

Using RNAi, Lu *et al.* (2018) downregulated both *AG* and *STICKSEED* (*STK*; formerly known as *AGAMOUS-LIKE11*) in *Populus alba* (clone 6K10). For *AG*, two RNAi constructs were used, PTG and MPG; the latter was flanked with MARs (Chapter 6, this volume). When six of the 22 lines transformed with PTG and 11 of the 12 MPG-transformed lines produced catkins in a field trial (near Corvallis, Oregon, USA), flowers exhibited "carpel-inside-carpel" phenotypes. Seed production was completely disrupted in seven of the transgenic lines, and sterile, anther-like organs were produced by 10 lines. Lines in which both *AG* and *STK* were strongly suppressed lacked both seeds and seed hairs. The modified floral phenotypes remained stable for 4 years under field conditions. The authors concluded that because the function of *AG* and *STK* is so strongly conserved, they are promising targets for genetic containment of exotic trees and transgenic confinement of genetically engineered trees.

A major challenge for engineering flowering control in poplar is avoiding negative impacts on vegetative growth. In addition, poplar flowering-gene homologs are known to control vegetative growth in response to abiotic signals (Böhlenius *et al.*, 2006; Hoenicka *et al.*, 2008; Hsu *et al.*, 2011; Azeez *et al.*, 2014). Klocko *et al.* (2016) recently reported that suppression of the *P. alba* ortholog of *LFY* (*PaLFY*) led to complete and stable reproductive sterility in field-grown trees. Transgenic lines in which *PaLFY* was highly suppressed had extremely small inflorescences that lacked functional sexual organs. The sterility phenotype was stable over two growing seasons, and the trees were normal with respect to survival, timing of dormancy, vegetative morphology, and, importantly, growth rate.

The broad functional conservation of *AG*, with respect to regulating male and female reproductive development, and the floral expression patterns of the poplar orthologs (Brunner *et al.*, 2000), make *AG* a good candidate for engineering sterility. Blocking expression of *AG* would likely not prevent the floral transition, but it is important to disrupt the floral-development process as early as possible to minimize the amount of carbohydrate diverted to reproductive effort. Moreover, knocking out genes that act at more than one stage of floral development would likely lead to more durable sterility; hence, it is desirable to identify additional targets. Because release from dormancy, resumption of vegetative growth, and the initiation of inflorescence development occur sequentially, studying these phenological events may enhance our ability to identify genes that control the transition to flowering but are not involved in regulating vegetative growth and development (Brunner *et al.*, 2014).

One example is the *FLOWERING LOCUS T1* (*FT1*) gene, which not only induces precocious flowering in poplar (Böhlenius *et al.*, 2006) but is also involved in controlling release from dormancy (Rinne *et al.*, 2011). Moreover, single-nucleotide polymorphisms (SNPs) of *FT1* affect the timing of bud flush (Evans *et al.*, 2014). Therefore, *FT1* appears to play a critical role in regulating seasonal vegetative growth, so it would not be a good candidate for engineering transgene confinement. In contrast, a transcription factor known as FD, which is preferentially expressed in the shoot apex and is needed for FT to promote flowering (Abe *et al.*, 2005), would be a good candidate. One of three *FD* orthologs, *FD2*, is expressed seasonally, mainly in initiating inflorescence buds (A.M. Brunner, 2017, Blacksburg, Virginia, personal communication), and *FD2* is the only ortholog that induces precocious flowering (Tylewicz *et al.*, 2015; Parmentier-Line and Coleman, 2016). Given this, blocking expression of both *AG* and *FD2* via CRISPR would likely lead to robust sterility.

Early flowering

Long juvenile periods are a significant impediment to the genetic improvement of trees. By causing trees to flower early, breeding cycles can be shortened, expediting genetic studies and leading to more rapid gains in productivity. A naturally occurring female *P. alba* genotype has been found to produce flowers as early as 1 year (Meilan *et al.*, 2004). It was thought that putative flowering-control constructs could be tested initially in this genotype. However, this approach proved to be impractical because of the need for a lengthy and cumbersome inductive treatment that was only partially effective.

Acceleration of flowering in poplar has also been explored through genetic engineering (Meilan *et al.*, 2001). Poplar orthologs of the flowering-time gene *LFY* have been expressed ectopically in poplar trees (Nilsson *et al.*, 1998; Rottmann *et al.*, 2000), but Rottmann *et al.* (2000) found that it was ineffective at modulating flowering time in many genotypes, and the flowers produced were largely abnormal and infertile.

Mohamed (2006) used RNAi to downregulate expression of the poplar ortholog of *CENTRORADIALIS*, *PtCENL1*, a gene that is involved in maintaining juvenility in trees. When transgenic poplars containing this RNAi vector were grown under field conditions, four of the most highly suppressed events produced inflorescences or floral buds within

2 years, which is several years earlier than is typically seen with wild-type trees. Unexpectedly, overexpression of *PtCENL1* also led to delayed vegetative bud break. Based on these results, it would seem that *PtCENL1* is involved in regulating release from dormancy and resumption of growth. It may be possible to use this gene to shorten breeding cycles, and possibly to inform further research on flowering control in trees.

At about the same time, Hsu *et al.* (2006) reported that *FT2*, the poplar ortholog of the *A. thaliana* flowering-time gene *FT*, controls first-time and seasonal flowering in poplar. *FT2* transcript levels are very low in juvenile poplar and are much higher during the transition from vegetative to reproductive growth, and in adult trees. They also showed that flowering is induced in poplar within 1 year of when *FT2* is over-expressed. Their results suggest that *FT2* is not only a part of the flower-initiation pathway in poplar but also is involved in integrating seasonal flower production in trees, which have a perennial growth habit.

Contemporaneously, Böhlenius *et al.* (2006) described how the *CONSTANS* (*CO*)/*FT* regulon, which controls flowering in response to variations in day length in annual plants, also regulates flowering in poplar. In *A. thaliana*, *FT* transcription is under the control of the *CO* gene (Searle and Coupland, 2004). Surprisingly, Böhlenius *et al.* (2006) found that the *CO*/*FT* regulatory module is also involved in regulating the critical day length for growth cessation and bud set in poplar during autumn, across a wide latitudinal range.

The *BpMADS4* gene has been shown to play a central role in inflorescence initiation in silver birch (Elo *et al.*, 2007), and its overexpression caused early flowering in apple (*Malus domestica*; Flachowsky *et al.*, 2007). Lemmetyinen *et al.* (2004) showed that 35S::*BpMADS6* caused early flowering in *A. thaliana*. In tobacco, both 35S::*BpMADS1* and 35S::*BpMADS6* promoted flowering, but *BpMADS6* overexpression led to developmental abnormalities in sepals and petals.

Hoenicka *et al.* (2016) recently reported that temperature, not photoperiod, is the key factor needed for inducing precocious flowering in poplar. They produced transgenic lines of poplar in which the *A. thaliana FT* gene was under the control of a heat-shock promoter. Fertile flowers were only achieved in these lines when a cold treatment preceded the inductive heat treatment. In addition, delivery of a hybrid *FT* gene with a plant viral vector resulted in very precocious flowering in apple and pear (*Pyrus communis*) (Yamagishi *et al.*, 2016). Alternatively, by constitutively expressing a splice variant of *FD2a*, a transgenic poplar has been produced (A.M. Brunner, 2017, Blacksburg, Virginia, personal communication) that flowers within 6 months under long day lengths, and can be regenerated and propagated readily. Both systems have potential for testing genetic constructs that are expected to affect floral development.

Excision Systems

A tremendous amount of time and resources have been spent trying to develop gene-excision systems that can target specific loci in the genome, and to remove transgenes. The "gene deletor," which is based on a novel combination of the bacterial phage CRE–loxP and yeast (*Saccharomyces cerevisiae*) FLP/FRT DNA recombination systems, may provide a tool to prevent gene flow from transgenic trees (Li, 2012; Luo *et al.*, 2007). As demonstrated in tobacco, the gene deletor is highly efficient at eliminating all transgenes from both pollen and seed; none of more than 25,000 progeny examined from self-pollination of T_0 transgenic plants contained transgenes. This technology, if successfully used in woody plants, could offer an advantage over sterility. Large-scale (e.g. thousands of acres) plantings of sterile trees with no flower, pollen, or seed production may negatively impact insects and other animals that depend on these tissues for nourishment, potentially affecting biodiversity in forests or forcing insect populations into nearby farmland, parks, and wilderness. Using this technology, it may be possible to engineer transgenic trees to produce pollen and seeds that develop normally but lack the

transgene that would still be present in the somatic cells (Ye *et al.*, 2011).

Excision systems may be useful for preventing transgene escape from vegetatively propagated trees, but nothing is known about their long-term effectiveness under field conditions. In addition, it will be challenging to predict whether transgenes will be absent from the gametes before they are produced by trees. Moreover, this approach would not interfere with fertility; it would only prevent the spread of the genetic material targeted for excision (i.e. the transgene). However, excision of a reproduction-related gene could be coupled with the insertion of a transgene that imparts sterility to provide a more durable confinement system (Brunner *et al.*, 2007).

Excision may also be needed when using CRISPR technology. The construct that encodes the guide RNA (Chapter 6, this volume) integrates in the genome at a location that is distal to the locus that is being targeted for manipulation. In rapidly cycling annual plants, once the desired editing has been confirmed, the CRISPR construct can be segregated away from the mutation(s) they cause. This is not so easily done with poplar, and regulatory restrictions may require that the CRISPR construct be removed from a transgenic plant before it can be commercialized.

Targeted Mutagenesis and Gene Replacement

For many years, researchers have tried to develop techniques for introducing specific deletions, insertions, or mutations in plant genes. Engineering reproductive sterility via gene targeting is complicated by redundancy; both alleles of two or more genes might need to be replaced or mutated. Substituting only one allele of a single gene with a dominant suppression transgene might be a more effective way of engineering sterility than using *Agrobacterium tumefaciens* to randomly insert a novel transgene, because it would avoid position effects. However, the feasibility of making specific changes in the plant genome is dependent on both the efficacy of gene targeting and the ease with which the species of interest can be transformed, regenerated, and screened. *Arabidopsis thaliana* can be transformed *in planta* with relative ease. Because of this capability, its size, and its short life cycle, comparatively little effort is required to produce and screen large numbers of transgenic plants. Until recently, no such system has existed for trees (Brunner *et al.*, 2007). With the advent of CRISPR technology, this is now possible with poplar (see Chapter 6).

Prospects for Flowering Control

Despite signs that one or more of the approaches that have been tested for flowering control has potential for engineering transgene confinement, no single method satisfies the basic requirements for regulatory approval, which will be needed before transgenic trees can be deployed commercially. To do so, a more thorough understanding of floral development in trees is needed. Comparative studies have revealed that similar genes regulate sexual reproduction in a variety of angiosperms and, to some extent, gymnosperms (Brunner *et al.*, 2007). However, gene duplications, losses, and diversification make it difficult to predict gene function based solely on sequence similarity or expression pattern (Irish and Litt, 2005). The Floral Genome Project (www.floralgenome.org/) led to a large collection of expressed-sequence tags (ESTs) for a wide range of species. This resource has helped to reveal patterns among genes that regulate flowering. With a steep decline in the cost of genome sequencing, there has been a rapid increase in the number of ESTs available for various trees species, which has expedited the discovery of genes that are orthologous to *A. thaliana* genes and regulate floral development in trees, especially poplar. In addition, the National Science Foundation has funded a Plant Genome Project that focused on studying floral transition in poplar (Brunner *et al.*, 2007). Collectively, these resources are contributing to the attainment of a robust sterility system for trees.

Summary

Plant domestication is the process by which a plant adapted for survival, growth, and reproduction in the wild is genetically altered by conventional means or through genetic engineering to yield products useful to humans and/or grow in a new environment. Because of their long juvenile periods, trees have not been domesticated to the same extent as annual, herbaceous plants. Due to the extent of their domestication, many agronomic crops have few or no wild relatives with which they are sexually compatible. This is not true for trees, many of which are wind pollinated and obligate out-crossers.

All cells produced by a genetically engineered plant contain the transgene, including its gametes. After a genetically engineered tree undergoes maturation, release of its sexual propagules can result in the establishment of progeny containing the transgene outside the plantation where the transgenic parent was grown. There is concern that if an introduced gene imparts a selective advantage to trees harboring it, they could become weedy or invasive. Therefore, a durable method for mitigating the risk of transgene spread and persistence in the environment is likely to be required before genetically engineered forest-tree species can be grown for commercial purposes. To achieve transgene confinement, researchers are attempting to genetically engineer trees that do not produce functional flowers.

During years that trees are investing heavily in reproductive effort, there is a measurable decrease in their vegetative growth. Because trees that are engineered for reproductive sterility will not be using some of their limited pool of carbohydrates for producing reproductive structures (e.g. pollen, seeds, catkins), they might grow faster than trees of the same genotype and age that are allowed to flower. Thus, genetically engineering trees for flowering control may have favorable consequences for yield and productivity.

Several approaches to engineering reproductive sterility have been tested in trees. These include cell ablation, dominant-negative mutations, and reducing or eliminating the expression of one or more genes involved in floral development. Tests are under way to determine which of these will cause sterility that is sufficiently specific, absolute, and lasting. In the end, a combination of approaches may be required to achieve this goal. Any approach that does not result in absolute, durable sterility would likely not alleviate regulatory and public concerns.

Through a more thorough understanding of the flower-development process, it may also be possible to alter the timing of flowering. In some cases, delaying the onset of initial flowering may be sufficient for preventing transgene escape. In addition to delaying flowering, it may also be possible to hasten its onset through genetic engineering, leading to briefer juvenile periods and, thus, shorter breeding cycles. Finally, it may be desirable to engineer sterility that is reversible, so that a tree can temporarily be used for sexual reproduction.

References

Abe, M., Kobayashi, Y., Yamamoto, S., Daimon, Y., Yamaguchi, A., *et al.* (2005) FD, a bZIP protein mediating signals from the floral pathway integrator FT at the shoot apex. *Science* 309, 1052–1056.

Azeez, A., Miskolczi, P., Tylewicz, S., and Bhalerao, R.P. (2014) A tree ortholog of *APETALA1* mediates photoperiodic control of seasonal growth. *Current Biology* 24, 717–724.

Beachy, R. (1993) Virus resistance through expression of coat protein genes. In: Chet, I. (ed.) *Biotechnology in Plant Disease Control*. Wiley-Liss, New York, pp. 89–104.

Böhlenius, H., Huang, T., Charbonnel-Campaa, L., Brunner, A.M., Jansson, S., *et al.* (2006) CO/FT regulatory module controls timing of flowering and seasonal growth cessation in trees. *Science* 312, 1040–1043.

Braatne, J.H., Rood, S.B., and Heilman, P.E. (1996) Life history, ecology, and conservation of riparian cottonwoods in North America. In: Stettler, R.F., Bradshaw, H.D. Jr., Heilman, P.E., and Hinkley, T.M. (eds.) *Biology of* Populus *and its Implications for Management and Conservation*. NRC Research Press, Ottawa, pp. 57–85.

Bradshaw, H.D. Jr., Ceulemans, R., Davis, J., and Stettler, R.F. (2000) Emerging model systems: poplar (*Populus*) as a model forest tree. *Journal of Plant Growth Regulation* 19, 306–313.

Brunner, A.M., Rottmann, W.H., Sheppard, L.A., Krutovskii, K., DiFazio, S.P., *et al.* (2000) Structure and expression of duplicate *AGAMOUS* orthologues in poplar. *Plant Molecular Biology* 44, 619–634.

Brunner, A.M, Li, J., DiFazio, S.P., Shevchenko, O., Montgomery, B.E., *et al.* (2007) Genetic containment of forest plantations. *Tree Genetics and Genomes* 3, 75–100.

Brunner, A.M., Evans, L.M., Hsu, C.Y., and Sheng, X. (2014) Vernalization and the chilling requirement to exit bud dormancy: shared or separate regulation? *Frontiers in Plant Science* 5, 732.

Dickmann, D.I. (2001) An overview of the genus *Populus*. In: Dickmann, D.I., Isebrands, J.G., Eckenwalder, J.E., and Richardson, J. (eds.) *Poplar Culture in North America*. NRC Research Press, Ottawa, pp. 1–42.

DiFazio, S.P., Leonardi, S., Slavov, G.T., Garman, S.L., Adams, W.T., and Strauss, S.H. (2012) Gene flow and simulation of transgene dispersal from hybrid poplar plantations. *New Phytologist* 193, 903–915.

Eis, S., Garman, E.H., and Ebell, L.F. (1965) Relation between cone production and diameter increment of Douglas-fir (*Pseudotsuga menziesii* (Mirb.) Franco), Grand fir (*Abies grandis* (Dougl.) Lindl.), and western white pine (*Pinus monticola* (Dougl.)). *Canadian Journal of Botany* 43, 1553–1559.

Ellis, D., Meilan, R., Pilate, G., and Skinner, J.S. (2001) Transgenic trees: Where are we now? In: Strauss, S.H. and Bradshaw, H.D. Jr. (eds.) *Proceedings of the First International Symposium on the Ecological and Societal Aspects of Transgenic Plantations*. College of Forestry, Oregon State University, Corvallis, Oregon, pp. 113–123.

Elo, A., Lemmetyinen, J., Novak, A., Keinonen, K., Porali, I., *et al.* (2007) *BpMADS4* has a central role in inflorescence initiation in silver birch (*Betula pendula*). *Physiologia Plantarum* 131, 149–158.

Evans, L.M., Slavov, G.T., Rodgers-Melnick, E., Martin, J., Ranjan, P., *et al.* (2014) Population genomics of *Populus trichocarpa* identifies signatures of selection and adaptive trait associations. *Nature Genetics* 46, 1089–1096.

Ferrario, S., Busscher, J., Franken, J., Gerats, T., Vandenbussche, M., *et al.* (2004) Ectopic expression of the petunia MADS box gene *UNSHAVEN* accelerates flowering and confers leaf-like characteristics to floral organs in a dominant-negative manner. *Plant Cell* 16, 1490–1505.

Flachowsky, H., Peil, A., Sopanen, T., Elo, A., and Hanke, V. (2007) Overexpression of *BpMADS4* from silver birch (*Betula pendula* Roth.) induces early-flowering in apple (*Malus* × *domestica* Borkh.). *Plant Breeding* 126, 137–145.

Fritsche, S., Klocko, A.L., Boron, A., Brunner, A.M., and Thorlby, G. (2018) Strategies for engineering reproductive sterility in plantation forests. *Frontiers in Plant Science* 9, 1671.

Greenfield, L., Bjorn, M.J., Horn, G., Fong, D., Buck, G.A., *et al.* (1983) Nucleotide sequence of the structural gene for diphtheria toxin carried by corynebacteriophage β. *Proceedings of the National Academy of Science USA* 80, 6853–6857.

Hackett, R.M., Lawrence, M.J., and Franklin, C.H. (1992) A *Brassica* S-locus related gene promoter directs expression in both pollen and pistil of tobacco. *Plant Journal* 2, 613–617.

Hartley, R.W. (1988) Barnase and barstar: expression of its cloned inhibitor permits expression of a cloned ribonuclease. *Journal of Molecular Biology* 202, 913–915.

Hoenicka, H., Nowitzki, O., Hanelt, D., and Fladung, M. (2008) Heterologous overexpression of the birch *FRUITFULL*-like MADS-box gene *BpMADS4* prevents normal senescence and winter dormancy in *Populus tremula* L. *Planta* 227, 1001–1011.

Hoenicka, H., Lehnhardt, D., Briones, V., Nilsson, O., and Fladung, M. (2016) Low temperatures are required to induce the development of fertile flowers in transgenic male and female early flowering poplar (*Populus tremula* L.). *Tree Physiology* 36, 667–677.

Hsu, C.-Y., Liu, Y., Luthe, D.S., and Yuceer, C. (2006) Poplar *FT2* shortens the juvenile phase and promotes seasonal flowering. *Plant Cell* 18, 1846–1861.

Hsu, C.-Y., Adams, J.P., Kim, H., No, K., Ma, C., *et al.* (2011) *FLOWERING LOCUS T* duplication coordinates reproductive and vegetative growth in perennial poplar. *Proceedings of the National Academy of Science USA* 108, 10756–10761.

Hughes, C.E. (1994) Risks of species introductions in tropical forestry. *Commonwealth Forestry Review* 73, 243–252.

Irish, V.F. and Litt, A. (2005) Flower development and evolution: gene duplication, diversification and redeployment. *Current Opinion in Genetics and Development* 15, 454–460.

Jansson, S. and Douglas, C.J. (2007) *Populus*: a model system for plant biology. *Annual Review of Plant Biology* 58, 435–458.

Jeon, J.-S., Jang, S., Lee, S., Nam, J., Kim, C., *et al.* (2000) *leafy hull sterile1* is a homeotic mutation in a rice MADS box gene affecting rice flower development. *Plant Cell* 12, 871–884.

Klocko, A.L., Brunner, A.M., Huang, J., Meilan, R., Lu, H., *et al.* (2016) Genetic containment of forest trees by RNAi suppression of *LEAFY*. *Nature Biotechnology* 34, 918–922.

Koltunow, A.M., Truettner, J., Cox, K.H., Wallroth, M., and Goldberg, R.B. (1990) Different temporal and spatial gene expression patterns occur during anther development. *Plant Cell* 2, 1201–1224.

Lännenpää, M., Hassinen, M., Ranki, A., Hölttä-Vuori, M., Lemmetyinen, J., *et al.* (2005) Prevention of flower development in birch and other plants using a *BpFULL1::BARNASE* construct. *Plant Cell Reports* 24, 69–78.

Lemmetyinen, J. and Sopanen, T. (2004) Modification of flowering in forest trees. In: Kumar, S. and Fladung, M. (eds.) *Genetics and Breeding of Forest Trees*. Molecular Haworth Press, New York, pp. 263–291.

Lemmetyinen, J., Hassinen, M., Elo, A., Porali, I., Keinonen, K., *et al.* (2004) Functional characterization of *SEPALLATA3* and *AGAMOUS* orthologues in silver birch. *Physiologia Plantarum* 121, 149–162.

Li, Y. (2012) Gene deletor: a new tool to address gene flow and food safety concerns over transgenic crop plants. *Frontiers in Biology* 7, 557–565.

Longman, K.A. and Wareing, P.F. (1959) Early induction of flowering in birch seedlings. *Nature* 184, 2037–2038.

Lu, H., Klocko, A.L., Brunner, A.M., Ma, C., Magnuson, A.C., *et al.* (2018) RNAi suppression of *AGAMOUS* and *SEEDSTICK* alters floral organ identity and impairs floral organ determinacy, ovule differentiation, and seed-hair development in *Populus*. *New Phytologist* 222, 923–937.

Luo, K., Duan, H., Zhao, D., Zheng, X., Deng, W., *et al.* (2007) 'GM-gene-deletor': fused *loxP-FRT* recognition sequences dramatically improve the efficiency of FLP or CRE recombinase on transgene excision from pollen and seed of tobacco plants. *Plant Biotechnology Journal* 5, 263–274.

Meilan, R. (2006) Challenges to commercial use of transgenic plants. *Journal of Crop Improvement* 18, 433–450.

Meilan, R., Brunner, A., Skinner, J., and Strauss, S. (2001) Modification of flowering in transgenic trees. In: Komamine, A. and Morohoshi, N. (eds.) *Molecular Breeding of Woody Plants*. Progress in Biotechnology series. Elsevier Science BV, Amsterdam, pp. 247–256.

Meilan, R., Sabatti, M., Ma, C., and Kuzminsky, E. (2004) An early-flowering genotype of *Populus*. *Journal of Plant Biology* 47, 52–56.

Mohamed, R. (2006) Expression and function of *Populus* homologs to *TERMINAL FLOWER 1* genes: roles in onset of flowering and shoot phenology. PhD thesis, Oregon State University, Corvallis, Oregon.

Nilsson, O., Lee, I., Blazquez, M.A., and Weigel, D. (1998) Flowering-time genes modulate the response to *LEAFY* activity. *Genetics* 150, 403–410.

Parmentier-Line, C.M. and Coleman, G.D. (2016) Constitutive expression of the Poplar FD-like basic leucine zipper transcription factor alters growth and bud development. *Plant Biotechnology Journal* 14, 260–270.

Richardson, D.M. (1998) Forestry trees as invasive aliens. *Conservation Biology* 12, 18–26.

Rinne, P.L., Welling, A., Vahala, J., Ripel, L., Ruonala, R., *et al.* (2011) Chilling of dormant buds hyper-induces *FLOWERING LOCUS T* and recruits GA-inducible 1,3-β-glucanases to reopen signal conduits and release dormancy in *Populus*. *Plant Cell* 23, 130–146.

Rottmann, W.H., Meilan, R., Sheppard, L.A., Brunner, A.M., Skinner, J.S., *et al.* (2000) Diverse effects of overexpression of *LEAFY* and *PTLF*, a poplar (*Populus*) homolog of *LEAFY/FLORICAULA*, in transgenic poplar and *Arabidopsis*. *Plant Journal* 22, 235–245.

Searle, I. and Coupland, G. (2004) Induction of flowering by seasonal changes in photoperiod. *EMBO Journal* 23, 1217–1222.

Sheppard, L.A., Brunner, A.M., Krutovskii, K.V., Rottmann, W.H., Skinner, J.S., *et al.* (2000) A *DEFICIENS* homolog from the dioecious tree *Populus trichocarpa* is expressed in both female and male floral meristems of its two-whorled, unisexual flowers. *Plant Physiology* 124, 627–639.

Skinner, J.S., Meilan, R., Brunner, A.M., and Strauss, S.H. (2000) Options for genetic engineering of floral sterility in forest trees. In: Jain, S.M. and Minocha, S.C. (eds.) *Molecular Biology of Woody Plants*. Kluwer Academic Publishers, Dordrecht, The Netherlands, pp. 135–153.

Skinner, J.S., Meilan, R., Ma, C.P., and Strauss, S.H. (2003) The *Populus* PTD promoter imparts floral-predominant expression and enables high levels of floral-organ ablation in *Populus, Nicotiana* and *Arabidopsis*. *Molecular Breeding* 12, 119–132.

Stein, W.I. and Owston, P.W. (2008) Douglas-fir. In: Bonner, F.T. and Karrfalt, R.P. (eds.) *The Woody Plant Seed Manual*. USDA-Forest Service Agricultural Handbook 727. US Department of Agriculture, Forest Service, Washington, DC, pp. 891–906.

Tappeiner, J.C. (1969) Effect of cone production on branch, needle and xylem ring growth of Sierra Nevada Douglas-fir. *Forest Science* 15, 171–174.

Teich, A.H. (1975) Growth reduction due to cone crops on precocious white spruce provenances. *Environment Canada Bi-monthly Research Notes* 31, 6.

Tennant, P.F., Gonsalves, C., Ling, K.S., Fitch, M., Manshardt, R., *et al.* (1994) Differential protection against papaya ringspot virus isolates in coat protein gene transgenic papaya and classically cross-protected papaya. *Phytopathology* 84, 1359–1366.

Tylewicz, S., Tsuji, H., Miskolczi, P., Petterle, A., Azeez, A., *et al.* (2015) Dual role of tree florigen activation complex component *FD* in photoperiodic growth control and adaptive response pathways. *Proceedings of the National Academy of Science USA* 112, 3140–3145.

Viherä-Aarnio, A. and Ryynänen, L. (1994) Seed production of micropropagated plants, grafts and seedlings of birch in a seed orchard. *Silva Fennica* 28, 257–263.

Wang, H., Wu, H.M., and Cheng, A.Y. (1993) Development and pollination regulated accumulation and glycosylation of a stylar transmitting tissue-specific proline-rich protein. *Plant Cell* 5, 1639–1650.

Watson, J.M., Fusaro, A.F., Wang, M., and Waterhouse, P.M. (2005) RNA silencing platforms in plants. *FEBS Letters* 579, 5982–5987.

Wei, H., Meilan, R., Brunner, A.M., Skinner, J.S., Ma, C., *et al.* (2007) Field trial detects incomplete *barstar* attenuation of vegetative cytotoxicity in *Populus* trees containing a poplar *LEAFY* promoter::*barnase* sterility transgene. *Molecular Breeding* 19, 69–85.

Weigel, D. and Nilsson, O. (1995) A developmental switch sufficient for flowering initiation in diverse plants. *Nature* 12, 495–500.

Wullschleger, S.D., Jansson, S., and Taylor, G. (2002) Genomics and forest biology: *Populus* emerges as the perennial favorite. *Plant Cell* 14, 2651–2655.

Yamagishi, N., Li, C., and Yoshikawa, N. (2016) Promotion of flowering by apple latent spherical virus vector and virus elimination at high temperature allow accelerated breeding of apple and pear. *Frontiers in Plant Science* 7, 171.

Ye, X., Busov, V., Zhao, N., Meilan, R., McDonnell, L.M., *et al.* (2011) Transgenic *Populus* trees for forest products, bioenergy, and functional genomics. *Critical Reviews in Plant Sciences* 30, 415–434.

8 Engineering for Bioenergy

Introduction

Photosynthesis is the process by which radiant energy from the sun is transformed into stored chemical energy. Energy stored in the products of photosynthesis can be liberated directly through oxidation. Alternatively, chemical constituents of plant biomass can be converted to various types of liquid fuel and biomaterials. In this chapter, we will focus on the former.

Some traits desired for most, if not all, bioenergy crops include: (i) rapid growth and biomass accumulation; (ii) flowering control, to avoid diversion of photosynthate away from vegetative organs, and for transgene containment; (iii) ease of harvesting (e.g. growth distributed across several stems rather than one central stem, so existing farm equipment may be used, with minor modification, to harvest the biomass); (iv) vegetative propagation (i.e. production of rooted plants from cuttings); and (v) greater conversion efficiency through the alteration of plant cell-wall chemical composition. Many of these same characteristics are sought in trees grown for other purposes. The first two topics were discussed in Chapter 7 (this volume) and the remainder will be discussed here.

The two primary criteria used to evaluate the potential of plant species as feedstocks for biofuel production are its growth rate (i.e. the amount of biomass it can produce annually) and conversion efficiency (i.e. the amount of fuel that can be produced from a defined amount of biomass). Ethanol is the main liquid biofuel being utilized in the USA today and is derived largely from maize (*Zea mays*). The starch in the kernels is extracted and enzymatically hydrolyzed to simple sugars (glucose units), which are fed to yeast (*Saccharomyces cerevisiae*) for fermentation. In the USA, maize is used extensively as a food additive (e.g. high-fructose corn syrup) and feed for livestock (cattle and pigs) and poultry (chicken and turkeys). Thus, diverting a significant portion of the maize crop to fuel production negatively impacts and reverberates throughout the economy and food-supply chain. In addition, the practice of using maize as a feedstock for producing bioethanol is not sustainable—ecologically, energetically, or economically (Patzek, 2004).

There are advocates for not only using the kernels for bioethanol production but also using maize stover—the residual plant material left in the field after harvesting the ears, including stems, leaves, and roots—as a cellulosic feedstock. One distinguishing difference between animal and plant cells is that the latter possess a cell wall, which is a rich source of polysaccharides, including cellulose and hemicelluloses (see below). These complex carbohydrates can be hydrolyzed, liberating simple sugars, which can be used to produce bioethanol, other liquid fuels, and industrial reagents.

There are problems associated with the use of maize stover as a feedstock. First, it needs to be collected in the fall, when soils are most susceptible to erosion and compaction. Second, the nutrients that would be returned to the soil via biodegradation, if left on the field, are removed from the site. Third, transporting stover bales is not very efficient because they are bulky and lack density. Finally, stover storage is problematic because of the space that is required and the instability of the material. Additionally, field-collected stover is heavily contaminated with microbial spores that germinate and utilize the stored plant biomass as a carbon source.

To optimize the production and supply stability of biofuels over the greatest area, a suite of biomass crops will, therefore, be required, and this will likely need to occur with geographic specificity (not all plants grow on all landscapes). Examples include temperate-zone perennial monocots, such as switchgrass (*Panicum virgatum*) and species within the genus *Miscanthus*, and woody perennials, including eucalypts (*Eucalyptus* spp.) and poplars (*Populus* spp.). This chapter will focus on poplars because they have been the object of a plethora of genomic and biotechnological research.

Poplars offer several advantages over traditional row crops. For example: (i) poplars redistribute minerals and nutrients on an annual basis, reducing the need for supplemental annual fertilizer applications; (ii) they are not harvested every year (the rotation length is dependent on the planting density), reducing the amount of annual soil compaction and disturbance; (iii) poplars create structural and ecological diversity within and among planted stands; and (iv) there is evidence that poplars harbor a nitrogen-fixing endophyte (Doty *et al.*, 2005), further reducing the need for fertilizer. In addition, there are extant genotypes adapted for growth across a wide geographical range. Commercial plantations of poplar have yielded 10 oven-dry tons (ODT)/ha/year in Minnesota and upwards of 20 ODT/ha/year in the north-western USA (Volk *et al.*, 2011; Jeong, 2018), and poplar offers harvesting and handling options that increase a landowner's choices and profitability (e.g. if feedstock prices are low, *Populus* fibers can be "stored on the stump"). In addition, trees can be harvested throughout the year, eliminating the need for long-term storage facilities.

Despite its rapid growth and adaptability to a wide range of conditions, the utility of *Populus* as a cellulosic feedstock for bioenergy production is limited by the chemical composition of its cell walls. This is also true of biomass from other plants used as cellulosic feedstocks. In order to realize the economic and environmental benefits of these energy crops and convert their constituents to liquid biofuel, tools will need to be developed to effectively and efficiently deconstruct the recalcitrant secondary walls of their cells (Mansfield, 2009). In addition, methods are

needed to effectively extract compounds, including lignin and pectin, deposited in the middle lamella that anchors cells to one another. This must also be done to liberate fibers that are used for making paper. To increase the utility of wood used for a variety of purposes, transgenesis has been used to alter its chemical and physical properties, including the content and composition of the walls of cells comprising it.

The three primary constituents of the plant cell wall are cellulose, hemicellulose, and lignin. As with starch, cellulose is a linear array of glucose molecules; the difference is in the type of bond used to connect adjoining glucose residues. For starch, it is an α-1,4 glycosidic linkage, while for cellulose it is a β-1,4 linkage (Fig. 8.1). The plant combines long strands of cellulose together to form microfibril bundles (Figs 8.2 and 8.3), and hemicellulose tethers are intercalated among the microfibril bundles. Hemicellulose is comprised of a mixture of five- and six-carbon sugars, the amounts and types of which vary by species. Xylose is the principle sugar in poplar hemicelluose. Lignin forms a matrix that anchors the cellulose and hemicellulose components (Fig. 8.4) (Jones *et al.*, 2001). In grasses, the linkages between lignin and the other cell-wall constituents must be broken in order to expose the underlying polysaccharides. In trees, there is no evidence for direct linkages between lignin and other cell-wall components, but they do form during the extraction process.

To maximize biofuel production, it is desirable for bioenergy crops to have cell walls with the highest possible polysaccharide content. In addition, these sugars need to be accessible so they can be hydrolyzed to monosaccharides, which subsequently can be utilized during fermentation. Pretreatment methods are used to solubilize and separate the four primary constituents of plant biomass (i.e. cellulose, hemicellulose, lignin, and extractives) to make the residual more susceptible to downstream chemical or biological treatments. Following pretreatment, hexoses and pentoses are liberated from the complex carbohydrates in the cell wall via hydrolysis, which is also known as saccharification. In general, there are two types of hydrolytic

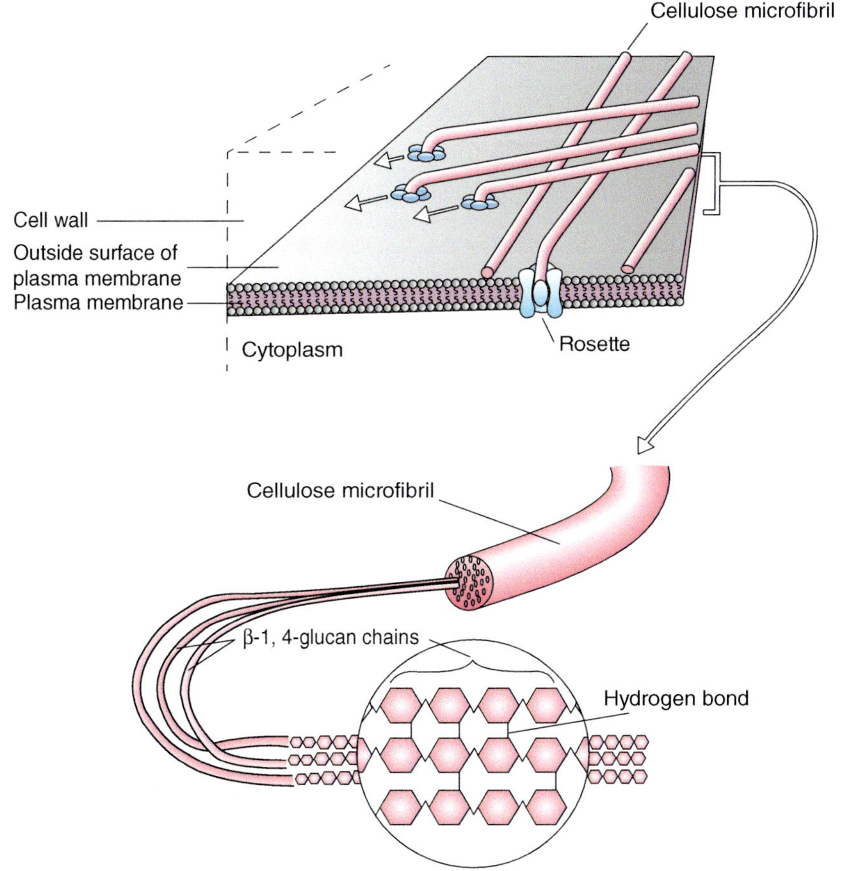

Fig. 8.1. Structures of cellulose and the linear form of starch, amylose.

Fig. 8.2. Synthesis of microfibril bundles in the plant cell wall. (From Cutler and Somerville, 1997.)

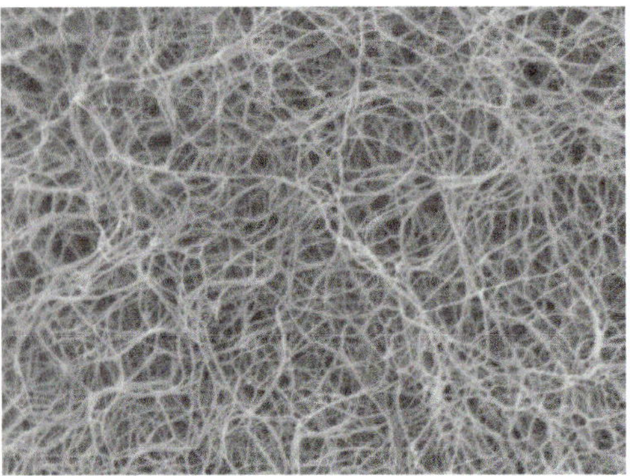

Fig. 8.3. Microfibril bundles in the wall of a beech parenchyma cell. Cellulose is synthesized by cellulose synthase or rosette complexes in the cell membrane. As the microfibrils are synthesized extracellularly, they push up against neighboring cells. These neighboring cells cannot move easily so the rosette complex is instead pushed around the cell through the fluid phospholipid membrane, which eventually results in the cell becoming wrapped in a microfibril layer. (Figure provided by Hiroyuki Yano, Laboratory of Active Bio-based Materials, Research Institute for Sustainable Humanosphere, Kyoto University, Japan.)

processes: chemical (e.g. dilute or concentrated acid, ammonia) and enzymatic. With respect to the former, pretreatment and hydrolysis can, in some cases, be carried out in a single step (Demirbas, 2005). Most pretreatment methods are not particularly effective at removing lignin, and they can affect the yield and integrity of the remaining polysaccharides. Researchers have altered the cell-wall structure in an attempt to improve saccharification rates.

Lignin Modification

Transgenesis

The term lignin is used to describe a polymer arising from the random coupling of aromatic compounds known as 4-hydroxyphenylpropanoids. Hydroxycinnamyl alcohols (i.e. monolignols), which are the principle building blocks of lignin, are products of the phenylpropanoid pathway. These include coniferyl, sinapyl, and *p*-coumaryl alcohols. Phenylalanine (Phe) is the first substrate in this pathway. Initially, Phe is deaminated by phenylalanine ammonia lyase (PAL), and is subsequently hydroxylated and methylated by a series of enzymes, ending in the production of the hydroxycinnamyl alcohols. When these monolignols are integrated into lignin, they are called guaiacyl (G), syringyl (S), and *p*-hydroxyphenyl (H) subunits, respectively. Lignins produced by angiosperms consist mainly of S and G subunits, with traces of H; in contrast, gymnosperm lignins are mainly made up of G, with low levels of H. The monomeric subunits are interconnected by various types of bonds. The relative abundance of these linkages depends on the monomeric composition of the lignin, and the ease with which the various bonds can be cleaved chemically varies. For example, G-rich lignins contain a higher percentage of resistant linkages than those with an abundance of S (Boerjan *et al.*, 2003; Vanholme *et al.*, 2010).

Other enzymes involved in the phenylpropanoid pathway include: cinnamic acid 4-hydroxylase (C4H), 4-(hydroxy)cinnamoyl CoA ligase (4CL), hydroxycinnamoyl-CoA:shikimate hydroxycinnamoyl transferase, p-coumaroylshikimate 3′-hydroxylase (C3′H), caffeoyl shikimate esterase (CSE), caffeoyl CoA O-methyltransferase (CCoAOMT), (hydroxy)cinnamoyl CoA reductase (CCR), ferulic acid 5-hydroxylase (F5H), caffeic acid O-methyltransferase (COMT), and (hydroxy)cinnamyl alcohol dehydrogenase (CAD) (Fig. 8.5) (Humphreys and Chapple, 2002; Boerjan *et al.*, 2003; Vanholme *et al.*, 2013). Following their synthesis, the monolignols are transported to the cell wall where they are polymerized into lignin. Although the mechanisms by which transport and polymerization occur are not well understood, it is known that the polymerization is not a template-driven process; it is the result of random coupling. Plants are able to utilize intermediates in the phenylpropanoid pathway for lignin assembly, but often do not because of the way in which the pathway is regulated.

Lignin is known to negatively impact biofuel production from cellulosic feedstocks in several ways. First and foremost, lignin blocks access of hydrolytic enzymes to the cell-wall

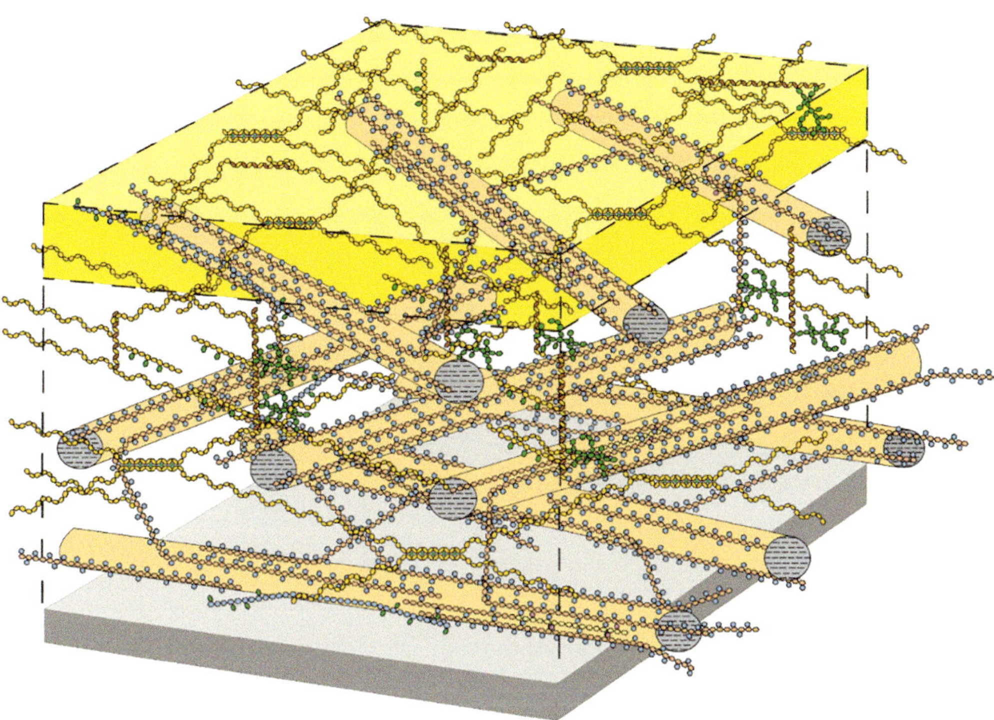

Fig. 8.4. Schematic representation of an idealized plant cell wall. The tan, tubular structures represent microfibril bundles, the yellow layer represents lignin, and the chains of small, circular structures represent hemicellulose and pectins. (After Carpita and McCann, 2008.)

polysaccharides; hence the need for pretreatment. In addition to simple sugars, a variety of compounds are generated during pretreatment of cellulosic feedstocks that can interfere with hydrolysis of the underlying carbohydrates. These inhibitors are divided in three chief classes: weak acids, phenolic compounds, and furan derivatives (Palmqvist and Hahn-Hägerdal, 2000; Keating *et al.*, 2006). Lignin can also adsorb hydrolytic enzymes that are involved in saccharification, and some lignin degradation products strongly inhibit fermentation (Olsson and Hahn-Hägerdal, 1996; Cullis and Mansfield, 2010).

Deriving sugars from plant cell walls (e.g. through acid hydrolysis or cellulase treatment) for fermentation is one of the most costly steps in bioethanol production. Modifying feedstocks to make them more responsive to pretreatments (i.e. to improve the access of cellulases to cellulose for its degradation) will reduce the overall cost of biofuel production. Pilate *et al.* (2002) demonstrated that downregulating the *CAD* gene in poplar led to enhanced lignin solubility in an alkaline medium. Similarly, Boudet *et al.* (2003) reported that cell-wall polysaccharides are more easily hydrolyzed when *CCR* is downregulated. Using a second-generation *Populus* cross, Davison *et al.* (2006) showed that a lower lignin content and lower S:G ratio led to higher xylose release from the hemicellulose fraction of the cell wall following hydrolysis with dilute acid.

Reducing lignin content in the cell wall is another way of improving the utility of cellulosic feedstocks, and downregulating *4CL* has been identified as a way to accomplish this objective. Lignin content of *Pinus radiata* was reduced to 36–50% of its wild-type (WT) levels by downregulating *4CL* expression using antisense suppression; however, these transgenic trees also exhibited a yield penalty (Wagner *et al.*, 2009). Voelker *et al.* (2010) examined wood chemistry and tree

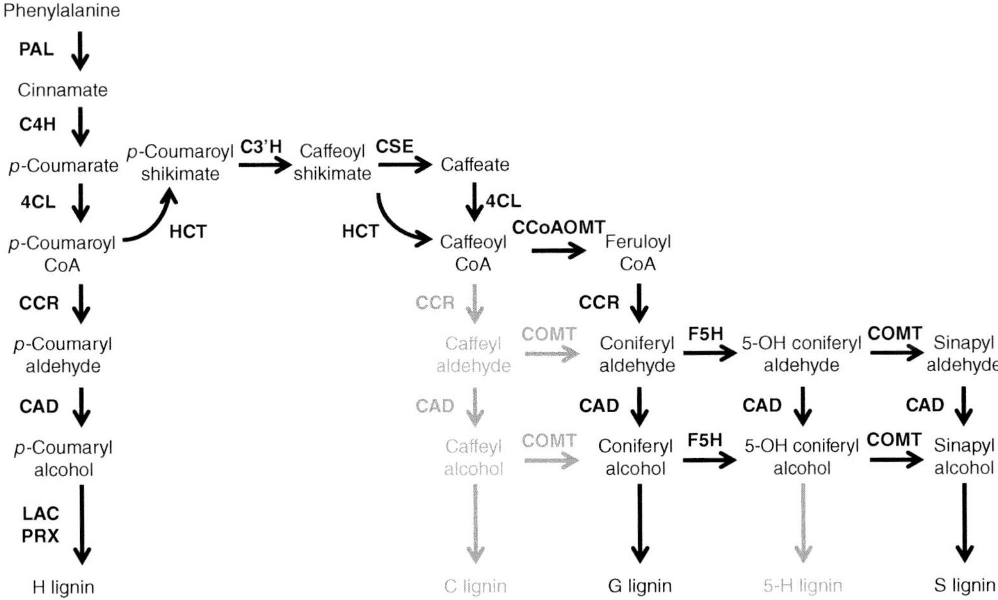

Fig. 8.5. A simplified pathway illustrating the enzymes and metabolites involved in lignin biosynthesis. Reactions in grey indicate that these steps are of only minor importance in wild-type plants. 4CL, 4-coumarate CoA ligase; C3′H, *p*-coumaroyl shikimate 3′-hydroxylase; C4H, cinnamate 4-hydroxylase; CAD, cinnamyl alcohol dehydrogenase; CCoAOMT, caffeoyl CoA *O*-methyltransferase; CCR, cinnamoyl CoA reductase;COMT, caffeic acid *O*-methyltransferase; CSE, caffeoyl shikimate esterase; F5H, ferulate 5-hydroxylase; HCT, hydroxycinnamoyl-CoA:shikimate hydroxycinnamoyl transferase; LAC, laccase; PAL, phenylalanine ammonia lyase; PRX, peroxidase. (Figure provided by Clint Chapple, Department of Biochemistry, Purdue University, West Lafayette, Indiana.)

productivity in field-grown hybrid aspen (*Populas tremula* × *P. alba*) in which *4CL* had been downregulated using an antisense approach. They observed that a sharp reduction in lignin content was correlated with *4CL* RNA expression. Moreover, with only small reductions in lignin (approximately 10%), reduced growth was apparent. They concluded that there are limits to reducing lignin content; below a certain threshold, alterations in plant metabolism and wood chemistry negatively affect biomass production and ethanol yield (Voelker *et al.*, 2010).

Using RNA interference (RNAi) to downregulate *C3′H* in a hybrid poplar (*Populus grandidentata* × *P. alba*) led to a nearly 55% reduction in lignin content in the most strongly suppressed line (Coleman *et al.*, 2008a). Relative to WT trees, the reduced-lignin trees had smaller stems, less root biomass, and altered leaf morphology. Inhibiting cell-wall lignification also led to reduced hydraulic conductivity

as a result of vascular collapse (Coleman *et al.*, 2008b). There is mounting evidence that lignin-modification-induced dwarfism (LMID) caused by downregulating *C3′H* is not simply due to mechanical failure of the plant cell wall but instead is the result of a biological process that can be mitigated by genetic manipulation. Suppressor genes important for LMID have recently been identified, and mutation of these genes has been shown to increase the growth of an *Arabidopsis thaliana* mutant that is defective in *C3′H* (X. Li, unpublished data).

Parallel studies have been performed in alfalfa (*Medicago sativa*). To evaluate the relationship between lignin content and composition and the rate of saccharification, Chen and Dixon (2007) evaluated lines of alfalfa transformed with antisense constructs that independently downregulated six different genes in the lignin biosynthetic pathway. The authors analyzed biomass digestibility following both acid pretreatment and enzymatic hydrolysis, and showed that the efficacy of both was proportional to lignin content.

The enzyme encoded by the *CCoAOMT* gene is bifunctional; it uses S-adenosyl-L-methionine as methyl source to catalyze the conversion of either caffeoyl CoA to feruloyl-CoA or 5-hydroxyferuloyl-CoA to sinapoyl-CoA. Antisense suppression of *CCoAOMT* led to a significant reduction in lignin content in *Populus* and, hence, a decrease in both the G and S subunits (Zhong *et al.*, 2000). In addition, the lignin that was produced was less condensed and had fewer cross-linkages. Using sense suppression, Meyermans *et al.* (2000) produced transgenic poplar with only 10% of the CCoAOMT protein levels found in the xylem of WT stems, leading to trees with 12% less lignin that concurrently displayed higher S:G ratios. In both studies, repression of *CCoAOMT* led to the incorporation of *p*-hydroxybenzoic acid into the cell wall (Meyermans *et al.*, 2000; Zhong *et al.*, 2000).

COMT is an enzyme that catalyzes the ortho-methylation of both caffeate to ferulate and 5-hydroxyferulate to sinapate, and may also be involved in methylation of other substrates. Expression of the *COMT* gene has been downregulated using both sense and antisense approaches (van Doorsselaere *et al.*, 1995; Tsai *et al.*, 1998; Lapierre *et al.*, 1999; Jouanin *et al.*, 2000). These manipulations have led to dramatic reductions in the S:G ratio, but antisense suppression of *COMT* did not result in significant changes in lignin content (van Doorsselaere *et al.*, 1995; Lapierre *et al.*, 1999; Tuskan *et al.*, 2006). The results of co-suppression studies have been inconsistent. In quaking aspen (*P. tremuloides*), no significant differences in lignin content were seen; however, the stems of their transgenic trees exhibited either a blotchy or complete red-brown discoloration (Tsai *et al.*, 1998). In contrast, sense suppression of *COMT* in hybrid poplar (*P. tremula* × *P. alba*) led to a 17% reduction in lignin content in 6-month-old trees (Jouanin *et al.*, 2000).

The expression product of the *CCR* gene catalyzes the reduction of hydroxycinnamoyl-CoA thioesters to their corresponding aldehydes, a reaction that is thought to be a control point in the phenylpropanoid pathway (Lacombe *et al.*, 1997). Leplé *et al.* (2007) downregulated *CCR* expression in *P. tremula* × *P. alba* using sense and antisense constructs, generating trees that displayed a 50% reduction in lignin content. The outer xylem of the transgenic plants was an orange-brown color, which was thought to be the result of the elevated levels of ferulic acid accumulating during lignification. Cohesion of the walls also was influenced, especially at sites that are typically richer in S subunits in WT poplar.

F5H is a cytochrome P450-dependent monooxygenase that catalyzes the hydroxylation of ferulic acid, coniferaldehyde, and coniferyl alcohol, leading to the synthesis of S monomers. To evaluate the regulatory role of F5H in plants that undergo secondary growth, Franke *et al.* (2000) expressed the *A. thaliana F5H* gene in *P. tremula* × *P. alba* under control of the *A. thaliana C4H* promoter. They reported a dramatic increase in S content (85 mol% in the highest expressing transgenic line versus 55 mol% in WT trees) but no change in lignin content. Further analysis of poplar lines containing the C4H::*F5H* construct revealed a significant increase in chemical (kraft) pulping efficiency of wood derived from plants grown in a greenhouse (Huntley *et al.*, 2003). A similar result was also reported when a sweetgum (*Liquidambar styraciflua*) F5H ortholog, controlled by a xylem-specific promoter (*Pt4CL1P*), was transformed into quaking aspen cells. Regenerated transgenic plants exhibited a 2.5-fold increase in the S:G ratio, but the lignin content remained unchanged (Li *et al.*, 2003).

More recently, Li *et al.* (2010) reported on the conversion efficiency for *A. thaliana* lines with altered S:G ratios. One was a *fah1-2* mutant that was defective in *F5H*, so that it did not synthesize S lignin; in the other, *F5H* was overexpressed. Following liquid hot-water pretreatment and enzymatic hydrolysis, a much greater glucose yield was achieved from the high-S plants than from either the WT or plants with G-rich lignin. Similarly, Mansfield *et al.* (2012) assessed the effect of steam-explosion pretreatment on saccharification rates for wood derived from one of the F5H-overexpressing poplar lines described by Franke *et al.* (2000). When subjected to enzymatic hydrolysis, following pretreatment, more of the available carbohydrate was liberated from the pretreated high-S (>93%) wood than from WT wood.

Depolymerization of lignin is restricted by the interunit carbon–carbon bonds present in native lignin, and by formation of these connections during lignin extraction. Shuai *et al.* (2016) reported that the addition of formaldehyde during pretreatment of biomass from high-S (*F5H*-overexpressing) poplar lines leads to a soluble lignin fraction that is readily converted to G and S monomers during the following hydrogenolysis.

CAD catalyzes the final step in monolignol biosynthesis. Baucher *et al.* (1996) used both antisense and co-suppression to downregulate *CAD* expression in transgenic *P. tremula* × *P. alba*. Several antisense and sense *CAD* transgenic plants showed nearly a 70% reduction in CAD activity. However, neither the lignin content nor its monomeric composition was changed significantly. Subsequently, Lapierre *et al.* (1999) evaluated 2-year-old *Populus* that had been engineered to suppress *CAD* via an antisense approach. The lignin content of the transgenic plants was only slightly lower than that of the controls. In addition, downregulation of *CAD* did not change the frequency of G subunits, but the lignin had a greater number of free phenolic groups, and the outer xylem exhibited a red-brown coloration. In the most suppressed line, the enhancement in free phenolic units facilitated lignin solubilization and disintegration during kraft pulping (Lapierre *et al.*, 1999).

CSE is the most recently discovered enzyme in the phenylpropanoid pathway (Vanholme *et al.*, 2013). *Arabidopsis thaliana* plants containing a mutant version of the gene encoding this enzyme deposited less lignin than their WT counterparts, and their lignin had a higher level of H subunits. The lignin biosynthetic pathway intermediate caffeoyl shikimate accumulated to a greater extent in *cse* mutants than in WT plants, suggesting that it is a substrate for CSE. Saccharification rates for the *cse* mutants was up to four times as great than was seen for the WT, in the absence of pretreatment.

Plants can readily incorporate novel monolignols into the lignin polymer, exemplifying the plasticity of its biosynthesis. The Zip-Lignin™ strategy, which involves the introduction of esters into the lignin backbone, illustrates the potential for using monolignol acyltransferases to engineer designer lignins (Mottiar *et al.*, 2016). Wilkerson *et al.* (2014) were the first to demonstrate that zip-lignins could be used to diminish lignin recalcitrance. The authors expressed a feruloyl-CoA:monolignol transferase gene from female ginseng (*Angelica sinensis*) under the control of a xylem-specific promoter in hybrid poplar, resulting in unique, chemically labile bonds being introduced into the lignin backbone. The lignin in these transgenic poplars was more easily digested, following mild alkaline pretreatment, than that in control plants. This novel approach reveals the potential for tailoring lignins to increase the utility of bioenergy crops as a feedstocks for biofuel production, without perturbing normal plant growth and development.

Naturally occurring mutants

Fahrenkrog *et al.* (2016) have genotyped and phenotyped around 600 unrelated genotypes of *Populus deltoides*, collected from 35 river systems in 15 states in central, southern, and eastern USA. Genotyping was carried out by targeted resequencing of over 18,000 genes, for which approximately 1.4 million single-nucleotide polymorphisms and almost 200,000 insertion or deletion variants were detected and tested for association with wood-quality traits. Genetic association analysis identified a few large-effect polymorphisms that impact wood chemistry, including a major locus that affects the lignin S:G ratio. A change in the coding sequence for the *F5H* gene led to an alteration in the amino acid sequence of the enzyme it encodes, resulting in elevated levels of S in the mutant genotype. Similarly, Studer *et al.* (2011) collected 1,100 samples from a natural population of undomesticated *P. trichocarpa* trees and quantified them for, among other traits, S:G ratio. The sampled trees had S:G ratios ranging from 1.0 to 3.0.

These examples demonstrate the power of using genomics, metabolomics, and bioinformatics in conjunction with genetic engineering. Through the characterization of genetically engineered plants, it has been shown that high-S poplar lines have higher conversion efficiencies and are, therefore, more desirable as a biofuel feedstock. Currently, this naturally occurring, high-S mutant can be vegetatively propagated and grown in plantations as a dedicated energy crop, unlike the genetically engineered high-S lines, which cannot be deployed commercially because of concerns about transgene escape (Chapter 7, this volume).

Transcriptional control of lignin biosynthesis

Monolignol biosynthesis is largely regulated transcriptionally, and considerable effort has been devoted to understanding this process so that it can be deliberately manipulated. The majority of the lignin biosynthetic genes contain the same *cis* element, designated AC, which is needed for their expression (Zhong and Ye, 2009). Much progress has been made in identifying the transcription factors that bind to these AC elements and coordinate the regulation of the phenylpropanoid pathway. In addition to the large family of MYB transcription factors that activate monolignol biosynthetic genes, members of the NAC family of transcription factors serve as master regulators of cell-wall formation (Mottiar *et al.*, 2016). Moreover, subunits of a complex transcriptional co-regulator known as Mediator have been identified as repressors of monolignol biosynthesis (Bonawitz *et al.*, 2014). For more information about transcriptional control of lignin biosynthesis, see Brunner and Beers (2011) and Zhong and Ye (2015).

Modifying expression of more than one gene

In contrast to these single-gene studies, Li *et al.* (2003) used *Agrobacterium tumefaciens* to simultaneously transfer antisense *4CL* and sense *F5H* constructs into quaking aspen. The effect was additive; transgenic plants had 52% less lignin and a 64% higher S:G ratio. This study suggests that a multi-gene, co-transformation approach may be effective for modifying lignin using genetic engineering.

Corngrass1 (*Cg1*) is a gene that encodes a miRNA (Chapter 6, this volume) that is a member of the *miR156* family. It was identified when a retrotransposon inserted into the *Cg1* promoter region, leading to its overexpression in maize. The *cg1* mutant exhibited adventitious root formation, facilitating vegetative propagation, pronounced shoot proliferation, and partial sterility, due to perturbations in the development of the floral meristem (Chuck *et al.*, 2007). Overexpression of *miR156* in *A. thaliana* caused a loss of apical dominance, resulting in numerous branches arising from axillary buds, and delayed flowering (Schwab *et al.*, 2005; Wu and Poethig, 2006). Tobacco (*Nicotiana tabacum*) and *A. thaliana* plants in which *Cg1* was expressed constitutively grew faster, produced multiple axillary branches, had lower lignin content, and were either sterile or exhibited delayed flowering (G. Chuck, 2011, Albany, California, personal communication). Rubinelli *et al.* (2012) overexpressed *Cg1* in *P. tremula* × *P. alba* under the control of the CaMV 35S promoter. Transgenic plants exhibited up to a 30% reduction in stem lignin content compared with that of WT plants. In addition, the S:G ratio was lower in 35S::*Cg1* lines than in WT poplar or a control transgenic line with a low abundance of *Cg1* transcript.

Carbohydrate Component of the Cell Wall

Cellulose

In contrast to lignin biosynthesis, which is reasonably well understood, much is unknown about cellulose production in woody plants (Joshi and Mansfield, 2007). There are only a few reports on genetic manipulation of cell-wall carbohydrates in trees. It has been shown that UDP-glucose pyrophosphorylase, an enzyme involved in the generation of UDP-glucose, is required for the biosynthesis of cellulose. Overexpression of an *Acetobacter xylinum* UDP-glucose pyrophosphorylase in *Populus* resulted in altered cell-wall chemistry, including increases in soluble sugar, starch, and cellulose content. In addition, the transgenic trees had slower growth, lower lignin content, and an increase in the S:G ratio (Coleman *et al.*, 2007). In a subsequent study, Coleman *et al.* (2009) showed that overexpression of a sucrose synthase gene (*SuSy*) from upland cotton (*Gossypium hirsutum*) in poplar also led to additional carbon being partitioned into cellulose, but not at the expense of growth. The authors concluded that SuSy is a key regulator of sink strength in poplar, and that cellulose synthesis and SuSy activity are carefully coordinated during formation of the secondary cell wall in poplar.

Park *et al.* (2009) differentially expressed an *A. thaliana* sucrose phosphate synthase (SPS) gene in hybrid poplar (*P. alba* × *P. grandidentata*) in an effort to manipulate the intracellular pools of sucrose. This was done to evaluate the role of SPS in controlling plant growth, phenology, and fiber development. There were no differences in tree height or stem diameter between WT and transgenic trees; however, there were significant differences in fiber length. In addition, elevated levels of intracellular sucrose in both leaf and stem tissue was associated with a delay in the onset of senescence in the fall, and with early vegetative bud flush in spring of the following year.

Sucrose 6-phosphate phosphatase (SPP) catalyzes the final reaction in the sucrose biosynthetic pathway, in which sucrose 6-phosphate, produced through the action of SPS, is irreversibly hydrolyzed to sucrose (Lunn and Rees, 1990). Maloney *et al.* (2015) reported evidence for an association between SPS and SPP, and showed that their interaction affects carbohydrate pools and the partitioning of carbon into starch. Moreover, they demonstrated that a fusion construct between the genes that encode SPS and SPP promotes the growth of both transgenic *A. thaliana* and hybrid poplar (*P. alba* × *P. grandidentata*, genotype P39). If an SPS–SPP enzyme complex does indeed exist *in planta*, it may provide an additional level of regulation for sucrose biosynthesis, leading to the direct metabolic channeling of sucrose 6-phosphate between these two enzymes.

Unda *et al.* (2017) examined the effects of altering carbon allocation on cell-wall characteristics and plant development in hybrid poplar (P39) through constitutive expression of an *A. thaliana* galactinol synthase gene (*AtGolS3*). Overexpression of *AtGolS3* had clear effects on poplar development. Lines with the highest transgene expression were stunted and had cell walls that resembled those in tension wood, while lines with moderate expression levels appeared to grow normally and had only modest changes in the composition and structure of their cell walls. Stem cross-sections from their transgenic poplars revealed an increase in the number of vessels in the developing xylem, and the xylem had elevated starch content. In addition, the *AtGolS3*-overexpressing lines displayed higher cellulose and lower lignin content, an increase in cellulose crystallinity, and significantly altered hemicellulose composition. Moreover, transcriptomic analysis revealed that genes involved in sucrose degradation were significantly upregulated in the transgenic plants. The authors concluded that the expression of *GolS* and the product synthesized by the enzyme it encodes, galactinol, could serve as a molecular signal that triggers metabolic changes that affect cell-wall development and possibly the formation of reaction wood.

Hemicellulose

After lignin is extracted from a cellulosic feedstock, cell-wall carbohydrates must be hydrolyzed to monosaccharides before they can be fermented into ethanol (Gray *et al.*, 2006). Most commonly employed WT yeast strains can only ferment six-carbon sugars, while the five-carbon sugars in the hemicellulose fraction cannot be utilized for ethanol production. Using metabolic engineering, Toon *et al.* (1997) developed a stable recombinant strain of yeast that could simultaneously ferment both glucose and xylose. This derived strain of yeast, LNH-ST, contains in its chromosome genes from the yeast *Pichia stipitis* that encode an enzyme capable of catabolizing xylose. With this development, the carbohydrates in the cell wall can be utilized more fully for producing biofuel.

By-products

Current methods for converting cellulosic feedstocks into ethanol and other liquid fuels rely principally on the use of the carbohydrate fraction of biomass. The lignin component (15–30% by weight) is underutilized and is often burned to produce electricity. A long-sought goal has been to develop processes for converting lignin to other high-energy products (Zakzeski *et al.*, 2010). Parsell *et al.* (2015) identified a bi-metallic (Zn/Pd) catalyst that permits the direct conversion of lignin from biomass into two value-added products, leaving the polysaccharide components of the cell wall largely intact for further transformation. Hydrodeoxygenation cleaves the lignin polymer C–O linkages, yielding the aromatic products dihydroeugenol and 2,6-dimethoxy-4-propylphenol, which are used in the fragrance and food-flavoring industries. Poplar that has been genetically engineered for enhanced S lignin content yields primarily a single product in this process. These aromatic products can be converted to propylcyclohexane, which can be upgraded easily to the high-octane propylbenzene. The biomass residue was shown to be composed of glucan and xylan. When this residue underwent fast pyrolysis, the results were comparable to those obtained with pure crystalline cellulose.

The addition of acyl substituents to the hydroxyl groups of monolignols occurs commonly in many plant species. For example, monocot lignins are adorned with *p*-coumarate moieties. An acyltransferase that catalyzes the synthesis of these conjugates has been identified in monocots (Withers *et al.*, 2012), but until recently the effect of conjugating monolignols with *p*-coumarate on the lignification process and on plant growth and development had not yet been evaluated in plants that do not normally incorporate them into their cell walls. Smith *et al.* (2015) transformed the rice *p-COUMAROYL-COENZYME A MONOLIGNOL TRANSFERASE* (*PMT*) gene, driven by the 35S promoter, into hybrid poplar (P39). Plants expressing the *PMT* transgene produced *p*-coumarate–monolignol conjugates without displacing other acylated monolignols, and without affecting usual monolignol production. *p*-Coumarates can be metabolized by a suite of microbes, resulting in the production of various high-value biomaterials (Pan *et al.*, 2008; Eudes *et al.*, 2014; Beckham *et al.*, 2016).

Side Effects of Cell-wall Modification

Susceptibility to pest attack

In addition to their involvement in plant growth and development, lignin and cellulose composition and content have an important function with respect to plant–environment interactions. Lignin influences cell-wall permeability, thus affecting the movement of water and solutes through the plant's vascular system; it also provides protection against pest attack (Sarkanen and Ludwig, 1971; Moerschbacher *et al.*, 1990). Very few studies have evaluated the performance of cell-wall-modified trees under field conditions. However, such information is essential before growth of transgenic trees in dedicated energy plantations will be feasible economically, and acceptable from a regulatory perspective. While cell-wall composition can be altered via transgenesis, it must be shown that these alterations do not have unintended consequences. This can only be accomplished through multi-year field trials.

In one such study, Buhl *et al.* (2017) evaluated field-grown, lignin-modified hybrid poplar (*P. tremula* × *P. alba*) over a 3-year period. Some of the transgenic trees had lower lignin content, while in others the monolignol composition was altered. The authors reasoned that the potential consequences of lignin modification included: (i) altered mechanical resistance because of structurally weakened lignin; (ii) altered chemical resistance traits due to resource trade-offs; and (iii) changes in arthropod communities in response to changes in herbivore resistance traits. Given this, they quantified mechanical and chemical resistance traits in the foliage of lignin-modified and WT trees. They demonstrated that the lignin modifications tested did not negatively affect herbivore resistance traits, arthropod abundance, or community composition.

Skyba *et al.* (2013) evaluated the susceptibility of wood with altered lignin composition to degradation by decay fungi. Wood samples from two transgenic poplar lines and a non-transgenic control were inoculated with six different species of wood-decaying fungi. The transgenic poplar harbored T-DNA containing the *A. thaliana F5H* gene, under the control of the *C4H* promoter, which led to elevated levels of S lignin (~94 mol%) relative to the WT (~65 mol%). Wood from the S-rich transgenic poplar lines exhibited greater resistance to decay by all of the fungi that were evaluated. The authors speculated that lignin composition and its distribution among cell types and within various cell layers are the key wood properties that affect its durability.

Interconnected pathways

Many transgenic plants that were engineered to modify lignin have exhibited additional and often unanticipated phenotypes. Qualitative and/or quantitative changes in one cell-wall constituent have often resulted in striking changes in others (Boudet *et al.*, 2003). For example, downregulation of *4CL* in *P. tremuloides* led to both a 45% decrease in lignin and a 15% increase in cellulose content (Hu *et al.*, 1999). Similarly, transgenic tobacco lines in which *CCR* expression was downregulated using antisense suppression not only displayed the expected decrease in lignin levels but also showed an increase in the content of cell-wall-associated xylose, glucose, and phenolic compounds, such as sinapic and ferulic acids (Chabannes *et al.*, 2001). Subsequently, Li *et al.* (2003) co-transformed antisense *4CL* and sense *CONIFERALDE-HYDE 5-HYDROXYLASE* (*CAld5H*) constructs into *P. tremuloides*. Co-expression of these transgenes led to plants having up to 52% less lignin, 64% higher S/G ratio, and 30% more cellulose.

Microfibril bundles are cross-linked by xyloglucans present in the hemicellulose component of the plant cell wall. Degradation of xyloglucan is thought to contribute to cell-wall loosening that must occur during cell expansion. Park *et al.* (2004) reported that increasing xyloglucanase activity in *Populus* leads to increased cellulose accumulation.

Leplé *et al.* (2007) analyzed the metabolome of poplar in which *CCR* had been downregulated. Maleate levels were 2.4–3.9 times higher in their transgenic lines than in WT trees. In addition, levels of fumaric and malic acids (Krebs cycle intermediates) increased 2–2.5-fold and 1.6–3.0-fold, respectively. The largest differential seen was in carbohydrate accumulation. Glucose, mannose, galactose, myo-inositol, raffinose, and melezitose were all 0.5–0.8-fold lower in the transgenic poplars than in the corresponding WT plants.

As stated above, Coleman *et al.* (2007) showed that increasing soluble sugar, starch, and cellulose content led to a reduction in growth and lignin content, and a change in the S:G ratio. In addition, when Coleman *et al.* (2008b) used RNAi to downregulate *C3'H* in *Populus*, they saw a striking decrease in lignin content and a change in its composition. The concentration of G subunits decreased, the S subunits were unchanged and the concentration of H subunits increased. In addition, a 13.5% increase in cell-wall carbohydrates—primarily xylose, glucose, and arabinose—was also observed in their RNAi lines.

The outer portion of the middle lamella and primary cell wall are, together, known as the compound middle lamella, which is rich in pectic substances, xyloglucan, and lignin. In contrast, the underlying secondary cell wall is: (i) thicker; (ii) comprised of abundant cellulose and hemicelluloses (e.g. xylans and mannans); and (iii) believed to contain less lignin (Biswal *et al.*, 2014). The multi-layered secondary wall makes up a majority of wood and has historically been thought to exert the most dominant influence on its characteristics. Surprisingly, hydrolysis of xyloglucan, a key component of hemicellulose in poplar, affected the properties of its wood (Park *et al.*, 2004) and made it easier to saccharify (Kaida *et al.*, 2009). In addition, Sexton *et al.* (2012)

reported that alleles of two pectin methyl esterase genes from *Eucalyptus*, *PME6* and *PME7*, are significantly associated with wood quality, inferring that the structure of pectin also influences wood properties.

The pectins found in wood are comprised chiefly of homogalacturonan (HG) and rhamnogalacturonan (RG), and are thought to be concentrated in the outermost layer of the compound middle lamella; however, they have not been shown to occur in the xylan-rich secondary walls (Mellerowicz and Gorshkova, 2012). The structure of the pectin network is determined, in part, by the activity of pectate lyases present in the cell wall (Fry, 2004). Plant pectate lyases, which are members of a family of polysaccharide lyases (PL1), cleave the α-1,4 glycosidic linkages between the galacturonic acid subunits of HG. In poplar, pectate lyase genes have been numbered sequentially as *PL1-1* to *-28* (Geisler-Lee *et al.*, 2006). Israelsson *et al.* (2003) showed that the *PL1-27* gene from hybrid aspen (*P. tremula* × *P. tremuloides*; Ptxt) was highly upregulated in a transgenic line that exhibited greater secondary growth as a result of overexpressing the gene encoding GA20 oxidase. Subsequently, Biswal *et al.* (2014) biochemically characterized the recombinant PtxtPL1-27 protein and analyzed the effects of its overexpression in hybrid aspen. Their results demonstrated that PtxtPL1-27 can enhance the extractability and saccharification of xylem cell-wall polysaccharides.

Rhamnogalacturonan-I (RG-I) is a pectic polymer consisting of alternating residues of rhamnose and galacturonic acid, which can be extensively substituted with arabinan and galactan side-chains (Carpita and Gibeaut, 1993). RG lyases are enzymes that cleave the RG-I backbone. To evaluate the role played by RG-I in cell–cell adhesion, Yang *et al.* (2019) generated transgenic poplar lines that expressed an *A. thaliana* RG lyase gene (*RGIL6*). These transgenic lines not only exhibited increased cell–cell separation in chemical, enzymatic, and mechanical assays but also yielded more glucose and xylose in saccharification assays, relative to non-transgenic controls. Their results demonstrated that it is possible to modulate cell–cell adhesion by modifying RG-I content without compromising biomass yield.

The *GAUT12/IRX8* gene encodes a putative glycosyltransferase that is thought to be involved in the production of glucuronoxylan and/or pectin in the secondary cell wall because *A. thaliana irx8* mutants show a reduction in both xylan and HG (Hao and Mohnen, 2014). There are two *GAUT12* orthologs in *P. trichocarpa*, *PtGAUT12.1* and *PtGAUT12.2*. Simultaneous suppression of both genes leads to a reduction in the xylan content of poplar wood (Li *et al.*, 2011). Biswal *et al.* (2015) used RNAi to downregulate only *GAUT12.1* to determine its effect on the recalcitrance of poplar wood. They generated 11 RNAi *P. deltoides* lines, 10 of which had 4–8% greater glucose release following enzymatic saccharification relative to control plants. In addition, downregulation of *PdGAUT12.1* led to 12–52% and 12–44% increases in plant height and stem diameter, respectively, compared with non-transgenic controls. Reducing the expression of *PdGAUT12.1* also resulted in 25–47% and 17–30% reductions in galacturonic acid and xylose, respectively, but there was no effect on lignin content. Thus, similar to what is observed in *A. thaliana*, *GAUT12* affects both pectin and xylan formation. Analysis of cell-wall extracts revealed a decrease in glucuronoxylan, HG, and RG in the *PdGAUT12.1* RNAi lines. In summary, these changes in poplar cell-wall composition led to both increased growth and reduced recalcitrance.

Together, these examples suggest that there is cross-talk between pathways responsible for the biosynthesis of the major cell-wall constituents. Such changes in mass balance may provide a way of maintaining mechanical strength in lignin-modified plants while improving the efficiency with which sugars can be released from their biomass.

Remodeling the Cell Wall

Endoglucanases

As cell walls undergo expansion, mechanical strength must be maintained to support the plant, while also allowing movement of construction materials and signaling molecules into the living cells (Levy *et al.*, 2002). Cell size and shape are governed by their elongation rates, which are, in turn, influenced by cell-wall plasticity and cell–cell adhesion. Growing regions are rich in cellulase, xylanase, and xyloglucanase activity, all of which seem to be involved in cell enlargement (Mellerowicz and Sundberg, 2008). In plants, there is a large family of enzymes known as endoglucanases, which are capable of hydrolyzing β-1,4 glucans (e.g. cellulose). Some of these enzymes are secreted into the extracellular space during growth, while others are believed to be associated with a complex that includes cellulose synthase, and they are involved in both cell-wall development and degradation. Cellulose strands are connected by a network of hydrogen bonds and van der Waals forces, leading to crystallization in certain regions of the microfibrils. The highly structured crystalline domains are separated by less organized, amorphous regions. The crystalline areas have limited accessibility, protecting the glycosidic bonds from enzymatic attack (Levy *et al.*, 2002). Endoglucanases are capable of hydrolyzing non-crystalline cellulose and xyloglucans, allowing cell-wall plasticity and the xyloglucan–cellulose matrix to be remodeled during plant growth and development, through a process known as cell-wall loosening (Carpita and Gibeaut, 1993; Shani and Shoseyov, 2006).

The protein encoded by an *A. thaliana* endo-(1-4)-β-glucanase gene (*AtCel1*) collects in young, rapidly growing plant tissues and appears to play a central role in controlling cell enlargement (Shani *et al.*, 2006). Overexpression of *Cel1* in poplar or of a poplar ortholog (*PopCel1*) in *A. thaliana* has resulted in longer internodes, greater cell elongation, and the accumulation of more biomass (Park *et al.*, 2003; Shani *et al.*, 2004). It is assumed that the cell walls in these transgenic plants contain less cross-linked material (Park *et al.*, 2003). It was proposed that non-crystalline glucan chains tethered by strands of hemicellulose become unraveled by the action of the endoglucanases, reducing the number of anchored xyloglucans and, as a result, increasing the plasticity of the cell wall, which is a prerequisite for cell-wall expansion and continued deposition (Hayashi, 1989).

A membrane-bound endoglucanase known as KORRIGAN (KOR) is needed for proper cellulose synthesis (Nicol *et al.*, 1998; Sato *et al.*, 2001). Maloney *et al.* (2012) characterized the white spruce (*Picea glauca*) ortholog of *KOR* (*PgKOR*), and showed that it can rescue the *A. thaliana kor1-1* mutant, suggesting that it is the functional equivalent. In white spruce, *KOR* expression is highest in young, developing tissues, corresponding to the period of primary cell-wall development. Downregulation of *PgKOR* via RNAi led to a reduction in growth and glucose content, but the cellulose ultrastructure remained unchanged. Conservation of *KOR* function in gymnosperms implies that its role in cell-wall synthesis evolved about 300 million years ago, prior to the emergence of angiosperms 130 million years ago, suggesting that enzymes involved in cellulose synthesis are conserved among vascular plants.

Glass *et al.* (2015) examined the role played by two orthologs of poplar and *A. thaliana* endoglucanase genes in secondary cell-wall development. The poplar and *A. thaliana* class B endoglucanases, PtGH9B5 and AtGH9B5, respectively, are secreted enzymes with a glycosylphosphatidylinositol anchor. Predicted roles for this anchor include: (i) serving as a bridge between the plasmalemma and the developing cell wall; (ii) targeting the protein near to where it functions, beneath the cell wall; and/or (iii) tethering proteins so they are in close proximity to one another and can function more efficiently (Borner *et al.*, 2002). Two class C endoglucanases, PtGH9C2 and AtGH9C2, are thought to be secreted, but they have a carbohydrate-binding moiety, which is assumed to facilitate attachment to crystalline cellulose (Urbanowicz *et al.*, 2007). These endoglucanase genes were expressed in *A. thaliana* under the control of both the 35S and the *A. thaliana CesA8* promoters to evaluate their effects on growth and development, cell-wall crystallinity, microfibril angle, and carbohydrate composition. Expression of *PtGH9B5* using the *AtCesA8* promoter, which targets the secondary cell wall, led to a decrease in plant stature and rosette diameter (see Fig. 8.2), implying that this gene might influence cell expansion. Altering the expression of *PtGH9B5* and *AtGH9B5* caused changes in xylose content. Overexpression of *PtGH9C2* and *AtGH9C2* affected crystallinity, which was negatively correlated with plant height and rosette diameter. Based on their results, the authors concluded that these endoglucanases alter secondary cell-wall development by influencing its crystallinity.

Intercalating iron

Thermal and chemical conversion of cellulosic feedstocks to high-value products requires large energy inputs, and conversion efficiency is limited by biomass recalcitrance. Ferrous ion (Fe^{2+}) enhancement during dilute acid pretreatment has led to an increase in the release of simple sugars (Wei *et al.*, 2011); however, the cost of adding metal catalysts to biomass during pretreatment is prohibitive operationally, and its efficacy is limited by rates of diffusion. Yang *et al.* (2016) engineered a binary vector to contain the sequence for a carbohydrate-binding module (CBM11), an iron-binding peptide, and a signal peptide that targets the cell wall. Expression of this fusion polypeptide in *A. thaliana* and rice (*Oryza sativa*) resulted in significant increases in iron content and an improvement in biomass conversion efficiency, relative to non-transgenic control plants. Introduction of this construct into a poplar hybrid (*P. tremula* × *P. alba*) led to a 30–40% increase in iron accumulation compared with WT plants (R. Meilan and M.C. McCann, unpublished data). It remains to be seen whether this sequestered iron will ultimately lead to higher conversion efficiencies.

Root Formation

Being able to clonally propagate desirable genotypes is an important attribute for all bioenergy crops. Species within the genus *Populus* can be divided into two groups, depending on the ease with which their stem segments and branches can be rooted, which is a simple and economical way to vegetatively propagate them (Dickmann and Stuart, 1983).

Generally, species within the sections *Turanga*, *Leucoides*, *Tacamahaca*, and *Aigeiros* are relatively easy to root, while those in the section *Populus* are recalcitrant. Dai *et al.* (2004) overexpressed the *rolB* gene in hybrid aspen (*Populus* × *canescens* × *P. grandidentata*) under the control of a heat-shock promoter. When exposed to elevated temperatures, hardwood cuttings taken from plants expressing the heat-shock construct had rooting frequencies of 70–80%. In contrast, the rate of rooting on cuttings taken from WT plants is normally in the range of 15–20%. Cheng *et al.* (2005) used *iaaM*, a gene involved in auxin biosynthesis, to enhance the formation of roots on aspen cuttings. Although the initial results seem promising, additional evaluation is needed (Ye *et al.*, 2011).

Using activation tagging in *Populus*, Trupiano *et al.* (2013) identified five tagged genes that influenced adventitious root development. Of the lines that were affected, three showed an increase and two a decrease in adventitious root production. They recovered the genomic sequences adjoining the left-hand border of the activation-tagging T-DNA insert and verified elevated transcription of the proximal genes. In two of the three lines with increased adventitious roots, the activated genes were found to negatively regulate either ethylene signaling or biosynthesis. This implies that ethylene is involved in controlling adventitious root formation in *Populus*.

Root development has also been shown to be improved in greenwood cuttings taken from a poplar hybrid (*P. tremula* × *P. alba*) whose pyruvate dehydrogenase kinase expression was downregulated via RNAi (Fig. 8.6; Harfouche *et al.*, 2011). Ordinarily, cuttings from this genotype will only root after being provided with an exogenous supply of auxin and extensive misting.

Lateral roots arise from the pericycle, a meristematic zone adjacent to the stele within the root. Digestive enzymes are needed to allow these lateral roots to emerge (Laskowski *et al.*, 2006). Pectins are complex polysaccharides deposited in plant cell walls and in the middle lamella that contribute to cell–cell adhesion. As described above, RG is a major component of the pectin network, and RG lyase is an enzyme involved in cleaving it. A family

Fig. 8.6. Rooting is a limiting factor for a significant number of woody species. The photograph shows the rooting ability of transgenic poplar (*P. tremula* × *P. alba*) in which an RNAi construct was used to downregulate pyruvate dehydrogenase kinase gene expression (left) compared with non-transgenic plants (right). The non-transgenic cuttings were dipped into a commercially available, auxin-containing root-induction product (Rootone®). The transgenic poplar produced more roots in medium without auxin than the difficult-to-root, non-transgenic cuttings that received an exogenous supply of auxin. (After Harfouche *et al.*, 2011.)

of *MYST* genes in *A. thaliana* has sequence similarity to microbial RG lyases. Overexpression of *MYST6* in *A. thaliana* led to increases in the length of the primary root, the numbers of lateral roots, and the lengths and density of root hairs (Zhou, 2010). Similar results have been observed when *MYST6* was ectopically expressed in *P. tremula* × *P. alba* (R. Meilan and M.C. McCann, unpublished data).

Dash *et al.* (2018) constructed a genetic network using time-series transcriptomic data collected from the roots of hybrid poplar (*P. tremula* × *P. alba*) plants that were treated with polyethylene glycol to simulate drought stress. The network was hierarchical in nature, the highest level of which was occupied by nine genes. These genes were designated "superhub genes" because they connected 18 hubs, which, in turn, linked 2,934 genes. The authors were only able to produce transgenic poplar lines overexpressing two superhub genes, which encoded poplar orthologs of the *A. thaliana* genes *JAZ3* and *RAP2.6*. Failure to recover transgenic lines using any of the other superhub genes suggests that they interfered with the regeneration process. Transgenic lines overexpressing *P. trichocarpa* orthologs of these genes, *PtaJAZ3* and *PtaRAP2.6*, exhibited significant increases in root length and lateral-root proliferation in response to drought stress. The authors presented evidence, suggesting that there were regulatory interactions between these two superhub genes. First, both were significantly induced by methyl-jasmonate treatment. Second, *PtaRAP2.6* expression was elevated in the lines overexpressing *PtaJAZ3*, but *PtaJAZ3* expression was unaffected in the *PtaRAP2.6*-overexpressing lines. A better understanding of the mechanism by which root architecture is regulated may allow us to tailor plants that are better adapted to growing under drought stress.

Summary

Ethanol is the main liquid biofuel being utilized in the USA today and is largely derived from maize. The starch in its kernels is extracted and enzymatically hydrolyzed to glucose, which is fermented by yeast (*S. cerevisiae*). There are advocates for not only using the kernels for bioethanol production but also using maize stover—the residual plant material left in the field after harvesting the ears, including stems, leaves, and roots—as a cellulosic feedstock.

One distinguishing feature of plant cells is that they possess a cell wall, which is a rich source of polysaccharides, including cellulose and hemicelluloses. These complex carbohydrates can be hydrolyzed, liberating simple sugars, which can be used to produce bioethanol, as well as other liquid fuels and industrial reagents. To optimize production and supply stability of biofuels, a suite of biomass crops will be needed. However, the use of any cellulosic feedstock for biofuel production is limited by the chemical composition of its cell walls.

To maximize biofuel production, it is not only desirable for bioenergy crops to have cell walls with the highest possible polysaccharide content, these sugars also need to be accessible so they can be hydrolyzed to monosaccharides, which subsequently can be utilized for fermentation or other conversion processes. Pretreatment methods are used to solubilize and separate the constituents of cell walls to make the residual more amenable to downstream chemical or biological treatments. However, most pretreatment methods are not particularly effective at removing lignin, and they can affect the yield and integrity of the remaining polysaccharides.

Researchers have altered cell-wall structure in an attempt to improve saccharification rates. Lignin is known to negatively impact biofuel production from cellulosic feedstocks in several ways. First and foremost, lignin blocks access of hydrolytic enzymes to the cell-wall polysaccharides; hence the need for pretreatment. Second, in addition to simple sugars, a variety of compounds are generated during pretreatment of cellulosic feedstocks that can interfere with hydrolysis of the underlying carbohydrates. Reducing lignin content can have deleterious consequences, so the focus has been on altering its composition. The latter affects the number and strength of the linkages connecting lignin to the other cell-wall components, thus affecting the ease with which it can be extracted. As a result, research on feedstock modification has emphasized genes encoding enzymes catalyzing various steps in the lignin biosynthetic pathway that contribute to the monomeric composition of lignin.

Through the characterization of genetically engineered plants, it has been shown that high-S poplar (*Populus* spp.) lines have higher conversion efficiencies than WT plants and are, therefore, more desirable as a biofuel feedstock. Subsequently, genomic sequencing identified a naturally occurring poplar mutant that produces high-S lignin. This mutant can now be vegetatively propagated and grown in plantations as a dedicated energy crop, unlike the genetically engineered high-S lines, which cannot be deployed commercially because of concerns about transgene escape (Chapter 7, this volume).

Many transgenic plants that were engineered to produce modified lignin have exhibited additional and often unanticipated phenotypes. Qualitative and/or quantitative alterations in one cell-wall constituent have often resulted in striking changes in others. In contrast to lignin deposition, which is now relatively well understood, cellulose synthesis in woody plants remains poorly characterized, but there are a few reports on genetic manipulation of cell-wall carbohydrates in plants. Through the limited work that has been done, it is now obvious that there is cross-talk between pathways responsible for the biosynthesis of the major cell-wall constituents. Understanding these interactions will undoubtedly provide a way to maintain mechanical strength and other desirable properties (e.g. pest resistance) in lignin-modified plants, while improving the efficiency with which sugars can be released from their biomass in order to improve their utility as a bioenergy crop.

References

Baucher, M., Chabbert, B., Pilate, G., van Doorsselaere, J., Tollier, M.T., *et al.* (1996). Red xylem and higher lignin extractability by down-regulating a cinnamyl alcohol dehydrogenase in poplar (*Populus tremula* × *Populus alba*). *Plant Physiology* 112, 1479–1490.

Beckham, G.T., Johnson, C.W., Karp, E.M., Salvachúa, D., and Vardon, D.R. (2016) Opportunities and challenges in biological lignin valorization. *Current Opinion in Biotechnology* 42, 40–53.

Biswal, A.K., Soeno, K., Gandla, M.L., Immerzeel, P., Pattathil, S., *et al.* (2014) Aspen pectate lyase

Ptxt PL1-27 mobilizes matrix polysaccharides from woody tissues and improves saccharification yield. *Biotechnology for Biofuels* 7, 11.

Biswal, A.K., Hao, Z., Pattathil, S., Yang, X., Winkeler, K., *et al.* (2015) Downregulation of *GAUT12* in *Populus deltoides* by RNA silencing results in reduced recalcitrance, increased growth and reduced xylan and pectin in a woody biofuel feedstock. *Biotechnology for Biofuels* 8, 41.

Boerjan, W., Ralph, J., and Baucher, M. (2003) Lignin biosynthesis. *Annual Review of Plant Biology* 54, 519–546.

Bonawitz, N.D., Kim, J.I., Tobimatsu, Y., Ciesielski, P.N., Anderson, N.A., *et al.* (2014) Disruption of Mediator rescues the stunted growth of a lignin-deficient *Arabidopsis* mutant. *Nature* 509, 376–380.

Borner, G.H.H., Sherrier, D.J., Stevens, T.J., Arkin, I.T., and Dupree, P. (2002) Prediction of glycosylphosphatidylinositol-anchored proteins in *Arabidopsis*. A genomic analysis. *Plant Physiology* 129, 486–499.

Boudet, A.M., Kajita, S., Grima-Pettenati, J., and Goffner, D. (2003) Lignins and lignocellulosics: a better control of synthesis for new and improved uses. *Trends in Plant Science* 8, 576–581.

Brunner, A.M. and Beers, E.P. (2011) Transcription factors in poplar growth and development. In: Joshi, C.P., DiFazio, S.P., and Kole, C. (eds.) *Genetics, Genomics and Breeding of Poplar*. CRC Press, Boca Raton, Florida, pp. 192–230.

Buhl, C., Meilan, R., and Lindroth, R.L. (2017) Genetic modification of lignin in hybrid poplar (*Populus alba × P. tremula*) does not substantially alter plant defense or arthropod communities. *Journal of Insect Science* 17, 1–8.

Carpita, N.C. and Gibeaut, D.M. (1993) Structural models of primary cell walls in flowering plants: consistency of molecular structure with the physical properties of the walls during growth. *Plant Journal* 3, 1–30.

Carpita, N.C. and McCann, M.C. (2008) Maize and sorghum: genetic resources for bioenergy grasses. *Trends in Plant Science* 13, 415–420.

Chabannes, M., Barakate, A., Lapierre, C., Marita, J.M., Ralph, J., *et al.* (2001) Strong decrease in lignin content without significant alteration of plant development is induced by simultaneous down-regulation of cinnamoyl CoA reductase (CCR) and cinnamyl alcohol dehydrogenase (CAD) in tobacco plants. *Plant Journal* 28, 257–270.

Chen, F. and Dixon, R.A. (2007) Lignin modification improves fermentable sugar yields for biofuel production. *Nature Biotechnology* 25, 759–761.

Cheng, Z.-M., Dai, W.H., Bosela, M.J., and Osburn, L.D. (2005) Genetic engineering approach to enhance adventitious root formation of hardwood cuttings. *Journal of Crop Improvement* 17, 211–225.

Chuck, G., Cigan, M., Saeteurn, K., and Hake, S. (2007) The heterochronic maize mutant *Corngrass1* results from overexpression of a tandem microRNA. *Nature Genetics* 39, 544–549.

Coleman, H.D., Canam, T., Kang, K.Y., Ellis, D.D., and Mansfield, S.D. (2007) Over-expression of UDP-glucose pyrophosphorylase in hybrid poplar affects carbon allocation. *Journal of Experimental Botany* 58, 4257–4268.

Coleman, H.D., Park, J.Y., Nair, R., Chapple, C., and Mansfield, S.D. (2008a) RNAi-mediated suppression of *p*-coumaroyl-CoA 3'-hydroxylase in hybrid poplar impacts lignin deposition and soluble secondary metabolism. *Proceedings of the National Academy of Science USA* 105, 4501–4506.

Coleman, H.D., Samuels, A.L., Guy, R.D., and Mansfield, S.D. (2008b) Perturbed lignification impacts tree growth in hybrid poplar—a function of sink strength, vascular integrity, and photosynthetic assimilation. *Plant Physiology* 148, 1229–1237.

Coleman, H.D., Yan, J., and Mansfield, S.D. (2009) Sucrose synthase affects carbon partitioning to increase cellulose production and altered cell wall ultrastructure. *Proceedings of the National Academy of Science USA* 106, 13118–13123.

Cullis, I.F. and Mansfield, S.D. (2010) Optimized delignification of wood-derived lignocellulosics for improved enzymatic hydrolysis. *Biotechnology and Bioengineering* 106, 884–893.

Cutler, S. and Somerville, C. (1997) Cellulose synthesis: cloning *in silico*. *Current Biology Dispatch* 7, R108–R111.

Dai, W.H., Cheng, Z.-M., and Sargent, W.A. (2004) Expression of the *rol*B gene enhances adventitious root formation in hardwood cuttings of aspen. *In Vitro Cellular & Developmental Biology—Plant* 40, 366–370.

Dash, M., Yordanov, Y.S., Georgieva, T., Wei, H., and Busov, V. (2018) Gene network analysis of poplar root transcriptome in response to drought stress identifies a *PtaJAZ3PtaRAP2.6*-centered hierarchical network. *PLoS One* 13, e0208560.

Davison, B.H., Drescher, S.R., Tuskan, G.A., Davis, M.F., and Nghiem, N.P. (2006) Variation of S/G ratio and lignin content in a *Populus* family influences the release of xylose by dilute acid hydrolysis. *Applied Biochemistry and Biotechnology* 130, 427–435.

Demirbas, A. (2005) Bioethanol from cellulosic materials: a renewable motor fuel from biomass. *Energy Sources* 27, 327–337.

Dickmann, D.I. and Stuart, K.W. (1983) *The Culture of Poplars in Eastern North America*. Michigan State University Press, East Lansing, Michigan.

Doty, S.L., Dosher, M.R., Singleton, G.L., Moore, A.L., van Aken, B., *et al.* (2005) Identification of an endophytic *Rhizobium* in stems of *Populus*. *Symbiosis* 39, 27–36.

Eudes, A., Liang, Y., Mitra, P., and Loqué, D. (2014) Lignin bioengineering. *Current Opinion in Biotechnology* 26, 189–198.

Fahrenkrog, A.M., Neves, L.G., Resende, M.F. Jr., Vazquez, A.I., de Los Campos, G., *et al.* (2016) Genome-wide association study reveals putative regulators of bioenergy traits in *Populus deltoides*. *New Phytologist* 213, 799–811.

Franke, R., McMichael, C.M., Meyer, K., Shirley, A.M., Cusumano, J.C., and Chapple, C. (2000) Modified lignin in tobacco and poplar plants overexpressing the *Arabidopsis* gene encoding ferulate 5-hydroxylase. *Plant Journal* 22, 223–234.

Fry, S.C. (2004) Primary cell wall metabolism: tracking the careers of wall polymers in living plant cells. *New Phytologist* 161, 641–675.

Geisler-Lee, J., Geisler, M., Coutinho, P.M., Segerman, B., Nishikubo, N., *et al.* (2006) Poplar carbohydrate-active enzymes (CAZymes). Gene identification and expression analyses. *Plant Physiology* 140, 946–962.

Glass, M., Barkwill, S., Unda, F., and Mansfield, S.D. (2015) Endo β-1,4-glucanases impact plant cell wall development by influencing cellulose crystallinity. *Journal of Integrative Plant Biology* 57, 396–410.

Gray, K.A., Zhao, L., and Emptage, M. (2006) Bioethanol. *Current Opinion in Chemical Biology* 10, 141–146.

Hao, Z. and Mohnen, D. (2014) A review of xylan and lignin biosynthesis: foundation for studying *Arabidopsis* irregular xylem mutants with pleiotropic phenotypes. *Critical Reviews in Biochemistry and Molecular Biology* 49, 212–241.

Harfouche, A., Meilan, R., and Altman, A. (2011) Tree genetic engineering and applications to sustainable forestry and biomass production. *Trends in Biotechnology* 29, 9–17.

Hayashi, T. (1989) Xyloglucans in the primary cell wall. *Annual Review of Plant Physiology* 40, 139–168.

Hu, W.-J., Harding, S.A., Lung, J., Popko, J.L., Ralph, J., *et al.* (1999) Repression of lignin biosynthesis promotes cellulose accumulation and growth in transgenic trees. *Nature Biotechnology* 17, 808–812.

Humphreys, J.M. and Chapple, C. (2002) Rewriting the lignin roadmap. *Current Opinion in Plant Biology* 5, 224–229.

Huntley, S.K., Ellis, D., Gilbert, M., Chapple, C., and Mansfield, S.D. (2003) Significant increases in pulping efficiency in C4H-F5H-transformed poplars: improved chemical savings and reduced environmental toxins. *Journal of Agricultural and Food Chemistry* 51, 6178–6183.

Israelsson, M., Eriksson, M.E., Hertzberg, M., Aspeborg, H., Nilsson, P., and Moritz, T. (2003) Changes in gene expression in the wood-forming tissue of transgenic hybrid aspen with increased secondary growth. *Plant Molecular Biology* 52, 893–903.

Jeong, D. (2018) Stochastic techno-economic analysis of electricity produced from poplar plantations in Indiana. MS thesis, Purdue University, West Lafayette, Indiana.

Jones, L., Ennos, A.R., and Turner, S.R. (2001) Cloning and characterization of *irregular xylem4* (*irx4*): a severely lignin-deficient mutant of *Arabidopsis*. *Plant Journal* 26, 205–216.

Joshi, C.P. and Mansfield, S.D. (2007) The cellulose paradox—simple molecule, complex biosynthesis. *Current Opinion in Plant Biology* 10, 220–226.

Jouanin, L., Goujon, T., de Nadai, V., Martin, M.T., Mila, I., *et al.* (2000) Lignification in transgenic poplars with extremely reduced caffeic acid O-methyltransferase activity. *Plant Physiology* 123, 1363–1373.

Kaida, R., Kaku, T., Baba, K., Oyadomari, M., Watanabe, T., *et al.* (2009) Loosening xyloglucan accelerates the enzymatic degradation of cellulose in wood. *Molecular Plant* 2, 904–909.

Keating, J.D., Panganiban, C., and Mansfield, S.D. (2006) Tolerance and adaptation of ethanologenic yeasts to lignocellulosic inhibitory compounds. *Biotechnology and Bioengineering* 93, 1196–2006.

Lacombe, E., Hawkins, S., van Doorsselaere, J., Piquemal, J., Goffner, D., *et al.* (1997) Cinnamoyl CoA reductase, the first committed enzyme of the lignin branch biosynthetic pathway: cloning, expression and phylogenetic relationships. *Plant Journal* 11, 429–441.

Lapierre, C., Pollet, B., Petit-Conil, M., Toval, G., Romero, J., *et al.* (1999) Structural alterations of lignins in transgenic poplars with depressed cinnamyl alcohol dehydrogenase or caffeic acid O-methyltransferase activity have an opposite impact on the efficiency of industrial kraft pulping. *Plant Physiology* 119, 153–163.

Laskowski, M., Biller, S., Stanley, K., Kajstura, T., and Prusty, R. (2006) Expression profiling of auxin-treated *Arabidopsis* roots: toward a molecular analysis of lateral root emergence. *Plant and Cell Physiology* 47, 788–792.

Leplé, J.C., Dauwe, R., Morreel, K., Storme, V., Lapierre, C., *et al.* (2007) Downregulation of cinnamoyl-coenzyme a reductase in poplar: multiple-level phenotyping reveals effects on cell wall polymer metabolism and structure. *Plant Cell* 19, 3669–3691.

Levy, I., Shani, Z., and Shoseyov, O. (2002) Modification of polysaccharides and plant cell wall by endo-1,4-β-glucanase and cellulose-binding domains. *Biomolecular Engineering* 19, 17–30.

Li, L., Zhou, Y.H., Cheng, X.F., Sun, J.Y., Marita, J.M., *et al.* (2003) Combinatorial modification of multiple lignin traits in trees through multigene cotransformation. *Proceedings of the National Academy of Science USA* 100, 4939–4944.

Li, Q., Min, D., Wang, J.P.-Y., Peszlen, I., Horvath, L., *et al.* (2011) Down-regulation of glycosyltransferase 8D genes in *Populus trichocarpa* caused reduced mechanical strength and xylan content in wood. *Tree Physiology* 31, 226–236.

Li, X., Ximenes, E., Kim, Y., Slininger, M., Meilan, R., *et al.* (2010) Lignin monomer composition affects *Arabidopsis* cell-wall degradability after liquid hot water pretreatment. *Biotechnology for Biofuels* 3, 27.

Lunn, J.E. and Rees, T.A. (1990) Apparent equilibrium constant and mass-action ratio for sucrose-phosphate synthase in seeds of *Pisum sativum*. *Biochemical Journal* 267, 739–743.

Maloney, V.J., Samuels, A.L., and Mansfield, S.D. (2012) Functional conservation of *Korrigan*, a putative membrane-bound endo-1,4-β-glucanase required for cellulose biosynthesis in vascular plants. *New Phytologist* 193, 1076–1087.

Maloney, V.J., Park, J.-Y., Unda, F., and Mansfield, S.D. (2015) Sucrose-phosphate synthase and sucrose-phosphate phosphatase interact *in planta* and promotes enhanced plant growth. *Journal of Experimental Botany* 66, 4383–4394.

Mansfield, S.D. (2009) Solutions for dissolution—engineering cell walls for deconstruction. *Current Opinions in Biotechnology* 20, 286–294.

Mansfield, S.D., Kang, K.-Y., and Chapple, C. (2012) Designed for deconstruction—poplar trees altered in cell wall lignification improve the efficacy of bioethanol production. *New Phytologist* 194, 91–101.

Mellerowicz, E.J. and Gorshkova, T.A. (2012) Tensional stress generation in gelatinous fibres: a review and possible mechanism based on cell wall structure and composition. *Journal of Experimental Botany* 63, 551–565.

Mellerowicz, E.J. and Sundberg, B. (2008) Wood cell walls: biosynthesis, developmental dynamics and their implications for wood properties. *Current Opinion in Plant Biology* 11, 293–300.

Meyermans, H., Morreel, K., Lapierre, C., Pollet, B., de Bruyn, A., *et al.* (2000) Modifications in lignin and accumulation of phenolic glucosides in poplar xylem upon down-regulation of caffeoyl-coenzyme A O-methyltransferase, an enzyme involved in lignin biosynthesis. *Journal of Biological Chemistry* 275, 36899–36909.

Moerschbacher, B.M., Noll, U., Gorrichon, L., and Reisener, H.J. (1990) Specific-inhibition of lignification breaks hypersensitive resistance of wheat to stem rust. *Plant Physiology* 93, 465–470.

Mottiar, Y., Vanholme, R., Boerjan, W., Ralph, J., and Mansfield, S.D. (2016) Designer lignins: harnessing the plasticity of lignification. *Current Opinion in Biotechnology* 37, 190–200.

Nicol, F., His, I., Jauneau, A., Vernhettes, S., Canut, H., and Höfte, H. (1998) A plasma membrane-bound putative endo-1,4-β-D-glucanase is required for normal wall assembly and cell elongation in *Arabidopsis*. *EMBO Journal* 17, 5563–5576.

Olsson, L. and Hahn-Hägerdal, B. (1996) Fermentation of lignocellulosic hydrolysates for ethanol production. *Enzyme and Microbial Technology* 18, 312–331.

Palmqvist, E. and Hahn-Hägerdal, B. (2000) Fermentation of lignocellulosic hydrolysates. I: inhibition and detoxification. *Bioresource Technology* 74, 17–24.

Pan, C., Oda, Y., Lankford, P.K., Zhang, B., Samatova, N.F., *et al.* (2008) Characterization of anaerobic catabolism of *p*-coumarate in *Rhodopseudomonas palustris* by integrating transcriptomics and quantitative proteomics. *Molecular and Cellular Proteomics* 7, 938–948.

Park, J.-Y., Canam, T., Kang, K.Y., Unda, F., and Mansfield, S.D. (2009) Sucrose phosphate synthase (SPS) expression influences phenology in poplar. *Tree Physiology* 29, 937–946.

Park, Y.W., Tominaga, R., Sugiyama, J., Furuta, Y., Tanimoto, E., *et al.* (2003) Enhancement of growth by expression of poplar cellulase in *Arabidopsis thaliana*. *Plant Journal* 33, 1099–1106.

Park, Y.W., Baba, K., Furuta, Y., Iida, I., Sameshima, K., *et al.* (2004) Enhancement of growth and cellulose accumulation by overexpression of xyloglucanase in poplar. *FEBS Letters* 564, 183–187.

Parsell, T., Jarrell, T., Klein, I., Degenstein, J., Hurt, M., *et al.* (2015) A synergistic biorefinery based on catalytic conversion of lignin prior to cellulose starting from lignocellulosic biomass. *Green Chemistry* 17, 1492–1499.

Patzek, T.W. (2004) Thermodynamics of the corn-ethanol biofuel cycle. *Critical Reviews in Plant Sciences* 23, 519–567.

Pilate, G., Guiney, E., Holt, K., Petit-Conil, M., Lapierre, C., *et al.* (2002) Field and pulping performances of transgenic trees with altered lignification. *Nature Biotechnology* 20, 607–612.

Rubinelli, P.M., Chuck, G., Li, X., and Meilan, R. (2012) Constitutive expression of microRNA *Corngrass1* in poplar affects axillary meristem outgrowth, internode length, and lignin quantity and composition. *Biomass and Bioenergy* 54, 312–321.

Sarkanen, K.V. and Ludwig, C.H. (eds.) (1971) *Lignins: Occurrence, Formation, Structure, and Reactions.* Wiley-Interscience, New York.

Sato, S., Kato, T., Kakegawa, K., Ishii, T., Liu, Y.G., *et al.* (2001) Role of the putative membrane-bound endo-1,4-β-glucanase KORRIGAN in cell elongation and cellulose synthesis in *Arabidopsis thaliana. Plant Cell and Physiology* 42, 251–263.

Schwab, R., Palatnik, J.F., Riester, M., Schommer, C., Schmid, M., and Weigel, D. (2005) Specific effects of microRNAs on the plant transcriptome. *Developmental Cell* 8, 517–527.

Sexton, T.R., Henry, R.J., Harwood, C.E., Thomas, D.S., McManus, L.J., *et al.* (2012) Pectin methylesterase genes influence solid wood properties of *Eucalyptus pilularis. Plant Physiology* 158, 531–541.

Shani, Z. and Shoseyov, O. (2006) Cell wall proteins as a tool for plant cell modification. In: Hayashi, T. (ed.) *The Science and Lore of the Plant Cell Wall: Biosynthesis, Structure and Function.* BrownWalker Press, Boca Raton, Florida, pp. 317–325.

Shani, Z., Dekel, M., Tsabary, G., Goren, R., and Shoseyov, O. (2004) Growth enhancement of transgenic poplar plants by overexpression of *Arabidopsis thaliana* endo-1,4-β-glucanase (*cel1*). *Molecular Breeding* 14, 321–330.

Shani, Z., Dekel, M., Roiz, L., Horowitz, M., Kolosovski, N., *et al.* (2006) Expression of endo-1,4-β-glucanase (*cel1*) in *Arabidopsis thaliana* is associated with plant growth, xylem development and cell wall thickening. *Plant Cell Reports* 25, 1067–1074.

Shuai, L., Amiri, M.T., Questell-Santiago, Y.M., Héroguel, F., Kim, H., *et al.* (2016) Stabilization with formaldehyde facilitates the high-yield production of monomers from lignin during integrated biomass depolymerization. *Science* 354, 329–333.

Skyba, O., Douglas, C.J., and Mansfield, S.D. (2013) Syringyl-rich lignin renders poplars more resistant to degradation by wood decay fungi. *Applied and Environmental Microbiology* 79, 2560–2571.

Smith, R., Gonzales-Vigil, E., Karlen, S., Park, J.-Y., Lu, F., *et al.* (2015) Engineering monolignol *p*-coumarate conjugates into poplar and *Arabidopsis* lignins. *Plant Physiology* 169, 2992–3001.

Studer, M.H., DeMartini, J.D., Davis, M.F., Sykes, R.W., Davison, B., *et al.* (2011) Lignin content in natural *Populus* variants affects sugar release. *Proceedings of the National Academy of Sciences USA* 108, 6300–6305.

Toon, S.T., Philippidis, G.P., Ho, N.W.Y., Chen, Z.D., Brainard, A., *et al.* (1997) Enhanced cofermentation of glucose and xylose by recombinant *Saccharomyces* yeast strains in batch and continuous operating modes. *Applied Biochemistry and Biotechnology* 63, 243.

Trupiano, D., Yordanov, Y., Regan, S., Meilan, R., Tschaplinski, T., *et al.* (2013) Identification and characterization of an AP2/ERF transcription factor that promotes adventitious and lateral root formation in *Populus. Planta* 238, 271–282.

Tsai, C.J., Popko, J.L., Mielke, M.R., Hu, W.J., Podila, G.K., and Chiang, V.L. (1998) Suppression of *O*-methyltransferase gene by homologous sense transgene in quaking aspen causes red-brown wood phenotypes. *Plant Physiology* 117, 101–112.

Unda, F., Kim, H., Hefer, C., Ralph, J., and Mansfield, S.D. (2017) Altering carbon allocation in hybrid poplar (*Populus alba* × *grandidentata*) impacts cell wall growth and development. *Plant Biotechnology Journal* 15, 865–878.

Urbanowicz, B.R., Catalá, C., Irwin, D., Wilson, D.B., Ripoll, D.R., and Rose, J.K.C. (2007) A tomato endo-β-1,4-glucanase, SlCel9C1, represents a distinct subclass with a new family of carbohydrate binding modules (CBM49). *Journal of Biological Chemistry* 282, 12066–12074.

van Doorsselaere, J., Baucher, M., Chognot, E., Chabbert, B., Tollier, M.T., *et al.* (1995) A novel lignin in poplar trees with a reduced caffeic acid 5-hydroxyferulic acid *O*-methyltransferase activity. *Plant Journal* 8, 855–864.

Vanholme, R., Demedts, B., Morreel, K., Ralph, J., and Boerjan, W. (2010) Lignin biosynthesis and structure. *Plant Physiology* 153, 895–905.

Vanholme, R., Cesarino, I., Rataj, K., Xiao, Y., Sundin, L., *et al.* (2013) Caffeoyl shikimate esterase (CSE) is an enzyme in the lignin biosynthetic pathway in *Arabidopsis. Science* 341, 1103–1106.

Voelker, S.L., Lachenbruch, B., Meinzer, F.C., Jourdes, M., Ki, C., *et al.* (2010) Antisense down-regulation of *4CL* expression alters lignification, tree growth, and saccharification potential of field-grown poplar. *Plant Physiology* 154, 874–886.

Volk, T.A., Buford, M.A., Berguson, B., Caputo, J., Eaton, J., *et al.* (2011) Woody feedstocks—management and regional differences. In: Braun, R., Karlen, D., and Johnson, D. (eds.) *Sustainable Alternative Feedstock Opportunities, Challenges and Roadmap for Six U.S. Regions. Proceedings of the Sustainable Feedstocks for Advance Biofuels Workshop.* Soil and Water Conservation Society, Ankeny, Iowa, pp. 120–141.

Wagner, A., Donaldson, L., Kim, H., Phillips, L., Flint, H., *et al.* (2009) Suppression of 4-coumarate-CoA ligase in the coniferous gymnosperm *Pinus radiata. Plant Physiology* 149, 370–383.

Wei, H., Donohoe, B.S., Vinzant, T.B., Ciesielski, P.N., Wang, W., *et al.* (2011) Elucidating the role of ferrous ion cocatalyst in enhancing dilute acid pretreatment of lignocellulosic biomass. *Biotechnology for Biofuels* 4, 48.

Wilkerson, C.G., Mansfield, S.D., Lu, F., Withers, S., Park, J.-Y., *et al.* (2014) Monolignol ferulate transferase introduces chemically labile linkages into the lignin backbone. *Science* 344, 90–93.

Withers, S., Lu, F., Kim, H., Zhu, Y., Ralph, J., and Wilkerson, C.G. (2012) Identification of grass-specific enzyme that acylates monolignols with *p*-coumarate. *Journal of Biological Chemistry* 287, 8347–8355.

Wu, G. and Poethig, R.S. (2006) Temporal regulation of shoot development in *Arabidopsis thaliana* by miR156 and its target SPL3. *Development* 133, 3539–3547.

Yang, H., Wei, H., Ma, G., Antunes, M.S., Vogt, S., *et al.* (2016) Cell wall targeted *in planta* iron accumulation enhances biomass conversion and seed iron concentration in Arabidopsis and rice. *Plant Biotechnology Journal* 14, 1998–2009.

Yang, H., Benatti, M.R., Karve, R.A., Fox, A., Meilan, R., *et al.* (2019) Rhamnogalacturonan-I is a determinant of cell–cell adhesion in poplar wood. *Plant Biotechnology Journal*.

Ye, X., Busov, V., Zhao, N., Meilan, R., McDonnell, L.M., *et al.* (2011) Transgenic *Populus* trees for forest products, bioenergy, and functional genomics. *Critical Reviews in Plant Sciences* 30, 415–434.

Zakzeski, J., Bruijnincx, P.C.A., Jongerius, A.L., and Weckhuysen, B.M. (2010) The catalytic valorization of lignin for the production of renewable chemicals. *Chemical Reviews* 110, 3552–3599.

Zhong, R. and Ye, Z.-H. (2009) Transcriptional regulation of lignin biosynthesis. *Plant Signaling and Behavior* 4, 1028–1034.

Zhong, R. and Ye, Z.-H. (2015) Transcriptional regulation of biosynthesis of cell wall components during xylem differentiation. In: Fukuda, H. (ed.) *Plant Cell Wall Patterning and Cell Shape.* Wiley, Hoboken, New Jersey, pp. 351–377.

Zhong, R., Morrison, W.H., Himmelsbach, D.S., Poole, F.L., and Ye, Z.-H. (2000) Essential role of caffeoyl coenzyme A *O*-methyltransferase in lignin biosynthesis in woody poplar plants. *Plant Physiology* 124, 563–578.

Zhou, S. (2010) *Arabidopsis MYST6* encodes a rhamnogalacturonan lyase that modulates root growth. PhD thesis, Purdue University, West Lafayette, Indiana.

Engineering Tolerance to Abiotic Stresses

Introduction

Forest trees account for over 80% of global biomass (Roy *et al.*, 2001) and forests provide more than half of terrestrial biodiversity (Neale and Kremer, 2011). Because of the carbon they sequester, forests help alleviate the severity of climate change, and they provide a vast array of ecosystem services and products that humans desire, including lumber, paper, fuel, and various biomaterial derivatives (Harfouche *et al.*, 2012, 2014).

Trees in our forests are regularly exposed to abiotic stresses, which are environmental conditions that can negatively impact their growth and development (e.g. drought, salinity, temperature extremes). Based on reports prepared by the Intergovernmental Panel on Climate Change (www.ipcc.ch, accessed July 15, 2019), these stresses are likely to increase in the short term due to global climate change. Anticipated changes in climatic conditions, along with greater reliance on marginal land to supply food, feed, fiber, biofuel, and other plant-derived products (Perlack *et al.*, 2005; Ragauskas *et al.*, 2006; Jordan *et al.*, 2007; Robertson *et al.*, 2008), as well as increased salinization of soil and water, pose serious threats to forest productivity globally (Ye *et al.*, 2011). Salinity and drought are already common in many parts of the world, and are predicted to adversely affect productivity on more than 50% of arable land across the globe by the year 2050 (Altman, 2003).

These dire forecasts have spurred the development of crop plants that have the potential to be high yielding, even under the influence of various stresses (Altman, 2003; Good *et al.*, 2004; Hirel *et al.*, 2007). These stresses elicit various physiological responses in plants, and their genetic composition determines whether they will be able to survive and remain productive under harsher environmental conditions. Historically, it has been challenging to genetically improve trees through conventional breeding because of their long juvenile periods. However, this impediment is being overcome with recent advances in marker-aided breeding (Chapter 5, this volume). In addition, genes controlling some sought-after traits are absent from the genome of the preferred plant, and therefore require alternative approaches for their introduction (Harfouche *et al.*, 2012). Using genetic engineering, it is possible to overcome these obstacles and hasten the domestication and commercial deployment of plants that can tolerate abiotic stresses (Altman, 2003) and meet societal demands for the products we derive from them.

A plant's molecular response to abiotic stress is complex and typically entails the integrated expression of many genes. In some cases, a given gene or assemblage of genes can provide protection against a variety of stresses. Progress in investigating stress tolerance in trees has not been as rapid as with other plants (Vinocur and Altman, 2005), but knowledge garnered through the study of herbaceous model species is allowing swifter progress. Recent developments in high-throughput sequencing, genomics, transcriptomics, metabolomics, and proteomics in several forest-tree species (Neale and Ingvarsson, 2008; Wilkins *et al.*, 2009; Tsai and Xue, 2015; Xue *et al.*, 2015) are providing tools and resources that will expedite the improvement of abiotic stress tolerance in trees (Harfouche *et al.*, 2012; Zimin *et al.*, 2014). For example, genome sequences are now available for several commercially important forest-tree species, including black cottonwood (*Populus trichocarpa*; Tuskan *et al.*, 2006); rose gum (*Eucalyptus grandis*; Myburg *et al.*, 2014); white spruce (*Picea glauca*; Birol *et al.*, 2013); Norway spruce (*Picea abies*; Nystedt *et al.*, 2013); loblolly pine (*Pinus taeda*; Neale *et al.*, 2014); Douglas-fir (*Pseudotsuga menziesii*; Neale *et al.*, 2017); and European ash (*Fraxinus excelsior*; Sollars *et al.*, 2017).

The rest of this chapter provides selected examples of how these advances can be utilized to introduce or enhance abiotic stress tolerance in trees.

Temperature

Temperature extremes can affect plant survival, productivity, and susceptibility to pests. Trees may exhibit cold-induced damage when exposed to low temperatures in spring, following their release from dormancy and expansion of their leaf primordia. Cold tolerance is a term used to describe the ability of plants to survive low temperatures, whereas freezing tolerance is the ability to withstand subzero temperatures, which can involve the formation of ice crystals in the xylem or in the apoplast (the space between cells). Warmer winters have led to early resumption of growth, and as a result, the prevalence of damage from frost in forest- and fruit-tree species has increased considerably (Gu *et al.*, 2008). Climate change is also likely to cause significantly higher temperatures during periods of active growth in many regions of the world. Genetic engineering can be used to provide protection for trees against both high and low temperatures. Some examples are provided below.

Heat and oxidative stress

Changes in their tertiary structure can render proteins non-functional when they are exposed to excessive heat. Many stress-related proteins were originally classified as heat-shock proteins because they prevented this denaturation from occurring. Chaperones are a class of stress-related polypeptides that aid in the folding and unfolding of other proteins but are not part of their functional structure. In addition to assisting in protein refolding under stressful conditions, chaperones are also involved in the production, movement, and disassembly of macromolecules involved in a plethora of standard cellular processes, such as the stabilization of membrane systems (Wang *et al.*, 2004). Chaperones and heat-shock proteins play a role in a plant's response to a variety of abiotic stresses, not just heat.

Abiotic stresses can lead to the production and build-up of reactive oxygen species (ROS), which can be harmful to plant cells. Kawaoka *et al.* (2003) reported that calli of hybrid aspen (*Populus sieboldii* × *P. grandidentata*) transformed with a horseradish (*Armoracia rusticana*) peroxidase gene, *prxC1a*, under the control of the promoter from the cauliflower mosaic virus *35S* gene, were resistant to the oxidative stress imposed by hydrogen peroxide (H_2O_2) treatment, and grew faster than control plants. Regenerated trees of the same genotype containing the same construct exhibited elevated peroxidase activity. In addition, leaf discs from transgenic plants were markedly more tolerant of H_2O_2 and methyl viologen treatment than control plants, and their stems were 25% longer than the corresponding non-transgenic plants, when grown in a greenhouse.

Kim *et al.* (2009) showed that expressing an *Arabidopsis thaliana* nucleoside diphosphate kinase gene (*AtNDPK2*) in sweet potato (*Ipomoea batatas*), under the control of a promoter that is induced by oxidative stress, led to greater tolerance to several environmental stresses. This same genetic construct was introduced into hybrid poplar (*P. alba* × *P. glandulosa*) to develop plants that are better able to tolerate oxidative stress. The activities of ascorbate peroxidase, catalase, and peroxidase (antioxidant enzymes) were increased in the transgenic plants in response to chemically induced (using methyl viologen) oxidative stress. Average branch number and stem diameter were significantly greater in two of the transgenic lines, relative to non-transgenic control plants, after one growing season in the field (Kim *et al.*, 2011).

Isoprene is one of the main biogenic volatile organic compounds emitted by plants, accounting for nearly half of all such compunds released into the atmosphere annually

$(412 \times 10^{12} - 601 \times 10^{12}$ g $= 4.5 \times 10^{8}$ tons; Laothawornkitkul *et al.*, 2009), which is nearly equivalent to the emission of methane from all sources globally (Guenther *et al.*, 2006; Sharkey *et al.*, 2008). Isoprene-production capacity is a highly variable trait; several plant families include species that can emit isoprene and others that cannot (Monson *et al.*, 2013), and it is released at the highest rate by trees, especially oaks (*Quercus* spp.) and poplars (*Populus* spp.) (Sharkey *et al.*, 2008). Isoprene is produced in plastids and is normally released in response to light and heat; it is not stored. Elevated isoprene synthesis in plants is also frequently associated with exposure to abiotic stresses (Sharkey *et al.*, 2008; Vickers *et al.*, 2009; Loreto and Schnitzler, 2010), and is thought to play a protective role by deactivating ROS produced in response to high temperatures, intense light, and other environmentally triggered oxidative stresses (e.g. ozone) (Vickers *et al.*, 2009; Jardine *et al.*, 2012). Moreover, it is thought to be involved in stabilizing thylakoid membranes (Velikova *et al.*, 2011; Harrison *et al.*, 2013).

To investigate the role isoprene plays in preventing the disruption of physiological processes in leaves exposed to thermal and oxidative stress, Behnke *et al.* (2007) used RNA interference (RNAi) to reduce expression of the isoprene synthase (*ISPS*) gene in grey poplar (*Populus* × *canescens*). Their transgenic plants were thermally stressed by exposing them to three heat-treatment cycles at between 38 and 40°C, followed by recovery at 30°C. In the transgenic plants, isoprene emissions were effectively blocked; however, the plants had reduced rates of photosynthetic electron transport and net assimilation during heat stress, which was not observed under control conditions. Photochemical efficiency was correlated with the amount of absorbed light energy that was dissipated, which was measured by non-photochemical quenching. Plants in which isoprene biosynthesis was suppressed also exhibited an increase in the formation of zeaxanthin in the absence of stress, which can enhance non-photochemical quenching. Production of this accessory pigment suggests an increased need for antioxidants. In contrast, overexpression of *ISPS* did not lead to increased isoprene production, which was attributed to post-transcriptional co-suppression. Overexpression of the *P. alba* ortholog of *ISPS* in *A. thaliana* led to elevated levels of isoprene. These transgenic plants were able to tolerate temperatures of up to 60°C, and exhibited a decrease in leaf surface temperature during heat-stress treatment. These data provide evidence that isoprene plays a crucial role in the protection of plants from heat stress (Sasaki *et al.*, 2007).

Subsequently, Behnke *et al.* (2013) reported on the response of transgenic poplars (*Populus* × *canescens*) that are incapable of emitting isoprene and wild-type (WT) plants of the same genotype to various stresses, both individually and in combination, including high salt, simulated sunflecks, and heat. They showed that salt stress together with sunflecks caused a modest decrease in the photosynthetic capacity of the transgenic lines. Following long-term exposure to heat stress (40°C), the leaves of the transgenic plants collapsed, demonstrating that disrupting isoprene production affects a plant's ability to tolerate elevated temperatures. These results confirmed an earlier report showing reduced heat tolerance of poplars in which isoprene production had been blocked (Behnke *et al.*, 2007).

Cold

To survive harsh winter conditions, temperate-zone trees normally enter a quiescent state before developing cold hardiness as they enter endodormancy during late fall. In response to shorter days and cooler weather, not only is there an increase in the production of storage proteins in vegetative tissues (Clausen and Apel, 1991), but other metabolic and structural modifications are also triggered, allowing plants to survive low temperatures (Weiser, 1970; Welling *et al.*, 2002). Plants synthesize various types of membranes, including the plasmalemma, the nuclear envelope, and those encapsulating the mitochondria and chloroplasts, all of which contain phospholipids (e.g. Fig. 9.1).

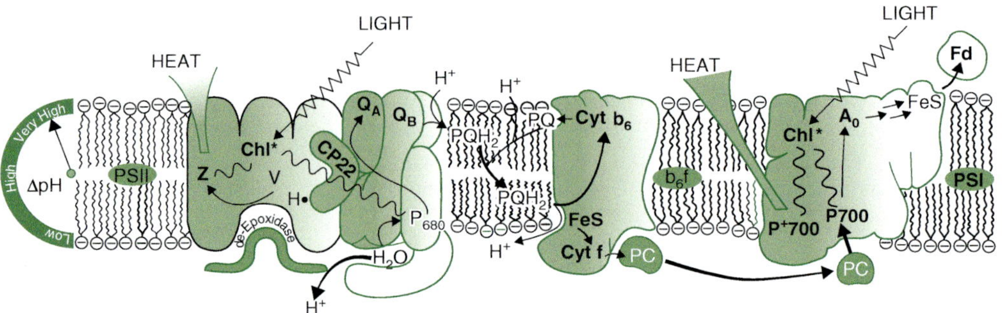

Fig. 9.1. Model of a thylakoid membrane, which is made up of phospholipids and has various macromolecules embedded in it. To perform their roles, many of these molecules need to be mobile in the membrane; thus, its fluidity must be maintained. Under high-light conditions, the membrane-bound epoxidase converts violaxanthin to zeaxanthin, ultimately leading to the dissipation of heat. A_0, primary acceptor of PSI; b6f, cytochrome b6f complex; Chl*, excited chlorophyll molecule; CP22, minor PSII pigment protein; Cyt, cytochrome; FeS, bound iron sulfur acceptors of PSI; Fd, soluble ferredoxin; P_{680}, reaction center chlorophyll of PSII; P700 and P$^+$700, reduced and oxidized forms of the reaction center chlorophyll of PSI; PC, plastocyanin; PQ and PQH_2, plastoquinone and reduced plastoquinone; PSI, photosystem I complex; PSII, photosystem II complex; Q_A and Q_B, quinone acceptors of PSII; V, violaxanthin; Z, zeaxanthin. (From: Ort, 2001; copyrighted by the American Society of Plant Biologists and reprinted with permission.)

These membrane systems are susceptible to damage if their liquid content freezes, which can lead to cell death. But plants can adjust membrane structure to avoid this disruption (Steponkus, 1984; Palta, 1990). For proper membrane function at low temperatures during periods of active growth, fluidity must be preserved (Beney and Gervais, 2001); therefore, while acquiring tolerance to freezing, plants modify the extent of saturation in the fatty-acid side chains of their membrane lipids (Kodama *et al.*, 1995; Falcone *et al.*, 2004). This adjustment lowers the temperature at which membranes begin to congeal (Table 1, Fig. 9.2).

Much of the early cold-response research was conducted in *A. thaliana*, which is a versatile model herbaceous plant because it is short statured, has a short life cycle and a small genome, and can be genetically modified with ease. *Populus* spp. and *A. thaliana* are related phylogenetically, so what is accomplished in the latter is often directly transferable to the former. The best-characterized cold-response pathway in *A. thaliana* includes the *C-REPEAT BINDING FACTOR* genes, also referred to as *DEHYDRATION RESPONSIVE ELEMENT BINDING (CBF/DREB1)* genes, which encode closely related transcription factors

belonging to the AP2/ERF family of proteins that bind to DNA (Thomashow, 1999; Yamaguchi-Shinozaki and Shinozaki, 2006). The CBF protein is an ethylene response factor (ERF) and an activator protein (AP2) that binds to a conserved *cis*-acting element, and is involved in signaling in *A. thaliana* in response to low temperatures (Jaglo-Ottosen *et al.*, 1998; Liu *et al.*, 1998).

Orthologs of *CBF* have been identified in *Populus* spp. (Benedict *et al.*, 2006) and *Eucalyptus gunnii* (El Kayal *et al.*, 2006), and the pathway within which CBF operates has been shown to function in trees (Puhakainen *et al.*, 2004). Some poplar *CBF* genes are expressed at higher levels only in leaves in response to cold (in a controlled environment), while others are upregulated in more persistent tissues (e.g. stems) (Benedict *et al.*, 2006), implying that individual CBF proteins might perform distinct functions in different tissue types. For example, poplar orthologs of some *A. thaliana CBF*-related genes (e.g. *CBF1*, *CBF3*, and *Inducer of CBF3 Expression 1 (ICE1)*) were upregulated nearly 400-fold in the stems of field-grown trees during the winter, while they are dormant (Ko *et al.*, 2011). The *ICE1* gene encodes a MYC-like basic helix–loop–helix transcription factor that is known to

Table 9.1. Fatty acids are long-chain hydrocarbons with a terminal carboxylic acid group. Naturally occurring fatty acids may be saturated (single bonds between carbon atoms in the chain) or unsaturated (double bonds between carbons). The longer the saturated fatty acid, the higher its melting point, whereas higher levels of unsaturation lead to lower melting points. Melting points derived from: www.chemicalbook.com (accessed July 15, 2019).

Formula	Common name	Melting point (°C)
Saturated		
$CH_3(CH_2)_{10}COOH$	Lauric acid	44–46
$CH_3(CH_2)_{12}COOH$	Myristic acid	52–54
$CH_3(CH_2)_{14}COOH$	Palmitic acid	61–62.5
$CH_3(CH_2)_{16}COOH$	Stearic acid	67–72
$CH_3(CH_2)_{18}COOH$	Arachidic acid	74–76
Unsaturated		
$CH_3(CH_2)_5CH{=}CH(CH_2)_7COOH$	Palmitoleic acid	0.5
$CH_3(CH_2)_7CH{=}CH(CH_2)_7COOH$	Oleic acid	13–14
$CH_3(CH_2)_4CH{=}CHCH_2CH{=}CH(CH_2)_7COOH$	Linoleic acid	−5
$CH_3CH_2CH{=}CHCH_2CH{=}CHCH_2CH{=}CH(CH_2)_7COOH$	Linolenic acid	−11
$CH_3(CH_2)_4(CH{=}CHCH_2)_4(CH_2)_2COOH$	Arachidonic acid	−49

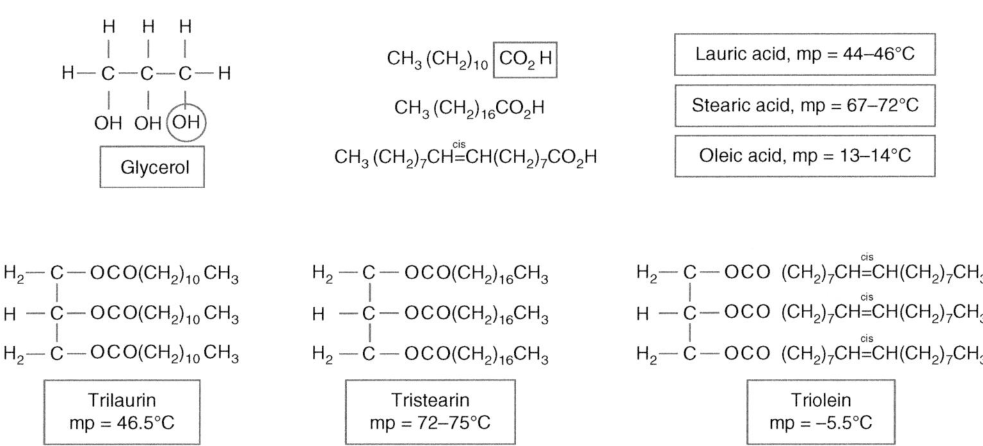

Fig. 9.2. Glycerol is a tri-alcohol that, together with lauric, stearic, and oleic acid, is commonly used by plants to synthesize the lipids (triglycerides) trilaurin, tristearin, and triolein, respectively. When the carboxyl acid group of a fatty acid (boxed) reacts with an alcohol group in glycerol (circled), the result is an ester linkage. If one of the alcohol groups is phosphorylated, a phospholipid is formed. Melting point (mp) values derived from: www.chemicalbook.com (accessed July 15, 2019).

activate *CBF* and cold-regulated (*COR*) genes in *A. thaliana* following exposure to low temperatures (Chinnusamy *et al.*, 2003). Constitutive expression of the *A. thaliana AtCBF1* gene in a hybrid aspen (*P. tremula* × *P. alba*) resulted in a significant increase in cold tolerance of the leaves and stems of unacclimated plants (Benedict *et al.*, 2006).

Welling and Palva (2008) investigated the molecular basis of freezing tolerance in silver birch (*Betula pendula*) in a growth-chamber study. They identified four birch orthologs of *CBF*, designated *BpCBF1* to -4, from an expressed-sequence-tag (EST) collection. Expression analysis revealed that, in addition to being responsive to low temperatures, these genes are also differentially regulated in dormant and actively growing plants; expression was retarded during dormancy. When freeze–thaw treatment was used to simulate winter conditions, thawing triggered a strong induction of *BpCBF* genes, which induced a *CBF* target gene, *BpLTI36*. Overexpression of birch *CBFs* in *A. thaliana* led to higher transcript levels for native *CBF* genes, and unacclimated transgenic plants were more cold tolerant. The authors concluded that, in addition to being involved in plants acclimating to low temperatures during the growing season, BpCBFs seem to play a role in regulating cold-hardiness in birch.

Expression analysis of four *CBF* genes (*CBF1a* to -*d*) in *E. gunnii* showed that they are all preferentially induced by cold, and that *CBF1c* is particularly responsive to salt stress. Because various *CBF* transcripts accumulate in response to different types of chilling treatments, the authors speculated that it might be important for winter survival of *E. gunnii*, which is a broad-leaved, evergreen tree (Navarro *et al.*, 2009).

Hinchee *et al.* (2011) introduced a genetic construct containing the *A. thaliana* stress-inducible *rd29A* promoter (Yamaguchi-Shinozaki and Shinozaki, 1993) to drive *CBF2* in an elite genotype of *E. grandis* × *E. urophylla* that exhibits fast growth and has desirable fiber and pulping characteristics, but is sensitive to winter freezing. The construct also included a gene to block pollen production, to address concerns about transgene escape (Hinchee *et al.*, 2009). Single-insertion events (lines) were evaluated in 21 replicated field trials at eight locations across three hardiness zones, as defined by the U.S. Department of Agriculture (8a, a zone with potentially killing winter temperatures; 8b, the zone targeted by the authors; and 9a, a freeze-free zone), in the south-eastern USA (Hinchee *et al.*, 2011). The best-performing transgenic lines showed growth that was comparable to that of WT plants in locales with mild conditions or that did not have freezing temperatures during the winter. In the zone of interest, which had a low temperature of −8.4°C during the fifth winter, the top-performing line exhibited about 10% dieback, whereas the control plants had 99% dieback.

Since the initial work by Hinchee and co-workers, improvements have been made in our understanding of the signaling and regulatory pathways involved in plant responses to cold stress. It had already been demonstrated that both *ICE1* and calmodulin-binding transcription activators upregulate CBFs (Chinnusamy *et al.*, 2003; Doherty *et al.*, 2009). Subsequently, a protein kinase encoded by *OPEN STOMATA1* (*OST1*), which is activated by cold stress, was shown to phosphorylate the ICE1 protein, thereby enhancing its ability to bind to the promoter of the *CBF3* gene, leading to an increase in its expression. When *OST1* was overexpressed in *A. thaliana*, it led to an improvement in freezing tolerance (Ding *et al.*, 2015). Moreover, Li *et al.* (2017) showed that *BZR1*, a gene encoding another transcription factor, is a critical element of a signaling pathway in brassinosteroids that acts upstream of *CBF1* and *CBF2* and directly regulates their expression; thus, it is a potential target for engineering cold tolerance. Unfortunately, none of these genes has yet been tested in a forest-tree species.

Dehydrins are a family of late embryogenesis abundant proteins that are thought to behave as chaperones in stressed plants. They are among the most widely studied gene families involved in cold tolerance in trees, as well as in drought and osmotic stress. As with *A. thaliana*, cold-inducible dehydrin genes in trees contain a C-repeat element (CRT) in their promoters (Puhakainen *et al.*, 2004; Benedict *et al.*, 2006), suggesting that CBF proteins are involved in controlling their expression at low temperatures. Wisniewski *et al.* (2006) identified a peach (*Prunus persica*) dehydrin gene (*PpDhn2*) containing two CRT/dehydration-responsive elements (DRE) that are responsive to low temperatures.

Ectopic expression of a pepper (*Capsicum annuum*) AP2/ERF transcription factor, *CaPF1*, in eastern white pine (*Pinus strobus*) resulted in a substantial improvement in its tolerance to cold, in addition to drought and salt stress (Tang *et al.*, 2007). The levels of putrescine, spermine, and spermidine remained unchanged in the *CaPF1*-overexpressing pine transgenics; however, the levels of these compounds decreased in the control plants, implying that the greater tolerance is related to polyamine biosynthesis. The *ThCAP* gene, which was isolated from a cDNA library constructed from salt-stressed *Tamarix hispida* roots, also encodes a cold-acclimation protein that is expressed in WT plants in response to a variety of stresses, and is organ-specific. When Guo *et al.* (2009) overexpressed *ThCAP* in poplar (*P. davidiana* × *P. bolleana*), the transgenic plants exhibited greater tolerance to cold temperatures than was observed in control plants.

Lu *et al.* (2008) examined the expression of micro-RNAs (miRNAs, Chapter 6, this volume) in response to chilling stress in poplar. They identified 19 miRNAs in *P. trichocarpa* that were responsive to cold; the expression of some was upregulated in response to low temperatures, while others were suppressed. Given that the genes predicted to be targets of these miRNAs have wide-ranging functions, these miRNAs are assumed to be involved in regulating more than one aspect of stress response in plants. Following in-depth sequencing, Chen *et al.* (2012) were able to identify additional miRNAs that are responsive to stress in *P. beijingensis*.

Drought and Salinity

Drought and salt stress are becoming major impediments to the survival, growth, and development of plants in many regions of the world; it has been predicted that by 2050, as much as 50% of all land that is currently suitable for crop production will no longer be arable (Wang *et al.*, 2003). Nearly 50% of China's land area is already classified as arid, semi-arid, or saline-alkali (Li *et al.*, 2009). While the world's human population is expanding, the land base from which we derive renewable resources is shrinking, yet the rate of resource consumption is increasing. To meet this growing demand, ways must be found not just to maintain but to improve yields, especially on marginal land, and in a sustainable manner. This will necessitate the development of plants that not only survive but thrive under stressful conditions.

Studies of plant molecular networks involved in sensing, transducing, and reacting to stress signals have revealed that water and salt stress are closely linked. Thus, plants genetically engineered to tolerate salt stress may have improved drought tolerance, as well as increased water-use efficiency (WUE).

Moreover, engineering plants for increased tolerance to abiotic stress cannot only permit survival under acute or chronic stress, but may also lead to sustained productivity under more modest levels of stress. It is thought that genes derived from halophytes might be helpful in this regard, but it remains to be seen whether they can be used to maintain growth under mild stress while also allowing survival under extreme conditions (Harfouche *et al.*, 2011).

Because of its ability to survive salt and drought stress, *P. euphratica* is used widely as a model system to study abiotic stress tolerance in woody perennials (see references in B. Li *et al.*, 2011) (Fig. 9.3). Using high-throughput sequencing of drought-stressed *P. euphratica* leaves, Li and colleagues found 197 miRNAs that were conserved between *P. euphratica* and *P. trichocarpa*, 58 of which were previously unknown, belonging to 38 families. By comparing high-throughput sequencing results with microarray profiling data, they showed that 104 of these miRNAs were upregulated and 27 were downregulated in response to drought stress. The authors speculated that these miRNAs are likely to play key roles in regulating stress resistance in *P. euphratica*, and that it may be possible to introduce this trait into other *Populus* spp. through conventional breeding. However, given the recent development of CRISPR technology, genome editing might be a more expedient approach (Chapter 6, this volume).

To help elucidate the mechanism by which *P. euphratica* is able to grow in arid regions, Tang *et al.* (2013) evaluated physiological and transcriptomic responses of leaves from control and drought-stressed plants. They found that stomatal closure was inhibited under moderate drought, presumably to maintain relatively high rates of photosynthesis and water transport under these conditions. They also presented evidence for transcriptional remodeling associated with stress perception and signaling, photoprotection systems, and oxidative stress detoxification (metabolism of ROS), among other stress-related responses. In addition, they found differential expression of genes involved with stomatal regulation, the ascorbate–glutathione pathway, and the

Fig. 9.3. Because of its ability to survive salinity and drought stress, *Populus euphratica* is commonly used as a model system for studying abiotic stress resistance in woody plants. (Photo provided by Arie Altman, The Hebrew University of Jerusalem, Faculty of Agricultural, Food and Environmental Quality Sciences, Rehovot, Israel.)

ubiquitin–proteasome system. Their analysis offers some insights into the molecular mechanisms that regulate drought response, and identifies potential targets for genetic modification of trees so they will be able to thrive in water-limiting environments.

Yoon *et al.* (2014) performed transcriptional analyses on hybrid poplar (*P. alba* × *P. glandulosa*) leaves exposed to drought and salt stress using a microarray comprised of 44,718 oligonucleotides derived from the *P. trichocarpa* genome. Overall, 1,604 and 1042 genes were upregulated more than twofold by drought and salt stress, respectively; expression of 765 genes was increased by both treatments. Moreover, 2,742 and 1,685 genes were downregulated in response to drought and salt stress, respectively, while

1,564 genes were downregulated by both treatments. The large number of genes whose expression is affected by both stresses suggests that cross-talk occurs between pathways that are responsive to these conditions. A majority of the upregulated genes were associated with functions related to subcellular localization, signal transduction, metabolism, and transcription. Because nearly 25% of the genes whose expression was significantly affected by these stresses had no known function, the authors speculated that poplar may provide an avenue for uncovering unique stress-related genes.

Barchet *et al.* (2014) used a non-targeted approach to profile the metabolites produced by four genotypes of *Populus* in response to drought stress. Their goal was to assess the relative levels of drought tolerance among genotypes and to elucidate the biochemical mechanism(s) underpinning this trait. The relative abundance of some metabolites increased following the imposition of drought, while that of others decreased. In general, amino acids; antioxidant phenolics, including catechin and kaempferol; and osmolytes such as raffinose and galactinol, increased in abundance in response to drought. In contrast, compounds associated with photorespiration, redox regulation, and carbon fixation were less abundant following drought stress. One genotype, Okanese, displayed a unique response to drought, relative to the other clones tested. Although it had higher leaf water potential, it had a slower rate of growth. In addition, Okanese had lower levels of various osmolytes (e.g. raffinose, galactinol, and proline) and higher levels of known antioxidants (e.g. catechin and dehydroascorbic acid). Given these results, the authors suggested that osmotic adjustment, as a means of drought tolerance, is not as well developed in this genotype, relative to the other genotypes that were investigated, and that Okanese may instead tolerate drought by reallocating resources or by retaining water better.

Osmotic Balance

Drought and salinity are interrelated in that both result in osmotic stress (Wang *et al.*,

2003). Hence, strategies for improving tolerance to either one often involve production of compounds (e.g. amino acids, amines, and sugars) that ameliorate stress by altering osmotic potential, which is involved in regulating water movement. For example, transgenic poplars expressing a bacterial mannitol 1-phosphate dehydrogenase gene, *mtlD* (Liu *et al.*, 2000), which encodes an enzyme that catalyzes the conversion of fructose to mannitol, showed increased tolerance to high levels of salt (Hu *et al.*, 2005). Hybrid poplars (*Populus* × *xiaozhannica* cv. Balizhuangyang) modified in this way survived up to 40 days when grown hydroponically in 75 mM NaCl, whereas the WT plants could only tolerate 25 mM NaCl. Because of its performance and its economic value, the Chinese government has granted a license for this genetically engineered variety to be grown operationally (Sun *et al.*, 2002; FAO, 2004; Ye *et al.*, 2011).

Overexpression of a vacuolar Na^+/H^+ antiporter gene from *A. thaliana*, *AtNHX1*, in *Populus* × *euramericana*, led to lines that were more resistant to NaCl treatment than WT plants (Jiang *et al.*, 2012). These transgenic poplars also accumulated more Na^+ and K^+ in their roots and leaves, exhibited greater photosynthetic capacity, and grew more rapidly than their WT counterparts.

Cross-tolerance

As described above, expression of various genes have allowed tolerance to a variety of abiotic stresses. For example, in *P. tremula*, the *STABLE PROTEIN 1* (*SP1*) gene encodes a chaperone-like, boiling-stable protein that is induced by salt, cold, heat, and desiccation stress (Wang *et al.*, 2002; Dgany *et al.*, 2004). The *PeSP1* gene that was cloned from *P. euphratica* is an ortholog of *SP1*, which was cloned from salt-sensitive *P. tremula*. Expression of *PeSP1* is upregulated in response to NaCl treatment, and in the absence of stress, its transcript levels were much higher in salt-tolerant than in salt-sensitive species. The PeSP1 protein undergoes sumoylation at specific sites following the imposition of stress. The sumoylation motifs are conserved in most of the orthologous SP1 proteins found in other species (e.g. *A. thaliana*, *Oryza sativa*, *P. tremula*, and the bacterium *Rhodopirellula baltica*) (Harfouche *et al.*, 2011).

Overexpression of the AP/ERF transcription factor from pepper, CaPF1, in Virginia pine (*Pinus virginiana*) imparted tolerance to heat, heavy metals (cadmium, copper, and zinc) and two bacterial plant pathogens, *Bacillus thuringiensis* and *Staphylococcus epidermidis*. Activities for ascorbate peroxidase, glutathione reductase, and superoxide dismutase (antioxidant enzymes) were higher in plants overexpressing *CaPF1* than in non-transgenic control plants. The authors suggested that these enzymes may prevent oxidative damage in cells exposed to various stresses. Overexpression of *CaPF1* also enhanced organ growth by increasing both cell number and organ size in transgenic plants, relative to the controls (Tang *et al.*, 2005).

When *ApDnaK*, a chaperone gene from *Aphanothece halophytica*, a halotolerant cyanobacterium, was overexpressed in *Populus alba*, the growth rate of transgenics was similar to that of WT plants, when they were grown under low-light conditions (Takabe *et al.*, 2008). However, under high-light conditions, the transgenic plants grew more rapidly and had higher levels of cellulose. The transgenics also recovered more rapidly from high salt, drought, and low temperature than control plants when grown under low light levels.

In an attempt to improve osmotic stress tolerance in paper mulberry (*Broussonetia papyrifera*), Li *et al.* (2011) overexpressed the *FaDREB2* gene from tall fescue (*Festuca arundinacea*). Plants expressing *FaDREB2* showed higher tolerance to salt and drought than WT controls. Although all WT plants died after water was withheld for 13 days or they were exposed to 250 mM NaCl for 15 days, the plants overexpressing *FaDREB2* survived these treatments. The transgenic plants also had higher chlorophyll and leaf water content, exhibited less membrane damage, and accumulated more proline and soluble sugars than the WT plants when both were exposed to dehydration or salt stress. Transgene expression was thought to promote the increased accumulation of osmolytes to counter the osmotic imbalance caused by these abiotic stresses. These results demonstrate that a single gene can be used to enhance tolerance to an assortment of environmental stresses.

Transcriptome Remodeling

Raj *et al.* (2011) wanted to determine whether a plant's environmental history affects its response to external stimuli. To test this hypothesis, they evaluated transcriptomic responses to drought in rooted cuttings (i.e. clonal material) taken from three *Populus* genotypes grown at nurseries in two Canadian provinces. Leaf tissue taken from two of the water-stressed clones showed dramatic differences in transcript abundance patterns, which were based on geographical origin. The extent of the transcriptome-level response was positively correlated with the time that had elapsed since the clones had been grown together in the same location. The same trend was seen with regard to genome-wide methylation: clones with the most divergent history (i.e. the length of time since they last shared a common environment) had the most striking differences with respect to the degree of total DNA methylation. Their evidence implies that there is an epigenetic basis for the observed differences at the transcriptome level. This effect that environmental preconditioning has on the transcriptome-level response of a clone to drought stress has practical implications: genetically identical planting stock obtained from two different sources can react differently to the conditions on a given site.

Membrane Protection

Choline is a head group in phosphatidylcholines, which is a class of phospholipids that plants use to assemble membranes. When Yu *et al.* (2009) overexpressed a choline oxidase gene from *Arthrobacter globiformis* (*codA*), it led to increased salt tolerance in several lines of *Eucalyptus globulus*. Superoxide (O_2^-) is a ROS that accumulates in plants in response to drought stress and can damage membranes (Selote *et al.*, 2004). Superoxide dismutases (SODs) are enzymes that catalyze the conversion of superoxide to diatomic oxygen (O_2). Overexpression of a manganese-dependent SOD from *Tamarix androssowii* (a halophyte) in a hybrid poplar (*P. davidiana* × *P. bolleana*) resulted in enhanced SOD activity following treatment with NaCl and a substantial increase in the growth of the transgenic poplars (Wang *et al.*, 2010).

Overexpression of an SK2-type dehydrin gene derived from *P. euphratica* in a *P. tremula* × *P. alba* genotype led to an increase in its drought tolerance. Apparently dehydrin-gene expression protected cellular membranes and/or macromolecules in engineered lines when water was deficient and led to improved water conservation and reduced water loss under drought conditions (Wang *et al.*, 2011). In a subsequent study, Han *et al.* (2013) produced drought- and salt-tolerant poplars by overexpressing the GSK3/shaggy-like kinase gene (*AtGSK1*) from *A. thaliana*. These transgenic lines exhibited increased stomatal conductance and higher rates of photosynthesis when stressed.

Zhang *et al.* (2013) showed that constitutive expression of the fatty acyl–acyl carrier protein thioesterase gene from wintersweet (*Chimonanthus praecox*), *CpFATB*, in poplar activated an oxidative signal cascade that led to drought tolerance. Their transgenic poplars maintained significantly higher rates of photosynthesis than controls when exposed to drought, suggesting that altering the composition and saturation of fatty acids affects dehydration tolerance.

Multi-gene Approaches

Due to the complexity and multitude of responses to abiotic stress, it has not been possible to simultaneously alter the expression of all associated genes. This hurdle has led to the development of regulon engineering (Umezawa *et al.*, 2006), a technique that involves regulating the co-expression of several genes using transcription factors that can control many aspects of plant growth, development, and responses to environmental signals. Transcription factors can trigger a reaction cascade, leading to activation or repression of several downstream processes or pathways. By strategically targeting key "master regulators", it may be possible to engineer protection against more than one abiotic stress. Using this approach, a gene encoding a tomato (*Solanum lycopersicum*) jasmonic acid-responsive AP2/ERF-domain transcription factor (*JERF*) that previously was shown to impart protection against drought, low temperature, and disease in *Nicotiana tabacum* (Li *et al.*, 2006) was transformed in *P. alba* × *P. berolinensis* (Li *et al.*, 2009). Expression of *JERF* in this hybrid poplar not only increased productivity in the presence of high salt but also affected other factors associated with stress tolerance, including the concentration of proline and chlorophyll. Transgenic, greenhouse-grown plants treated with 200 and 300 mM NaCl were 128.9% and 98.8% taller, respectively, than control plants after 60 days. Three-year-old, field-grown transgenic plants were 14.5–33.6% taller than the corresponding control plants at two test sites where the total soil salt concentration was 0.3%.

Su *et al.* (2011) described how overexpression of several genes in poplar improves its tolerance to various environmental stresses. The tested genes were: *vgb*, which encodes a bacterial hemoglobin; *SacB*, a gene from *Bacillus subtilis* that encodes an enzyme involved in fructan biosynthesis; and *JERF36* from tomato. Enhanced growth of the transgenics under drought and salinity stress was attributed primarily to higher fructan levels and water-use efficiency (WUE), and improved root architecture. Gene pyramiding such as this has revealed that expressing genes from various pathways can lead to levels of stress tolerance that might not be achieved via a monogenic approach (Su *et al.*, 2011). However, to better understand the interactions among genes and pathways, comparing plants transformed with several genes with those containing a single transgene and various combinations will help elucidate the relationships between phenotypes and gene function.

Water-use Efficiency

When water loss exceeds absorption and plant water potential declines, plants limit transpirational losses by reducing the aperture of their stomata; however, restricted water loss occurs at the expense of carbon fixation (Jarvis and Jarvis, 1963; Cowan, 1977). Severe water stress can lead to cavitation, and the embolisms that form reduce the plant's capacity to transport water (Tyree *et al.*, 1994; Rice *et al.*, 2004). The extent to which a plant is able to avoid drought stress is related to its ability to minimize water loss and maximize its absorption (Chaves *et al.*, 2003). Some forest-tree species increase water uptake by having deeper and more expansive root systems (Nguyen and Lamant, 1989) and/or by modifying various foliar characteristics, including cuticular wax deposition (Hadley and Smith, 1990); leaf area (Munné-Bosch and Alegre, 2004); and stomatal conductance. Active uptake of solutes into vacuoles (i.e. osmotic adjustment) is a typical physiological response to water deficit. The success of this strategy depends, among other things, on the extent to which photosynthate is diverted from growth to pathways involved in acclimation and/or tolerance. This metabolic cost of stress tolerance is compounded over time and can negatively impact vegetative growth.

However, it may be possible to reduce water loss without compromising growth. The rate of transpiration is positively correlated with stomatal conductance, but CO_2 assimilation increases more rapidly, and it reaches saturation at higher conductance rates in many agronomic crops (Jones, 1976; Farquhar and Sharkey, 1982; Martin and Ruiz-Torres, 1992; Radin *et al.*, 1994) and in poplar (Voltas *et al.*, 2006). Because of this differential response, it is theoretically possible to improve WUE without compromising net CO_2 assimilation or yield (Fig. 9.4) (Avola *et al.*, 2008; Yoo *et al.*, 2009). Small daily reductions in water loss can lead to substantial water savings over the entire growing season (Sinclair *et al.*, 2005). Thus, altering stomatal development to induce even slight reductions in midday water loss has the potential to improve plant WUE in the long term.

Hamanishi *et al.* (2012) investigated the impact of drought on stomatal characteristics during leaf formation in *P. balsamifera*. Drought resulted in reduced stomatal conductance and altered stomatal-development patterns in two genotypes. Leaves that matured under water deficiency had fewer stomata than those that developed under well-watered conditions. In addition, transcript-abundance patterns of poplar orthologs of

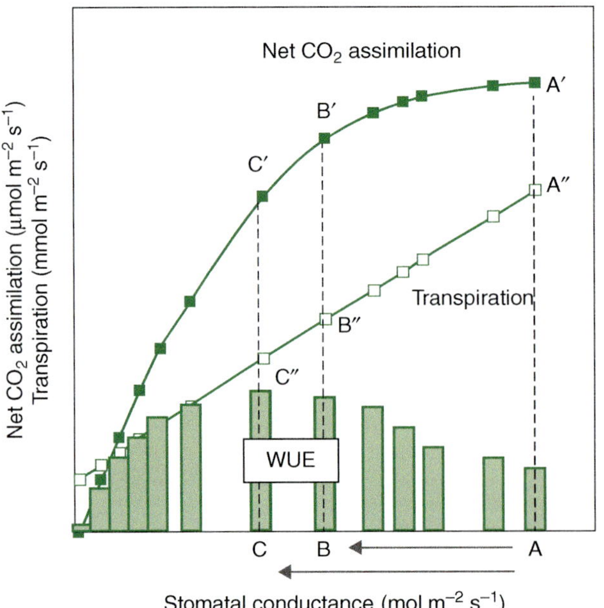

Fig. 9.4. Relationship between transpiration, net CO_2 assimilation, and WUE. Water-use efficiency is influenced by the differential effects of stomatal conductance on the rates of CO_2 assimilation and transpiration. The rate of transpiration increases linearly with stomatal conductance. While net CO_2 assimilation also increases with stomatal conductance, it does so more rapidly and eventually becomes saturated. A reduction in stomatal conductance from A to B will lead to an increase in WUE because transpiration is reduced to a greater extent (A″ to B″) than CO_2 assimilation (A′ to B′). (After Yoo *et al.*, 2009.)

STOMAGEN, ERECTA (ER), STOMATA DENS-ITY AND DISTRIBUTION 1 (SDD1), and *FAMA* (a basic helix–loop–helix transcription factor) were consistent with their having a role in the control of stomatal development in response to drought. In contrast, there were no significant differences in transcript abundance among genotypes or in response to water stress for two other genes that are involved in regulating stomatal development, *YODA (YDA)* and *TOO MANY MOUTHS (TMM)*. These findings demonstrated the potential for reducing leaf water loss by varying patterns of stomatal development; however, the long-term effects of altered stomatal development still need to be evaluated.

RECEPTOR-LIKE KINASES (RLKs) have been associated with plant responses to a variety of abiotic stresses, including drought (Marshall *et al.*, 2012). Xing *et al.* (2011) cloned the *P. nigra* × (*P. deltoides* × *P. nigra*) ortholog of *ERECTA (PdERECTA)*, which is a member of the RLK gene family. They showed that *PdERECTA* can recapitulate the WT phenotype of the *A. thaliana erecta* mutant, which suggests that there is sequence conservation in the stomatal-development genes of these two species. Overexpression of *PdERECTA* in *A. thaliana* resulted in early seedling establishment, longer primary roots, and greater leaf area. Plants overexpressing *PdERECTA* also exhibited an increase in the rate of photosynthesis, a decrease in the rate of transpiration, and lower stomatal density. Although the biochemical basis for the effects of *PdERECTA* is not known, the lower stomatal density, and the resulting decrease in transpiration, led to an improvement in WUE.

Summary

Abiotic stresses are environmental conditions that can negatively impact plant growth and development; they include, among others, drought, salinity, and temperature extremes. The predicted changes in climatic conditions, along with greater reliance on marginal land for the production of food, feed, biofuel, and other plant-derived products, as well as increased salinization of soil and water, pose serious threats to forest productivity globally. These stresses will affect the ability of our forests to continue providing the wide range of services and products that we expect, such as wood, biomass, paper, fuel, and biomaterial derivatives.

Abiotic stresses elicit various physiological responses in plants, and their genetic composition determines whether they will be able to survive and remain productive under adverse environmental conditions. At the molecular level, a plant's response to abiotic stress is complex and usually involves the coordinated expression of many genes. In some cases, a given gene or assemblage of genes can provide protection against a variety of stresses. While progress in investigating the molecular basis for stress tolerance in trees has not been as rapid as with other plants, the knowledge gathered through the study of herbaceous model species is allowing faster progress. With the knowledge of molecular targets, genetic engineering has the potential to hasten the domestication and commercial deployment of trees that are better able to tolerate abiotic stresses and meet societal demands for the products we derive from them.

Temperature extremes can affect plant survival, productivity, and susceptibility to pests. Trees may exhibit cold-induced damage when exposed to low temperatures in spring following their release from dormancy and expansion of their leaf primordia. Warmer winters have led to early resumption of growth and, as a result, the prevalence of frost damage in forest trees has increased dramatically. Climate change is also likely to cause significantly higher temperatures during the growing season in many regions of the world. Genetic engineering has been used extensively to develop ways of protecting trees from both high and low temperatures.

Abiotic stresses can lead to accumulation of ROS, which are damaging to plant cells. Genes that encode or trigger the production of antioxidant enzymes have been used to enhance tolerance to oxidative stress. The production of isoprene is associated with resistance to abiotic stresses, possibly by quenching ROS. Overexpression of an isoprene synthase gene has led to elevated levels of isoprene and plants that were able to tolerate temperatures of up to 60°C.

When proteins are exposed to excessive heat, changes in their tertiary structure can render them non-functional. Chaperones are a class of stress-related polypeptides that aid in the folding and unfolding of other proteins but are not part of the functional structure. In addition to assisting in protein refolding under elevated temperatures, chaperones are also involved in regulating a plant's response to a variety of other stresses and to a wide range of cellular processes under normal growing conditions.

To survive harsh winter conditions, temperate-zone trees typically enter a quiescent state before entering dormancy during late fall. In response to shorter days and cooler weather, biochemical and physical changes are triggered, allowing plants to survive low temperatures. The plant's various membrane systems are susceptible to damage if the cytosol freezes, which can lead to cell death. For proper membrane function at low temperatures during periods of active growth, fluidity must be maintained. During the acquisition of cold tolerance, plants can modify the saturation level of the fatty-acid side chains in membrane lipids to lower the temperature at which they begin to congeal. Various genes encoding transcription factors that are members of the AP2/ERF family have been targeted for improving cold tolerance via this mechanism.

In addition to temperature-related stresses, drought and salinity are also becoming major impediments to plant growth and development in many parts of the world. Because of its ability to survive salinity and drought stress, *Populus euphratica* is widely considered a suitable model system for studying abiotic stress resistance in woody plants. Studies of plant molecular networks involved in sensing, transducing, and responding to stress signals have revealed that salinity and water stress are closely linked. Thus, engineering tolerance to salinity stress may improve both drought tolerance and WUE. Drought and salinity are also interrelated in that both result in osmotic stress. Approaches to increasing tolerance to either

one often involve production of compounds that alleviate stress by altering osmotic potential, which determines water movement.

Because of the complexity and multitude of molecular responses to abiotic stress, it has previously not been possible to simultaneously alter the expression of all of the associated genes. Techniques are being developed to regulate the co-expression of several genes using transcription factors that can control many aspects of plant growth, development, and responses to various environmental conditions. Researchers have also simultaneously overexpressed several transgenes to enhance tolerance to environmental stressors. However, to better understand cross-talk among genes and pathways, comparing plants transformed with several genes with those containing a single transgene and various combinations thereof may help to elucidate the relationships between traits and gene function.

References

Altman, A. (2003) From plant tissue culture to biotechnology: scientific revolutions, abiotic stress tolerance, and forestry. *In Vitro Cellular & Developmental Biology—Plant* 39, 75–84.

Avola, G., Cavallaro, V., Patané, C., and Riggi, E. (2008) Gas exchange and photosynthetic water use efficiency in response to light, CO_2 concentration and temperature in *Vicia faba*. *Journal of Plant Physiology* 165, 796–804.

Barchet, G.L.H., Dauwe, R., Guy, R.D., Schroeder, W.R., Soolanayakanahally, R.Y., *et al.* (2014) Investigating the drought stress response of hybrid poplar genotypes by metabolite profiling. *Tree Physiology* 34, 1203–1219.

Behnke, K., Ehlting, B., Teuber, M., Bauerfeind, M., Louis, S., *et al.* (2007) Transgenic, non-isoprene emitting poplars don't like it hot. *Plant Journal* 51, 485–499.

Behnke, K., Kleist, E., Uerlings, R., Wildt, J., Rennenberg, H., and Schnitzler, J.-P. (2013) RNAi-mediated suppression of isoprene biosynthesis in hybrid poplar impacts ozone tolerance. *Tree Physiology* 33, 562–578.

Benedict, C., Skinner, J.S., Meng, R., Chang, Y., Bhalerao, R., *et al.* (2006) The CBF1-dependent low temperature signaling pathway, regulon and increase in freeze tolerance are conserved in *Populus* spp. *Plant, Cell & Environment* 29, 1259–1272.

Beney, L. and Gervais, P. (2001) Influence of the fluidity of the membrane on the response of microorganisms to environmental stresses. *Applied Microbiology and Biotechnology* 57, 34–42.

Birol, I., Raymond, A., Jackman, S.D., Pleasance, S., Coope, R., *et al.* (2013) Assembling the 20 Gb white spruce (*Picea glauca*) genome from whole-genome shotgun sequencing data. *Bioinformatics* 29, 1492–1497.

Chaves, M.M., Maroco, J.P., and Pereira, J.S. (2003) Understanding plant responses to drought—from genes to the whole plant. *Functional Plant Biology* 30, 239–264.

Chen, L., Zhang, Y., Ren, Y., Xu, J., Zhang, Z., and Wang, Y. (2012) Genome-wide identification of cold-responsive and new microRNAs in *Populus tomentosa* by high-throughput sequencing. *Biochemical and Biophysical Research Communications* 417, 892–896.

Chinnusamy, V., Ohta, M., Kanrar, S., Lee, B.-H., Hong, X., *et al.* (2003) ICE1: a regulator of cold-induced transcriptome and freezing tolerance in *Arabidopsis*. *Genes and Development* 17, 1043–1054.

Clausen, S. and Apel, K. (1991) Seasonal changes in the concentration of the major storage protein and its mRNA in xylem ray cells of poplar trees. *Plant Molecular Biology* 17, 669–678.

Cowan, I.R. (1977) Stomatal behaviour and environment. *Advances in Botanical Research* 4, 117–228.

Dgany, O., Gonzalez, A., Sofer, O., Wang, W., Zolotnitsky, G., *et al.* (2004) The structural basis of the thermostability of SP1, a novel plant (*Populus tremula*) boiling stable protein. *Journal of Biological Chemistry* 279, 51516–51523.

Ding, Y., Li, H., Zhang, X., Xie, Q., Gong, Z., and Yang, S. (2015) OST1 kinase modulates freezing tolerance by enhancing ICE1 stability in *Arabidopsis*. *Developmental Cell* 32, 278–289.

Doherty, C.J., van Buskirk, H.A., Myers, S.J., and Thomashow, M.F. (2009) Roles for *Arabidopsis* CAMTA transcription factors in cold-regulated gene expression and freezing tolerance. *Plant Cell* 21, 972–984.

El Kayal, W., Navarro, M., Marque, G., Keller, G., Marque, C., and Teulieres, C. (2006) Expression profile of *CBF*-like transcriptional factor genes from *Eucalyptus* in response to cold. *Journal of Experimental Botany* 57, 2455–2469.

Falcone, D.L., Ogas, J.P., and Somerville, C. (2004) Regulation of membrane fatty acid composition by temperature in mutants of *Arabidopsis* with alterations in membrane lipid composition. *BMC Plant Biology* 4, 17–31.

FAO (2004) *Preliminary Review of Biotechnology in Forestry, Including Genetic Modification*. Forest Genetic Resources Working Paper FGR/59E. Forest Resources Development Service, Forest Resources Division, FOA, Rome, Italy. Available at: www.fao.org/3/ae574e/ae574e00.htm (accessed June 27, 2019).

Farquhar, G.D. and Sharkey, T. (1982) Stomatal conductance and photosynthesis. *Annual Review of Plant Physiology* 33, 317–345.

Good, A.G., Shrawat, A.K., and Muench, D.G. (2004) Can less yield more? Is reducing nutrient input into the environment compatible with maintaining crop production? *Trends in Plant Science* 9, 597–605.

Gu, L., Hanson, P.J., Post, W.M., Kaiser, D.P., Yang, B., *et al.* (2008) The 2007 eastern US Spring freeze: increased cold damage in a warming world? *BioScience* 58, 253–262.

Guenther, A., Karl, T., Harley, P., Wiedinmyer, C., Palmer, P.I., and Geron, C. (2006) Estimates of global terrestrial isoprene emissions using MEGAN (Model of Emissions of Gases and Aerosols from Nature). *Atmospheric Chemistry and Physics* 6, 3181–3210.

Guo, X.-H., Jiang, J., Lin, S.J., Wang, B.C., Wang, Y.C., *et al.* (2009) A *ThCAP* gene from *Tamarix hispida* confers cold tolerance in transgenic *Populus* (*P. davidiana* × *P. bolleana*). *Biotechnology Letters* 31, 1079–1087.

Hadley, J. and Smith, W. (1990) Influence of leaf surface wax and leaf area to water content ratio on cuticular transpiration in western conifers, U.S.A. *Canadian Journal of Forest Research* 20, 1306–1311.

Hamanishi, E.T., Thomas, B.R., and Campbell, M.M. (2012) Drought induces alterations in the stomatal development program in *Populus*. *Journal of Experimental Botany* 63, 4959–4971.

Han, M.S., Noh, E.W., and Han, S.H. (2013) Enhanced drought and salt tolerance by expression of *AtGSK1* gene in poplar. *Plant Biotechnology Reports* 7, 39–47.

Harfouche, A., Meilan, R., and Altman, A. (2011) Tree genetic engineering and applications to sustainable forestry and biomass production. *Trends in Biotechnology* 29, 9–17.

Harfouche, A., Meilan, R., Kirst, M., Morgante, M., Boerjan, W., *et al.* (2012) Accelerating the domestication of forest trees in a changing world. *Trends in Plant Science* 17, 64–72.

Harfouche, A., Meilan, R., and Altman, A. (2014) Molecular and physiological responses to abiotic stress in forest trees and their relevance to tree improvement. *Tree Physiology* 34, 1181–1198.

Harrison, S.P., Morfopoulos, C., Dani, K.G.S., Prentice, I.C., Arneth, A., *et al.* (2013) Volatile isoprenoid emissions from plastid to planet. *New Phytologist* 197, 49–57.

Hinchee, M., Rottmann, W., Mullinax, L., Zhang, C., Chang, S., *et al.* (2009) Short-rotation woody crops for bioenergy and biofuels applications. *In Vitro Cellular & Developmental Biology — Plant* 45, 619–629.

Hinchee, M., Zhang, C., Chang, S., Cunningham, M., Hammond, W., and Nehra, N. (2011) Biotech *Eucalyptus* can sustainably address society's need for wood: the example of freeze tolerant *Eucalyptus* in the southeastern U.S. *BMC Proceedings* 5, 124.

Hirel, B., Le Gouis, J., Ney, B., and Gallais, A. (2007) The challenge of improving nitrogen use efficiency in crop plants: towards a more central role for genetic variability and quantitative genetics within integrated approaches. *Journal of Experimental Botany* 58, 2369–2387.

Hu, L., Lu, H., Liu, Q., Chen, X., and Jiang, X. (2005) Overexpression of *mtl*D gene in transgenic *Populus tomentosa* improves salt tolerance through accumulation of mannitol. *Tree Physiology* 25, 1273–1281.

Jaglo-Ottosen, K.R., Gilmour, S.J., Zarka, D.G., Schabenberger, O., and Thomashow, M.F. (1998) *Arabidopsis CBF1* overexpression induces *COR* genes and enhances freezing tolerance. *Science* 280, 104–106.

Jardine, K.J., Monson, R.K., Abrell, L., Saleska, S.R., Arneth, A., *et al.* (2012) Within-plant isoprene oxidation confirmed by direct emissions of oxidation products methyl vinyl ketone and methacrolein. *Global Change Biology* 18, 973–984.

Jarvis, P.G. and Jarvis, M.S. (1963) The water relations of tree seedlings. IV. Some aspects of the tissue water relations and drought resistance. *Physiologia Plantarum* 16, 501–516.

Jiang, C., Zheng, Q., Liu, Z., Xu, W., Liu, L., *et al.* (2012) Overexpression of *Arabidopsis thaliana* Na$^+$/H$^+$ antiporter gene enhanced salt resistance in transgenic poplar (*Populus* × *euramericana* 'Neva'). *Trees — Structure and Function* 26, 685–694.

Jones, H.G. (1976) Crop characteristics and the ratio between assimilation and transpiration. *Journal of Applied Ecology* 13, 605–622.

Jordan, N., Boody, G., Broussard, W., Glover, J.D., Keeney, D., *et al.* (2007) Sustainable development of the agricultural bio-economy. *Science* 316, 1570–1571.

Kawaoka, A., Matsunaga, E., Endo, S., Kondo, S., Yoshida, K., *et al.* (2003) Ectopic expression of a horseradish peroxidase enhances growth rate and increases oxidative stress resistance in hybrid aspen. *Plant Physiology* 132, 1177–1185.

Kim, Y.-H., Lim, S., Yang, K.-S., Kim, C.Y., Kwon, S.-Y., *et al.* (2009) Expression of *Arabidopsis NDPK2* increases antioxidant enzyme activities and enhances tolerance to multiple environmental stresses in transgenic sweetpotato plants. *Molecular Breeding* 24, 233–244.

Kim, Y.-H., Kim, M.D., Choi, Y.I., Park, S.-C., Yun, D.-J., *et al.* (2011) Transgenic poplar expressing *Arabidopsis NDPK2* enhances growth as well as oxidative stress tolerance. *Plant Biotechnology Journal* 9, 334–347.

Ko, J.H., Prassinos, C., Keathley, D., Han, K.-H., and Li, C. (2011) Novel aspects of transcriptional regulation in the winter survival and maintenance mechanism of poplar. *Tree Physiology* 31, 208–225.

Kodama, H., Horiguchi, G., Nishiuchi, T., Nishimura, M., and Iba, K. (1995) Fatty acid desaturation during chilling acclimation is one of the factors involved in conferring low-temperature tolerance to young tobacco leaves. *Plant Physiology* 107, 1177–1185.

Laothawornkitkul, J., Taylor, J.E., Paul, N.D., and Hewitt, C.N. (2009) Biogenic volatile organic compounds in the Earth system. *New Phytologist* 183, 27–51.

Li, B., Qin, Y., Duan, H., Yin W., and Xia, X. (2011) Genome-wide characterization of new and drought stress responsive microRNAs in *Populus euphratica*. *Journal of Experimental Botany* 62, 3765–3779.

Li, H., Ye, K., Shi, Y., Cheng, J., Zhang, X., and Yang, S. (2017) BZR1 positively regulates freeze tolerance via CBF-dependent and CBF-independent pathways in *Arabidopsis*. *Molecular Plant* 10, 545–559.

Li, M.-R., Li, Y., Li, H.-Q., and Wu, G.-J. (2011) Ectopic expression of *FaDREB2* enhances osmotic tolerance in paper mulberry. *Journal of Integrative Plant Biology* 53, 951–960.

Li, W.Z., Zhang, H.W., Wang, J.Y., and Huang, R.F. (2006) Ethylene responsive factors and their application in improving tobacco tolerance to biotic and abiotic stresses. *Biotechnology Bulletin* 4, 30–34.

Li, Y., Su, X., Zhang, B., Huang, Q., Zhang, X., and Huang, R. (2009) Expression of jasmonic ethylene responsive factor gene in transgenic poplar tree leads to increased salt tolerance. *Tree Physiology* 29, 273–279.

Liu, F.H., Sun, Z.X., Cui, D.C., Du, B.X., Wang, C.R., and Chen, S.Y. (2000) [Cloning of *E. coli mtl-D* gene and its expression in transgenic Balizhuangyang (*Populus*)]. *Acta Genetica Sinica* 27, 428–433 (article in Chinese, abstract in English).

Liu, Q., Kasuga, M., Sakuma, Y., Abe, H., Miura, S., *et al.* (1998) Two transcription factors, DREB1 and DREB2, with an EREBP/AP2 DNA binding domain separate two cellular signal transduction pathways in drought- and low-temperature-responsive gene expression, respectively, in *Arabidopsis*. *Plant Cell* 10, 1391–1406.

Loreto, F. and Schnitzler, J.P. (2010) Abiotic stresses and induced BVOCs. *Trends in Plant Science* 15, 154–166.

Lu, S., Sun, Y.H., and Chiang, V.L. (2008) Stress-responsive micro-RNAs in *Populus*. *Plant Journal* 55, 131–151.

Marshall, A., Aalen, R.B., Audenaert, D., Beeckman, T., Broadley, M.R., *et al.* (2012) Tackling drought stress: RECEPTOR-LIKE KINASES present new approaches. *Plant Cell* 24, 2262–2278.

Martin, B. and Ruiz-Torres, N.A. (1992) Effects of water-deficit stress on photosynthesis, its components and component limitations, and on water use efficiency in wheat (*Triticum aestivum* L.). *Plant Physiology* 100, 733–739.

Monson, R.K., Jones, R.T., Rosenstiel, T.N., and Schnitzler, J. (2013) Why only some plants emit isoprene. *Plant, Cell & Environment* 36, 503–516.

Munné-Bosch, S. and Alegre, L. (2004) Die and let live: leaf senescence contributes to plant survival under drought stress. *Functional Plant Biology* 31, 203–216.

Myburg, A.A., Grattapaglia, D., Tuskan, G.A., Hellsten, U., Hayes, R.D., *et al.* (2014) The genome of *Eucalyptus grandis*. *Nature* 510, 356–362.

Navarro, M., Marque, G., Ayax, C., Keller, G., Borges, J.P., *et al.* (2009) Complementary regulation of four *Eucalyptus CBF* genes under various cold conditions. *Journal of Experimental Botany* 60, 2713–2724.

Neale, D.B. and Ingvarsson, P.K. (2008) Population, quantitative and comparative genomics of adaptation in forest trees. *Current Opinion in Plant Biology* 11, 1–7.

Neale, D.B. and Kremer, A. (2011) Forest tree genomics: growing resources and applications. *Nature Reviews Genetics* 12, 111–122.

Neale, D.B., Wegrzyn, J.L., Stevens, K.A., Zimin, A.V., Puiu, D., *et al.* (2014) Decoding the massive genome of loblolly pine using haploid DNA and novel assembly strategies. *Genome Biology* 15, R59.

Neale, D.B., McGuire, P.E., Wheeler, V.C., Stevens, K.A., Crepeau, M.W., *et al.* (2017) The Douglas-fir genome sequence reveals specialization of the photosynthetic apparatus in Pinaceae. *G3: Genes Genomes Genetics* 7, 3157–3167.

Nguyen, A. and Lamant, A. (1989) Variation in growth and osmotic regulation of roots of water-stressed maritime pine (*Pinus pinaster* Ait.) provenances. *Tree Physiology* 5, 123–133.

Nystedt, B., Street, N.R., Wetterbom, A., Zuccolo, A., Lin, Y.-C., *et al.* (2013) The Norway spruce genome sequence and conifer genome evolution. *Nature* 497, 579–584.

Ort, D.R. (2001) When there is too much light. *Plant Physiology* 125, 29–32.

Palta, J.P. (1990) Stress interactions at membrane and cellular levels. *HortScience* 25, 1377–1381.

Perlack, R.D., Wright, L.L., Turhollow, A.F., and Graham, R.L. (2005) *Biomass as Feedstock for Bioenergy and Bioproducts Industry: the Technical Feasibility of a Billion-ton Annual Supply*. Report no. ORNL/TM-2005/66. Oak Ridge National Laboratory, Oak Ridge, Tennessee.

Puhakainen, T., Li, C., Boije-Malm, M., Kangasjarvi, J., Heino, P., and Palva, E.T. (2004) Short-day potentiation of low temperature-induced gene expression of a C-repeat-binding factor-controlled gene during cold acclimation in silver birch. *Plant Physiology* 136, 4299–4307.

Radin, J.W., Lu, Z.M., Percy, R.G., and Zeiger, E. (1994) Genetic variation for stomatal conductance in Pima cotton and its relation to improvements of heat adaptation. *Proceedings of the National Academy of Science USA* 94, 7217–7221.

Ragauskas, A.J., Williams, C.K., Davison, B.H., Britovsek, G., Cairney, J., *et al.* (2006) The path forward for biofuels and biomaterials. *Science* 311, 484–489.

Raj, S., Bräutigama, K., Hamanishi, E.T., Wilkins, O., Thomas, B.R., *et al.* (2011) Clone history shapes *Populus* drought responses. *Proceedings of the National Academy of Science USA* 108, 12521–12526.

Rice, K., Matzner, S., Byer, W., and Brown, J. (2004) Patterns of tree dieback in Queensland, Australia: the importance of drought stress and the role of resistance to cavitation. *Oecologia* 139, 190–198.

Robertson, G.P., Dale, V.H., Doering, O.C., Hamburg, S.P., Melillo, J.M., *et al.* (2008) Sustainable biofuels redux. *Science* 322, 49–50.

Roy, J., Saugier, B., and Mooney, H.A. (eds.) (2001) *Terrestrial Global Productivity*. Academic Press, London.

Sasaki, K., Saito, T., Lamsa, M., Oksman-Caldentey, K.-M., Suzuki, M., *et al.* (2007) Plants utilize isoprene emission as a thermotolerance mechanism. *Plant Cell Physiology* 48, 1254–1262.

Selote, D.S., Bharti, S., and Khanna-Chopra, R. (2004) Drought acclimation reduces $O_2^{\cdot-}$ accumulation and lipid peroxidation in wheat seedlings. *Biochemical and Biophysical Research Communications* 314, 724–729.

Sharkey, T.D., Wiberley, A.E., and Donohue, A.R. (2008) Isoprene emission from plants: why and how. *Annals of Botany* 101, 5–18.

Sinclair, T.R., Hammer, G.L., and van Oosterom, E.J. (2005) Potential yield and water-use efficiency benefits in sorghum from limited maximum transpiration rate. *Functional Plant Biology* 32, 945–952.

Sollars, E.S.A., Harper, A.L., Kelly, L.J., Sambles, C.M., Ramirez-Gonzalez, R.H., *et al.* (2017) Genome sequence and genetic diversity of European ash trees. *Nature* 541, 212–216.

Steponkus, P.L. (1984) Role of the plasma membrane in freezing injury and cold acclimation. *Annual Review of Plant Physiology* 35, 543–584.

Su, X., Chu, Y., Li, H., Hou, Y., Zhang, B., *et al.* (2011) Expression of multiple resistance genes enhances tolerance to environmental stressors in transgenic poplar (*Populus* × *euramericana* 'Guariento'). *PLoS One* 6, e24614.

Sun, Z.X., Yang, H.H., Cui, D.C., Zhao, C.Z., and Zhao, S.P. (2002) [Analysis of salt resistance on the poplar transferred with salt tolerance gene.] *Chinese Journal of Biotechnology* 18, 481–485 (article in Chinese, abstract in English).

Takabe, T., Uchida, A., Shinagawa, F., Terada, Y., Kajita, H., *et al.* (2008) Overexpression of DnaK from a halotolerant cyanobacterium *Aphanothece halophytica* enhances growth rate as well as abiotic stress tolerance of poplar plants. *Plant Growth Regulation* 56, 265–273.

Tang, S., Liang, H., Yan, D., Zhao, Y., Han, X., *et al.* (2013) *Populus euphratica*: the transcriptomic response to drought stress. *Plant Molecular Biology* 83, 539–557.

Tang, W., Charles, T.M., and Newton, R.J. (2005) Overexpression of the pepper transcription factor CaPF1 in transgenic Virginia pine (*Pinus virginiana* Mill.) confers multiple stress tolerance and enhances organ growth. *Plant Molecular Biology* 59, 603–617.

Tang, W., Newton, R.J., Li, C., and Charles, T.M. (2007) Enhanced stress tolerance in transgenic pine expressing the pepper *CaPF1* gene is associated with the polyamine biosynthesis. *Plant Cell Reports* 26, 115–124.

Thomashow, M.F. (1999) Plant cold acclimation, freezing tolerance genes and regulatory mechanisms. *Annual Review of Plant Physiology and Plant Molecular Biology* 50, 571–599.

Tsai, C.-J. and Xue, L.-J. (2015) CRISPRing into the woods. *GM Crops and Food* 6, 206–215.

Tuskan, G.A., DiFazio, S., Jansson, S., Bohlmann, J., Grigoriev, I., *et al.* (2006) The genome of black cottonwood, *Populus trichocarpa* (Torr. & Gray). *Science* 313, 1596–1604.

Tyree, M.T., Kolb, K.J., Rood, S.B., and Patino, S. (1994) Vulnerability to drought-induced cavitation of riparian cottonwoods in Alberta: a possible factor in the decline of the ecosystem? *Tree Physiology* 14, 455–466.

Umezawa, T., Fujita, M., Fujita, Y., Yamaguchi-Shinozaki, K., and Shinozaki, K. (2006) Engineering drought tolerance in plants: discovering and tailoring genes to unlock the future. *Current Opinion in Biotechnology* 17, 113–122.

Velikova, V., Várkonyi, Z., Szabó, M., Maslenkova, L., Nogues, I., *et al.* (2011) Increased thermostability of thylakoid membranes in isoprene-emitting leaves probed with three biophysical techniques. *Plant Physiology* 157, 905–916.

Vickers, C.E., Possell, M., Cojocariu, C.I., Velikova, V.B., Laothawornkitkul, J., *et al.* (2009) Isoprene synthesis protects transgenic tobacco plants from oxidative stress. *Plant, Cell & Environment* 32, 520–531.

Vinocur, B. and Altman, A. (2005) Recent advances in engineering plant tolerance to abiotic stress: achievements and limitations. *Current Opinion in Biotechnology* 16, 123–132.

Voltas, J., Serrano, L., Hernandez, M., and Peman, J. (2006) Carbon isotope discrimination, gas exchange and stem growth of four Euramerican hybrid poplars under different watering regimes. *New Forests* 31, 435–451.

Wang, W.-X., Pelah, D., Alergand, T., Shoseyov, O., and Altman, A. (2002) Characterization of SP1, a stress-responsive, boiling-soluble, homo-oligomeric protein from aspen. *Plant Physiology* 130, 865–875.

Wang, W., Vinocur, B., and Altman, A. (2003) Plant responses to drought, salinity and extreme temperatures: towards genetic engineering for stress tolerance. *Planta* 218, 1–14.

Wang, W., Vinocur, B., Shoseyov, O., and Altman, A. (2004) Role of plant heat-shock proteins and molecular chaperones in the abiotic stress response. *Trends in Plant Science* 9, 244–252.

Wang, Y., Wang, H., Li, R., Ma, Y., and Wei, J. (2011) Expression of a SK2-type dehydrin gene from *Populus euphratica* in a *Populus tremula* × *Populus alba* hybrid increased drought tolerance. *African Journal of Biotechnology* 10, 9225–9232.

Wang, Y.C., Qu, G.Z., Li, H.Y., Wu, Y.J., Wang, C., *et al.* (2010) Enhanced salt tolerance of transgenic poplar plants expressing a manganese superoxide dismutase from *Tamarix androssowii*. *Molecular Biology Reports* 37, 1119–1124.

Weiser, C.J. (1970) Cold resistance and injury in woody plants. *Science* 169, 1269–1278.

Welling, A. and Palva, E.T. (2008) Involvement of CBF transcription factors in winter hardiness in birch. *Plant Physiology* 147, 1199–1211.

Welling, A., Moritz, T., Palva, E.T., and Junttila, O. (2002) Independent activation of cold acclimation by low temperature and short photoperiod in hybrid aspen. *Plant Physiology* 129, 1633–1641.

Wilkins, O., Waldron, L., Nahal, H., Provart, N.J., and Campbell, M.M. (2009) Genotype and time of day shape the *Populus* drought response. *Plant Journal* 60, 703–715.

Wisniewski, M.E., Bassett, C.L., Renaut, J., Farrell, R. Jr., Tworkoski, T., and Artlip, T.S. (2006) Differential regulation of two dehydrin genes from peach (*Prunus persica*) by photoperiod, low temperature and water deficit. *Tree Physiology* 26, 575–584.

Xing, H.T., Guo, P., Xia, X.L., and Yin, W.L. (2011) *Pd*ERECTA, a leucine-rich repeat receptor-like kinase of poplar, confers enhanced water use efficiency in *Arabidopsis*. *Planta* 234, 229–241.

Xue, L.-J., Alabady, M.S., Mohebbi, M., and Tsai, C.-J. (2015) Exploiting genome variation to improve next-generation sequencing data analysis and genome editing efficiency in *Populus tremula* × *alba* 717-1B4. *Tree Genetics and Genomes* 11, 82.

Yamaguchi-Shinozaki, K. and Shinozaki, K. (1993) Characterization of the expression of a desiccation-responsive *rd29* gene of *Arabidopsis thaliana* and analysis of its promoter in transgenic plants. *Molecular and General Genetics* 236, 331–340.

Yamaguchi-Shinozaki, K. and Shinozaki, K. (2006) Transcriptional regulatory networks in cellular responses and tolerance to dehydration and cold stresses. *Annual Review of Plant Biology* 57, 781–803.

Ye, X., Busov, V., Zhao, N., Meilan, R., McDonnell, L.M., *et al.* (2011) Transgenic *Populus* trees for forest products, bioenergy and functional genomics. *Critical Reviews in Plant Sciences* 30, 415–434.

Yoo, C.Y., Pence, H.E., Hasegawa, P.M., and Mickelbart, M.V. (2009) Regulation of transpiration to improve crop water use. *Critical Reviews in Plant Sciences* 28, 410–431.

Yoon, S.K., Park, E.J., Choi, Y.I., Bae, E.K., Kim, J.H., *et al.* (2014) Response to drought and salt stress in leaves of poplar (*Populus alba* × *Populus glandulosa*): expression profiling by oligonucleotide microarray analysis. *Plant Physiology and Biochemistry* 84, 158–168.

Yu, X., Kikuchi, A., Matsunaga, E., Morishita, Y., Nanto, K., *et al.* (2009) Establishment of the evaluation system of salt tolerance on transgenic woody plants in the special netted-house. *Plant Biotechnology* 26, 135–141.

Zhang, L., Liu, M., Qiao, G., Jiang, J., Jiang, Y., and Zhuo, R. (2013) Transgenic poplar "NL895" expressing *CpFATB* gene shows enhanced tolerance to drought stress. *Acta Physiologiae Plantarum* 35, 603–613.

Zimin, A., Stevens, K.A., Crepeau, M.W., Holtz-Morris, A., Koriabine, M., *et al.* (2014) Sequencing and assembly of the 22-gb loblolly pine genome. *Genetics* 196, 875–890.

10 Engineering Pest Tolerance

Introduction

Genetically engineered pest (generally insects and diseases) resistance is not only commercially important but also can be environmentally beneficial because it will likely lead to reduced reliance on synthetic pesticides. Commonly used pesticides kill a broad spectrum of organisms, many of which are beneficial, not just the intended target. Competing vegetation is widely viewed as a type of pest, so herbicide tolerance is also discussed in this chapter. There is clear evidence that adoption of herbicide-tolerant agronomic crops has led not only to a reduction in the use of herbicides but also to the utilization of more environmentally benign active ingredients (Ault *et al.*, 2016). This chapter summarizes the ways in which genetic engineering has been used to improve the survival, growth, and defense against various pests of poplar (*Populus* spp.).

Insects

Insect pests are highly detrimental to productivity in poplar plantations globally. To date, much of the tree-related research on engineering pest resistance has focused on poplar because of its commercial importance and its amenability to genetic improvement (both conventional breeding and genetic engineering). Only a few of the approximately 150 insect species that feed on poplars in North America are considered serious pests. Nevertheless, this small number of insect species can hinder establishment, reduce growth, and lead to mortality in managed populations (Dickmann and Stuart, 1983; Coyle *et al.*, 2005; Nordman *et al.*, 2005). The cottonwood leaf beetle (*Chrysomela scripta*) is widespread and is the most destructive defoliating insect in North American poplar plantations (Harrell *et al.*, 1981, 1982; Coyle *et al.*, 2005). Outbreaks of this polyvoltine insect can cause severe defoliation, with young trees being especially vulnerable (Hart *et al.*, 1996).

Gruppe *et al.* (1999) reported up to 50% defoliation on various poplar genotypes in Europe; most of the damage was caused by a few of the most prevalent species of herbivorous insects, such as the aspen leaf beetle (*Chrysomela tremulae*), the brassy willow beetle (*Phratora vitellinae*), and the leaf-rolling weevil (*Byctiscus populi*).

In China, there are in excess of 700,000 ha of poplar plantations, accounting for 19% of the total artificially reforested area (Lu, 2008). Damage to poplar plantations by the poplar looper (*Apochemia cineraria*) and the gypsy moth (*Lymantria dispar*) has resulted in up to 40% tree mortality (Hu *et al.*, 2001). In 2002, genetically engineered insect-resistant black poplar (*Populus nigra*) was approved for commercial deployment by the Chinese Gene Security Committee (Su *et al.*, 2003). China is the first country to allow fertile, genetically engineered forest-tree species to be grown commercially.

Two methods have been used to genetically engineer insect resistance in trees: (i) upregulating the expression of a native resistance gene and (ii) introducing a resistance gene from a heterologous source (Hjältén and Axelsson, 2015). A disadvantage of the former approach is that some specialist insects that coevolved with their hosts have developed mechanisms whereby they use plant-derived compounds to synthesize defense chemicals of their own (Pasteels *et al.*, 1983). In addition, coevolution of herbivorous insects and their hosts could cause compounds that previously served as deterrents to become feeding stimulants (Hjältén *et al.*, 2007). Because the target insects have typically not been exposed to compounds produced as the result of heterologous gene expression, their ability to overcome their effects may be restricted. Therefore, the products of non-native gene expression tend to result in higher mortality of the target insects. However, there are indications that toxins encoded by heterologous transgenes can interact antagonistically with innate defense compounds (Bauce *et al.*, 2006). Moreover, effective control of the most abundant pest species can be advantageous to less dominant species, leading to the outbreak of a secondary pest (Raffa, 1989).

Introducing heterologous genes

Bacillus thuringiensis *(Bt) endotoxins*

Bacillus thuringiensis is a gram-positive bacterium that produces insecticidal proteins in the form of parasporal crystals. These crystals, which include one or more Cry and Cyt proteins, are also known as δ-endotoxins (Bravo *et al.*, 2007). These endotoxins are activated within the gut of the target insect by proteolytic cleavage, and operate by various modes of action. They commonly cause the lysis of epithelial cells in the midgut, resulting in the development of pores, eventually leading to insect death (Knowles and Dow, 1993; Bravo *et al.*, 2007). These naturally occurring endotoxins have been used safely as insecticides for many years in and on numerous plant species (Carozzi and Koziel, 1997).

Bacillus thuringiensis toxins are selective in two ways: they only affect insects and each toxin is only effective against a narrow group of closely related species; hence, they have few non-target effects (James, 1997). Coleopteran beetles and lepidopteran caterpillars are the two major groups of poplar insect pest; both are susceptible to Bt (see below). More than 150 different Cry proteins have been classified (Schnepf *et al.*, 1998); thus, trees genetically engineered to produce various types of Bt toxin can provide desirable and beneficial alternatives for protecting tree plantations from a wide array of insect pests.

Endogenous expression of transgenes encoding Bt toxins may be preferable to topical applications of purified Bt proteins for many reasons. First, plants, soil, and water adjacent to the crop would not be contaminated by spray drift and/or runoff. Therefore, non-target insects present in and around areas where transgenic plants are grown would not be exposed, reducing the risk of them developing resistance to Bt toxins. Second, exogenously applied Bt degrades quickly, often within a few days (Thompson *et al.*, 1995; James *et al.*, 1999). Because genetically engineered plants produce the toxin constitutively, it is not necessary to carefully time the applications, or bear the cost of repeated treatments. Moreover, because transgenic trees produce the toxin *in planta*, insects residing in the stem (e.g. wood borers) are vulnerable. For many of these pests, there are no insecticides available to effectively target the developmental stage that is most damaging to the host plant.

McCown *et al.* (1991) were the first to stably express an introduced Bt toxin gene in poplar. One of their transgenic trees exhibited high levels of protection against both the forest tent caterpillar (*Malacosoma disstria*) and the gypsy moth. Subsequently, James *et al.* (1999) demonstrated that a protein encoded by the *Cry3A* gene was highly effective against the cottonwood leaf beetle. In a related study, a binary vector containing a codon-optimized *Cry3Aa* gene, driven by the promoter from the cauliflower mosaic virus (CaMV) 35S gene, was introduced into four genotypes of hybrid cottonwood (*P. trichocarpa* × *P. deltoides*, clones 24-305, 50-197, and 198-434; and *P. deltoides* × *P. nigra*, clone OP-367) (Meilan *et al.*, 2000). The transgene was flanked by matrix-attachment regions (MARs), elements that had previously been shown to elevate transgene expression (Han *et al.*, 1997). In addition, the level of expression enhancement is affected by transgene copy number (Li *et al.*, 2013). In the field study referred to above, dead adult cottonwood leaf beetles were found on the transgenic trees (R. Meilan, unpublished results). This is unusual because adult insect are less susceptible to Bt than the larvae, and the adults consume much less plant tissue than the larvae.

An initial screening of 502 trees derived from 51 lines (i.e. independent transformation events) led to the identification of transgenic trees with reduced insect damage that grew similarly to the non-transgenic controls (Klocko *et al.*, 2014). A large-scale field trial, which included 402 trees from nine lines, conducted over two growing seasons, revealed

Fig. 10.1. Insect damage seen in a large-scale field trial of poplar expressing a *Cry3Aa* gene. (A) Control (non-transgenic) tree with severe insect herbivory. (B) An adjacent transgenic tree with little damage in the initial screening. (C) The top of a control tree on which there has been significant feeding. (D) A large transgenic tree with little whole-tree damage. (E) The view down two rows of trees in the study showing control trees (left) and transgenic trees (right); note that the foliage is denser in the latter than in the former. (F) Cottonwood leaf beetle larvae observed feeding on leaves of non-transgenic trees. (After Klocko *et al.*, 2014.)

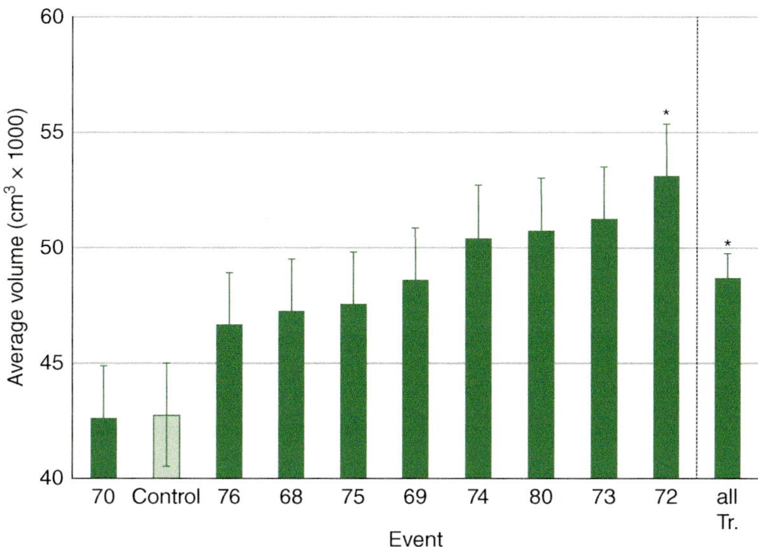

Fig. 10.2. Average volumetric growth of transgenic and control poplar (clone OP-367) at the end of the second growing season in a large-scale field trial conducted in eastern Oregon. The "all Tr." bar shows the average growth of all transgenic trees. Asterisks indicate significant differences ($p < 0.05$). Bars indicate standard error of the mean, based on trees within a line. (After Klocko *et al.*, 2014.)

reduced tree damage, and significantly increased volumetric growth (Figs 10.1 and 10.2). Measurement of the Cry3Aa protein demonstrated high levels of transgene expression continued after 14 years of annual or bi-annual coppicing (Klocko *et al.*, 2014).

In China, the field performance of black poplar lines expressing the Bt *Cry1Ac* gene was evaluated at the Manas Forest Station in the Xinjiang Uygur Autonomous Region between 1994 and 1997 (Hu *et al.*, 2001). The insect pests that cause severe damage on poplar plantations in this region are *Apocheima cinerarius* and *Orthosia incerta*. Only 10% of the leaves on the transgenic trees were highly damaged, while a similar level of herbivory was seen on 80–90% of the leaves on control trees. The average number of pupae per m^2 at a soil depth of 20 cm in a plantation where these transgenic poplars were grown was 18, which was only 20% of the levels seen in a field where corresponding non-transgenic trees were grown. The authors noted that the number of pupae and the amount of herbivory on the engineered trees was far below the threshold set for standard, chemical-control treatments. Non-transgenic poplars grown near the insect-resistant trees exhibited a re-

duced level of insect herbivory, indicating that cross-protection was imparted by the engineered trees.

Génissel *et al.* (2003) reported that leaves of hybrid aspen (*P. tremula* × *P. tremuloides*) expressing a *Cry3Aa* transgene were highly resistant to damage by the phytophagous beetle *Chrysomela tremulae*. They used a synthetic Bt gene in which codon usage had been modified to remove AT-rich regions. Insects feeding on leaves of high- and low-expressing transgenic lines died within 2–13 days of exposure, depending on their stage of development; older instars and adult beetles survived longer but eventually were killed. A high level of resistance is important for the development of a successful resistance-management program (Estruch *et al.*, 1997). It was subsequently shown that the transgenic aspen described by Génissel *et al.* (2003) was also resistant to the leaf beetle *Phratora vitellinae* under greenhouse conditions, but the observed growth benefit may have been context dependent (Hjältén *et al.*, 2012).

Axelsson *et al.* (2012) went on to evaluate the performance of two of the transgenic aspen lines (high- and low-expressing) described by Génissel *et al.* (2003) under semi-natural field conditions (potted plants grown in the field). In general, leaf damage was shown to be higher on the non-transgenic plants than on those with high transgene expression levels. However, biomass accumulation was not significantly greater for the transgenic plants. Moreover, feeding damage from *Byctiscus populi* was similar on the transgenic and non-transgenic plants. The authors concluded that more spatially and temporally elaborate field trials are needed.

Resistance management

There is concern that, through their widespread use, insect populations could develop resistance to Bt toxins (DiCosty and Whalon, 1997; James, 1997; Roush and Shelton, 1997). Before transgenic plants can be grown commercially in the USA, the developer must submit a petition for non-regulated status to the Animal and Plant Health Inspection Service (Meilan, 2006; Chapter 11, this volume). The U.S. Environmental Protection Agency (EPA) requires that petitions for plants containing a Bt gene (referred to as a plant-incorporated protectant, or PIP) must include a resistance-management plan, which describes steps that will be taken to reduce the risk of insects developing resistance to the PIP.

Based on prior experience with synthetic pesticides, several management strategies have been proposed to reduce the risk of insects becoming resistant to PIPs (e.g. Luttrell and Caprio, 1996; Roush, 1997; Gould, 1998; McGaughey *et al.*, 1998). Gene stacking is one approach to reducing the risk of insects becoming resistant to Bt proteins (Nwanze *et al.*, 1995; Maredia, 1997; Roush, 1997). This method involves engineering plants with more than one endotoxin gene, each of which kills the target insect via a different mode of action. The likelihood that an insect will become resistant to more than one toxin concurrently is the product of the probabilities of it becoming resistant to each toxin separately. Thus, for example, if the probability of an insect in a population becoming resistant to one Bt toxin is one in 10^6 and the probability of it becoming resistant to a second toxin alone is one in 10^5, the probability of it simultaneously becoming resistant to both toxins is one in 10^{11}. This strategy has been effective for managing resistance in a variety of insect species, including the cotton bollworm (*Helicoverpa armigera*; Zhao *et al.*, 1997).

Resistance-management plans often specify the establishment of refugia, which are adjoining areas on which non-transgenic plants of the same species must be grown. Insects cannot become resistant unless they are exposed to the toxin, and the greater the sublethal exposure, the greater the chance of resistance developing. Insects feeding on non-transgenic plants will remain vulnerable to the toxin. Mating of these susceptible insects with those that have become resistant by feeding on transgenic plants will prevent the fixation of resistance (Estruch *et al.*, 1997).

Some growers spray insect-infested trees with a liquid formulation containing one or more Bt toxins. The drawback to this approach is that non-target insects are much more likely to encounter exogenously applied Bt toxin than a PIP (where the insect actually has to consume some transgenic plant material in order to become exposed), which also can contribute to the development of resistant insects. However, both exogenous and endogenous applications of Bt are preferable to the use of broad-spectrum insecticides, which indiscriminately kill a wide range of insects, both targeted and beneficial.

Other toxin genes

Although the most commonly used approach for genetically engineering insect resistance involves the use of genes encoding insecticidal proteins from Bt, a variety of plants have also been engineered to express plant-derived transgenes that encode other insecticidal proteins. These include, among others, enzyme inhibitors, lectins, and hydrolytic enzymes (Jongsma and Bolter, 1997; Gatehouse and Gatehouse, 1998; Jouanin *et al.*, 1998).

For example, Dowd *et al.* (1998) tried expressing a tobacco (*Nicotiana tabacum*) anionic peroxidase in sweetgum (*Liquidambar styraciflua*). Overall, leaves from the engineered trees were more resistant to herbivory by caterpillars and beetles than leaves from non-transgenic control plants. However, the level of resistance depended on insect size and leaf ontogeny. The authors concluded that the tobacco anionic peroxidase might be more effective if it was expressed in a plant more closely related to the species from which the gene was derived.

A cowpea (*Vigna unguiculata*) trypsin inhibitor (CpTI) was the first heterologous-gene alternative to Bt to be expressed successfully and to impart useful levels of insect resistance (Bell *et al.*, 2005). The CpTI polypeptide belongs to the Bowman–Birk class of double-headed serine proteinase inhibitors (Hammond *et al.*, 1984; Hilder *et al.*, 1989; Ryan, 1989, 1990; Liu *et al.*, 1993). It has been highly effective at modifying the growth and development of a broad array of insects within the orders Coleoptera, Lepidoptera, and Orthoptera (Hilder *et al.*, 1990; Bell *et al.*, 2001). It is thought that insects are unlikely to become resistant to this toxin given that it binds competitively to an active site that is conserved among various digestive enzymes. Because it is very susceptible to pepsin, it is not likely to be harmful to mammals (Zhang *et al.*, 2005). However, while expression of the gene encoding CpTI conferred resistance to the bright-line brown-eye moth (*Lacanobia oleracea*), it adversely affected the ability of the ectoparasitic wasp *Eulophus pennicornis* to parasitize this lepidopteran plant pest. These results demonstrate how transgene expression can have unintended consequences for beneficial biological control agents (Bell *et al.*, 2005).

Chinese white poplar (*Populus tomentosa*) is indigenous to the middle-eastern region of China, where it occupies approximately 1 million km^2 (approximately one-ninth of the total area of China) (Zhang *et al.*, 2005). It is valued for its rapid growth, high yields, short rotations, and superior wood. It is also widely used for afforestation and landscaping in the north of China (Zhu and Zhang, 1997). Triploids produced using it ((*P. tomentosa* × *P. bolleana*) × *P. tomentosa*) (Zhang *et al.*, 1992, 1997) have greatly improved photosynthetic rates (Li and Zhang, 2000) and wood quality (Pu *et al.*, 2002; Xing and Zhang, 2002) but are very susceptible to insect herbivory. These triploids are good candidates for genetic engineering because of an established *in vitro* regeneration system (Hao *et al.*, 1999; Lu *et al.*, 2001) and their reduced fertility (Zhang *et al.*, 2000), decreasing the possibility of transgene spread. When transformed with the *CpTI* gene, one of these triploid poplars acquired resistance to three important insect pests: forest tent caterpillar, gypsy moth, and willow moth (*Stilpnotia candida*) (Zhang *et al.*, 2005).

Lines of the hybrid poplar clone 741 (*P. alba* × (*P. davidiana* × *P. simonii*) × *P. tomentosa*) containing both an arrowhead proteinase inhibitor (*API*) gene and a Bt *Cry1Ac* gene have exhibited high levels of resistance to the scarce chocolate-tip moth (*Clostera anachoreta*) and *Lymantria dispar*, and are currently being grown commercially in China (Zheng *et al.*, 2000). Similarly, Zhang *et al.* (2011) expressed in poplar (clone BOGA-5) genes encoding two proteins to which coleopteran insects are sensitive: the δ-endotoxin *Cry3A* gene from Bt and the proteinase inhibitor oryzacystatin I gene (*OC-I*) from rice (*Oryza sativa*). When larvae from the target pest, the leaf beetle *Plagiodera versicolora*, were fed leaves from transgenic plants, their mortality was 76.7% higher than for controls. However, these authors also showed that there were no significant differences in the mortality, exuviation index, pupation rate, or adult eclosion rate when *Clostera anachoreta*, a non-target insect, was fed leaves from transgenic and non-transgenic poplars. Furthermore, field-grown transgenic poplar remained resistant to coleopteran insects for all 3 years of this study, demonstrating the potential for using a two-gene strategy for pest management in commercial plantations.

Various other genes have been evaluated for their effect on insect pests of poplar. For example, Confalonieri *et al.* (1998) transformed black poplar (cv. Jean Pourret) with a soybean (*Glycine max*) gene encoding a Kunitz trypsin proteinase inhibitor (*KTi$_3$*). An *in vitro* assay revealed that trypsin-like digestive proteinases from the polyphagous gypsy moth and the black-back prominent moth (*Clostera anastomosis*) were inhibited by the KTi$_3$ protein produced in selected transgenic plants. Bioassays were performed on both insect species using leaf material from engineered and WT plants. There were no significant differences in larval mortality and growth, or pupal weight, for either species when reared on leaves from transgenic and control plants. It is possible that higher expression levels or a different proteinase inhibitor would have been more effective against these pests.

More promising results were obtained with white poplar (*P. alba*) expressing an *Arabidopsis thaliana* cysteine proteinase inhibitor gene, *Atcys*, under the control of the CaMV 35S promoter (Delledonne *et al.*, 2001). Up to 100% mortality was observed for larvae of the poplar leaf beetle (*Chrysomela populi*) after feeding on leaf tissue from selected transgenic lines for 16 days. Using an *in vitro* assay, papain activity was inhibited by extracts from these lines. In all transgenic lines that were evaluated, a correlation was seen between the level of papain inhibition and resistance to this leaf beetle.

Transforming hybrid poplar (*P. deltoides* × *P. simonii*) with a gene encoding an insect-specific neurotoxin from scorpion, *AaIT*, led to tolerance of the gypsy moth. Insects reared on leaves from these transgenic poplars exhibited slower growth and higher mortality (Wu *et al.*, 2000, as reported in Lin *et al.*, 2006). Finally, expression of an agglutinin gene in hybrid poplar led to lines that were highly resistant to longhorn beetles (*Apriona germari*) (Wang *et al.*, 2002, as reported in Lin *et al.*, 2006).

Upregulating native genes

Because condensed tannins (CTs) are known to deter insect feeding (Miranda *et al.*, 2007; Barbehenn and Constabel, 2011), Boeckler *et al.* (2014) upregulated a native tannin regulatory gene, *MYB134*, in hybrid aspen (*P. tremula* × *P. tremuloides*). Overexpression of this *MYB* gene led to a dramatic increase in the content of the CT proanthocyanidin, along with changes in the levels of other phenolic metabolites. The authors conducted a series of bioassays with gypsy moth and forest tent caterpillars to determine how these metabolic perturbations affected feeding preference and the performance of generalist, tree-feeding insects. Surprisingly, they found that both insect species showed an obvious preference for trees with high tannin levels, and gypsy moth performance was enhanced (greater pupal weight and reduced time to pupation), relative to control plants, when it was reared on the transgenics. The authors attributed these unexpected results to decreases in the levels of one or more of the other metabolites (e.g. salicinoids, which are phenolic glycosides, salicortin, and tremulacin) present in their CT-rich transgenic plants. This outcome demonstrates that overexpression of just one regulatory gene can have a dramatic effect on insect herbivory.

There are many possible roles for polyphenol oxidase (PPO) in plants (Wang and Constabel, 2004), but there is evidence that it is involved in defense against various pests (Constabel *et al.*, 1996). Wang and Constabel (2004) overexpressed a *Populus* gene encoding PPO, *PtdPPO1*, in hybrid aspen (*P. tremula* × *P. alba*) to investigate its putative defensive role. Transgenic plants contained high levels of *PtdPPO1* mRNA and its corresponding protein, and elevated PPO activity. Forest tent caterpillar larvae that were fed leaves from the *PtdPPO1*-overexpressing plants had significantly lower weight gain and higher mortality relative to larvae that were fed leaves from control plants. Although proteolytic cleavage is required for PPO activity, the enzyme extracted from caterpillar frass was in its fully activated form. These researchers also showed that PPO activity increased with leaf age. They believed that PPO latency and its activation during digestion argue for it imparting an adaptive and defense-related trait.

Trichomes are small hairs or outgrowths of leaf epidermal cells that are thought to be involved in reducing transpirational water loss and discouraging herbivory. In a screen of a population of *P. tremula* × *P. alba* (INRA 717-1B4) that was mutagenized via activation tagging, Plett *et al.* (2010) identified a mutant line with an overabundance of foliar trichomes, hence the designation *fuzzy*. Relative to its non-transgenic progenitor, the mutant line also showed a 35% increase in its rate of growth and a 200% increase in its rate of photosynthesis. In addition, the mutant was significantly less susceptible to herbivory by larvae of the white-spotted tussock moth (*Orgyia leucostigma*), an insect pest of poplar. The profusion of trichomes was attributed to increased expression of the *PtaMYB186* gene, which is an ortholog of a gene known to control trichome initiation in *A. thaliana*, *MYB106*. The enhanced-trichome phenotype was reproduced in poplar by overexpressing *PtaMYB186*. The *fuzzy* mutant also exhibited changes in the transcript abundance of several other genes related to trichome differentiation. As the authors pointed out, it is noteworthy that mis-expression of a gene involved in trichome differentiation affected both plant growth rate and insert herbivory. This unconventional approach to plant improvement may be more acceptable to the public and governmental regulators than the incorporation of heterologous genes that encode various types of toxins.

Diseases

Forest trees are susceptible to attack by a wide variety of fungal, bacterial, and viral pathogens. Trees genetically engineered for disease resistance can provide both ecological and economic benefits. *De novo* or elevated disease resistance has been achieved through the use of an assortment of genes obtained from heterologous sources, with varying amounts of success.

Fungi

Poplar

Fungal infections can severely damage trees, particularly poplars. Historically, hybrid poplars have been grown for fuel, pulpwood, and various solid-wood products (Ostry and McNabb, 1985). Increasingly, they are being grown as a bioenergy crop. However, use of hybrid poplar is limited in the eastern USA because of its susceptibility to the fungal pathogen *Sphaerulina musiva* (previously known as *Septoria musiva*). This fungus causes necrotic lesions on leaves and cankers on stems and branches (Moore and Wilson, 1983). Infection can reduce photosynthetic capacity and lead to premature defoliation, resulting in decreasing annual growth. Cankers on the bole can make the tree vulnerable to secondary infection and stem breakage. In addition, the wood of infected trees is unsuitable for pulp production (Ostry and McNabb, 1983). Growth losses due to this pathogen vary by genotype, but reductions as high as 63% have been reported in plantations (McNabb *et al.*, 1982). Although chemical and cultural treatments have been tried to control *Septoria* leaf spot and canker, they have not been completely effective (Ostry and McNabb, 1983; Yang *et al.* 1994).

Several plant-derived genes have been evaluated for their ability to impart fungal resistance, but results have been mixed. Expression of the bacterio-opsin gene (*bO*) from *Halobacterium halobium* in tobacco resulted in what resembled a hypersensitive response (Mittler *et al.*, 1995). Unfortunately, the expression of *bO* in hybrid poplar did not lead to protection against a variety of fungal pathogens, including leaf rust, leaf and shoot blight, and stem canker (Mohamed *et al.*, 2001). Similarly, expression of grapevine (*Vitis vinifera*) stilbene synthase (*StSy*), which has been associated with the production of resveratrol compounds, did not significantly reduce the susceptibility of white poplar to *Melampsora pulcherrima*, a fungal rust disease (Giorcelli *et al.*, 2004).

Defensins are a family of small, cysteine-rich, cationic peptides that possess antimicrobial and cytotoxic properties. One defensin, encoded by *NP-1*, which is expressed primarily in rabbit neutrophils, provides protection against many gram-positive and -negative bacteria, fungi, and viruses (Ting *et al.*, 2011). Extracts from *P. tomentosa* transformed with the *NP-1* gene inhibited the growth of *Bacillus subtilis* and *Agrobacterium tumefaciens* (Zhao *et al.*, 1999; Ting *et al.*, 2011).

Many fungal and bacterial pathogens produce compounds that are involved in pathogenicity and influence plant metabolism. One such toxin that is not host-specific is oxalic acid (OA), which can cause infected plants to wilt (Dutton and Evans, 1996). It is released soon after the plant–pathogen interaction is initiated and accumulates as the infection progresses (Maxwell and Lumsden, 1970), resulting in a decrease in apoplastic pH and a weakening of the plant cell wall (Dutton and Evans, 1996). It is thought that chelation of calcium by the oxalate anion compromises the function of calcium-dependent defense responses and is responsible for cell-wall breakdown (Bateman and Beer, 1965). Studies with cell cultures have shown that OA inhibits the oxidative burst that occurs in plants soon after pathogen invasion. This defense response involves the controlled release of O_2^- and H_2O_2 at the site of entry (Cessna *et al.*, 2000).

The enzyme encoded by a wheat (*Triticum aestivum*) germin-like oxalate oxidase gene, *OxO*, is capable of metabolizing OA, producing H_2O_2 and CO_2 in the process. The H_2O_2 liberated by OxOs may have a direct antimicrobial effect or trigger other defense responses in the plant (Lane, 1994; Lane *et al.*, 1993). When the *OxO* gene was expressed in hybrid poplar (*Populus* × *euramericana*) under the control of the CaMV 35S promoter, its transcript was detected in leaves, stems, and roots. Leaves from these transgenic plants were more tolerant of OA and had a greater capacity to elevate the pH of an OA solution relative to control plants. In addition, when leaf discs were inoculated with conidia from *S. musiva*, an important poplar pathogen that produces OA, the *OxO*-expressing plants were more resistant than non-transgenic control plants (Liang *et al.*, 2001).

Chitin is an important structural component of fungal cell walls but is absent from higher plants (Raikhel and Lee, 1993). *Ac*-AMP1, a short polypeptide (29 amino acids), is a small chitin-binding protein that is produced in the coat of *Amaranthus caudatus* seed. This polypeptide inhibits the growth of several plant pathogens at much lower levels than other known antifungal proteins. In addition, it does not agglutinate erythrocytes, suggesting low toxicity to mammalian cells (Broekaert *et al.*, 1992).

Liang *et al.* (2002) investigated antimicrobial peptides in hybrid poplar in an attempt to alter resistance to *S. musiva*. The Ogy (*Populus* × *euramericana*) and NM6 (*P. nigra* × *P. maximowiczii*) hybrid poplar genotypes were transformed with genes encoding *Ac*-AMP1.2 and ESF12; *Ac*-AMP1.2 is an analog of *Ac*-AMP1 in which the amino-terminal valine was replaced with methionine to facilitate expression (Powell and Maynard, 1997). ESF12 is a synthetic polypeptide that can mimic the amphipathic α-helix present in magainins, which are found in skin secretions of the African clawed frog (*Xenopus laevis*). The inhibitory effect of ESF12 on *S. musiva* and other pathogenic fungi has been demonstrated *in vitro* (Powell *et al.*, 1995). Two-year-old transgenic trees expressing a combination of these two antimicrobial peptides showed greater resistance than nontransgenic control plants in leaf-disc assays, and fewer *S. musiva* cankers were found on transgenic trees grown in field trials (unpublished data, as reported in Powell *et al.*, 2006).

Using an alternative strategy, Jia *et al.* (2010) ectopically expressed a chitinase gene derived from *Beauveria bassiana* (*Bbchit1*) in Chinese white poplar. The resulting transgenic trees exhibited greater resistance to a pathogenic fungus (*Cytospora chrysosperma*) than the control trees.

Matias Kirst and J.A. Smith (unpublished data, School of Forest Resources and Conservation, University of Florida, Florida, USA) have identified significant genetic variation for resistance to the main fungal diseases of *P. deltoides*. A genome-wide search for associations between DNA variants and field-measured disease traits detected a suite of loci that contribute to resistance to these pathogens, including several previously characterized disease-resistance genes. Unfortunately, most of the elite poplar genotypes selected for biomass productivity and biofuel conversion efficiency do not carry these favorable alleles. Although conventional breeding could be used to introduce these alleles into desirable germplasm, it would take decades to achieve this outcome because of the relatively long juvenile periods for poplar (5–7 years). Moreover, it is unknown how useful these resistance loci would be in some of the areas where commercial poplar plantations are likely to be established. Efforts are currently under way to verify the effectiveness of these disease-resistance loci across the geographic range of *P. deltoides* and to incorporate those that are beneficial into elite poplar germplasm through genome editing.

Hardwood restoration

American chestnut (*Castanea dentata*) was virtually eliminated as an overstory species by the chestnut blight fungus (*Cryphonectria parasitica*), a disease that was inadvertently introduced into the USA in the early 1900s. Likewise, American elm (*Ulmus americana*) was effectively eliminated from the northeastern USA in the late 1920s following the accidental release of the fungus that causes Dutch elm disease (*Ophiostoma ulmi*). Progress has been made toward restoration of both tree species via conventional means, such as identifying and propagating tolerant cultivars or breeding with sexually compatible, tolerant species. However, in cases where resistance alleles are absent from native populations, inserting genes from heterologous species may provide the only means of restoring them. In other cases, transgenesis may be used as part of an integrated strategy, complementing conventional breeding and/or biocontrol agents (Merkle *et al.*, 2007).

Until recently, significant progress with transformation and *in vitro* regeneration of forest-tree species has been restricted almost exclusively to those within the genus *Populus*. In addition, compared with other hardwood species, most progress in the areas of genomics and genetic modification has been with *Populus* spp. However, advances in *in vitro* propagation, particularly somatic embryogenesis, have led to the development of clonal propagation methods for commercially important hardwood tree species, especially *Eucalyptus* spp. Moreover, transformation/regeneration systems have been developed for a number of hardwood tree species (Merkle and Nairn, 2005; Wang and Pijut, 2014).

For example, elm and chestnut have been genetically engineered with antifungal genes to impart resistance to the fungi causing Dutch elm disease and chestnut blight (Merkle and Nairn, 2005). English elm (*Ulmus procera*) plantlets transformed via biolistics with (unspecified) antifungal genes have been challenged with Dutch elm disease (Gartland *et al.*, 2005). Notwithstanding these advances, the lack of field-test results suggests that durable resistance remains elusive (Harfouche *et al.*, 2011). American elm trees have been transformed with a gene encoding the synthetic antimicrobial peptide ESF39A under the control of a vascular-specific promoter. Four independent, single-copy lines were regenerated into rooted plants. These lines exhibited less wilting and sapwood staining than non-transgenic control plants, following inoculation with *Ophiostoma novo-ulmi*, but it has not been shown that the resistance is stable. However, initial evaluations revealed that there was no significant difference in mycorrhizal colonization between the genetically engineered and WT trees (Newhouse *et al.*, 2007).

Polin *et al.* (2006) transformed cells in somatic embryos of American chestnut and regenerated phenotypically normal rooted plants that expressed the *OxO* gene. Vascular tissue-specific transgene expression was detected in one of their transgenic lines. Welch *et al.* (2007) cultured American chestnut callus derived from WT and tissue transformed with the *OxO* gene on media containing various concentrations of OA. In response to this treatment, WT tissues exhibited a significant decrease in lignin and an increase in cellulose content, while transformed tissues did not. This led the authors to conclude that overexpressing *OxO* may prove to be useful for preventing changes in the cell-wall composition caused by the oxalate produced during *Cryphonectria parasitica* infection of American chestnut, and might possibly confer tolerance to chestnut blight.

Zhang *et al.* (2013) regenerated transgenic lines of American chestnut expressing the wheat *OxO* gene under the control of promoters from the CaMV 35S gene and the *VspB* gene, which is predominantly expressed in vascular tissue. *OxO* expression levels were approximately 3,000 times higher in the 35S::*OxO* lines than in for those in which *OxO* is under the control of a predominantly vascular promoter (VspB). The authors measured the length of necrotic lesions in leaf assays that were conducted to evaluate blight tolerance *in planta*. Lesion lengths in five of the 35S::*OxO* lines were similar to those seen in blight-tolerant Chinese chestnut (*Castanea mollissima*). Newhouse *et al.* (2014) demonstrated that this enhanced blight tolerance can be transmitted sexually to T_1 seedlings. This was confirmed via stem inoculations of 3–4-year-old, field-grown T_1 trees. The authors emphasized the importance of out-crossing for maintaining genetic diversity during the process of restoring this keystone species.

Bacteria

Reports of poplars being genetically engineered for resistance to bacterial pathogens are not as plentiful as those for fungi. Although severe bacterial outbreaks are unusual, serious infections do occur and have been reported for *Xanthomonas* spp. (de Kam, 1984; Haworth and Spiers, 1988). A sequence encoding a synthetic antimicrobial polypeptide, D4E1, which consists of 17 amino acids, was introduced into hybrid poplar (*P. tremula* × *P. alba*). The resulting transgenic lines were evaluated for resistance to *A. tumefaciens*, *Xanthomonas populi* pv. *populi*, and *Hypoxylon mammatum*. The line with the highest transgene expression, Tr23, exhibited a significant reduction in susceptibility to *A. tumefaciens* and *X. populi*. However, resistance to one strain of *A. tumefaciens*, C58, was not improved, suggesting that D4E1 lacks generality. Interestingly, none of the transgenic lines showed a significant response when challenged with *H. mammatum*, a fungal pathogen (Mentag *et al.*, 2003).

Competing Vegetation

In addition to insect and disease resistance, another major objective of genetic modification in trees is herbicide tolerance (HT). Genes imparting HT are not thought to be present in the genome of any tree, so they must be derived from heterologous sources. Having HT greatly reduces the cost of controlling competing vegetation in plantations, which is vital to achieving maximal tree growth rates (Fig. 10.3). Below we describe the various ways in which poplar has been engineered for HT.

Glyphosate

Fillatti *et al.* (1987) were first to report on genetically engineered HT in poplar (*P. alba* × *P. grandidentata*). They employed the *Salmonella typhimurium aroA* gene, which encodes 5-enolpyruvyl-3-phosphoshikimate synthase (EPSPS), a key enzyme in the shikimic acid pathway. Unlike the plant enzyme, activity of the bacterial analog is not blocked by glyphosate, the active ingredient in the herbicide Roundup® (Comai *et al.*, 1983). Transformants containing *aroA* under the control of the promoter for the gene encoding mannopine synthase (MAS) resulted in disappointing levels of HT (Riemenschneider *et al.*, 1988), which was attributed to low cytosolic expression of the transgene. Poplar expressing the *aroA* gene fused to a chloroplast transit peptide and driven by the CaMV 35S promoter exhibited somewhat higher levels of glyphosate tolerance (Riemenschneider and Haissig, 1991; Donahue *et al.*, 1994).

Although use of the 35S promoter led to higher transgene expression levels than were seen with MAS, and the transit peptide guided the translation product to the chloroplast, performance of these *aroA*-expressing lines was still lower than desired (Karnosky *et al.*, 1997). During greenhouse trials, the chlorophyll content in the transgenic lines was inversely related to glyphosate concentration, height growth was halted following herbicide treatment, and only one line retained live leaves 6 weeks post-treatment (Donahue *et al.*, 1994).

Subsequently, Meilan *et al.* (2002a) evaluated a genetic construct containing *CP4*, another gene that imparts glyphosate tolerance in plants. This gene, which is derived from *A. tumefaciens* strain CP4, encodes an alternative form of EPSPS that has a low affinity

Fig. 10.3. A trial demonstrating the importance of controlling competing vegetation in poplar plantings (Hibbs *et al.*, 2003). This photo was taken 4 years after the trial was established, near Corvallis, Oregon. The same *Populus* genotypes were planted in the plot depicted in the background as in the one shown in the foreground. In the rear plot, competing vegetation was controlled for the first 2 years of the study, until canopy closure was achieved. In the front plot, no vegetation management was done. The trees in the rear plot were, on average, about 20 m tall at the end of 4 1years; the trees in the foreground were (flagged blue) no more than 2 m tall, if they survived at all.

for glyphosate. Using an *Agrobacterium*-mediated transformation protocol (Leple *et al.*, 1992; Han *et al.*, 2000), Meilan *et al.* (2000) produced transgenic plants in 12 hybrid poplar genotypes. They also transformed the same genotypes with a construct containing both *CP4* and *GOX*. The latter gene encodes an enzyme that inactivates glyphosate by metabolizing it. The resulting lines were field-tested for 2 years on the west and east sides of the Cascade Mountains in northern Oregon. The lines expressing only *CP4* appeared to exhibit less damage in response to glyphosate treatment than those transformed with both genes. Action of the GOX enzyme on glyphosate leads to the production of aminomethylphosphonic acid, a metabolic by-product that causes the development of leaf chlorosis, which can be confused with herbicide damage. Growth of lines expressing only *CP4* was significantly better than those expressing both *CP4* and *GOX*. This was the first report of transgenic poplars displaying high levels of glyphosate tolerance when grown and treated under field conditions (Fig. 10.4). HT remained stable under field conditions for at least four growing seasons (Meilan *et al.*, 2002b).

Ault *et al.* (2016) studied the effect of engineered glyphosate tolerance on the control of competing vegetation and the growth of poplar under field conditions. Using an *Agrobacterium*-mediated transformation protocol, they regenerated 94 lines in four clones of hybrid poplar (three *P. trichocarpa* × *P. deltoides* and one *P. trichocarpa* × *P. nigra*). The transgenic lines were scored for their level of tolerance at two locations in Oregon. Based on the results of this screening, the authors propagated four lines from two hybrid genotypes to study their utility for facile weed control and growth in a 2-year management trial conducted in eastern Oregon. In this study, they compared conventional weed-control practices to foliar applications of Roundup® during the growing season. Herbicide tolerance remained stable in all of the transgenic trees during the course of the study. Control of competing vegetation was substantially improved using broadcast herbicide applications. Transgenic trees subjected to glyphosate treatment grew nearly 20% faster than both the transgenic and non-transgenic control trees that were grown under conventional weed-control regimes.

As a part of this project, these authors also conducted an exploratory life cycle analysis to evaluate the potential benefits of dedicated woody-crop energy plantations. They demonstrated that over a 6-year rotation, which included three coppice cycles, the growth benefit resulting from more effective weed control could provide a nearly 8% saving in greenhouse-gas emissions per unit of wood produced.

Chlorsulfuron

The sulfonylurea herbicide chlorsulfuron blocks the biosynthesis of valine and isoleucine by acting on acetolactate synthase (Ray, 1984). A mutant acetolactate synthase gene from *A. thaliana*, *crs1-1*, confers tolerance to chlorsulfuron. Brasileiro *et al.* (1992) expressed *crs1-1* in hybrid poplar (*P. tremula* × *P. alba*) under the control of either its own promoter or a dual CaMV 35S promoter (2×35S). Both genetic constructs resulted in transgenic plants that, in a greenhouse study, were completely resistant to chlorsulfuron applied at rates that were higher than those typically used in field applications. Control trees died within 2–3 weeks of herbicide application. Despite an initial delay in shoot growth and root development, the transgenic plants subsequently exhibited normal growth and development.

Chloroacetanilides

Glutathione *S*-transferase (GST) catalyzes the attachment of glutathione (GSH) to a variety of xenobiotic molecules, including various herbicides. These GSH–herbicide conjugates generally have lower toxicity and are more water soluble than the free active ingredients (Edwards *et al.*, 2000; Pascal *et al.*, 2000). Acetochlor and metolachlor are herbicides within the chloroacetanilide family whose active ingredients, which are aniline derivatives, can be deactivated through the action of GST (Gullner *et al.*, 2001). The bacterial γ-glutamylcysteine synthetase (γ-ECS) gene encodes an enzyme that catalyzes the rate-limiting step in the biosynthesis of the GSH. When poplar overexpressing γ-ECS in either their cell's cytoplasm or chloroplasts were exposed to acetochlor and metolachlor, the growth of all lines was greatly

Fig. 10.4. Morphology of transgenic lines after glyphosate application. (A) A transgenic hybrid cottonwood (clone 184-402), grown in western Oregon, with no apparent damage. Spray rate: 4.7 l Roundup Pro™/ha (1×). (B) A transgenic hybrid cottonwood (clone 189-434), grown in western Oregon, exhibiting minor damage. Spray rate: 9.3 l Roundup Pro™/ha (2×). (C) A non-transgenic hybrid cottonwood (clone 24-305) that died in response to treatment. (D) A typical transgenic hybrid aspen (clone 717-1B4), grown in western Oregon, free of damage, in response to a 1× spray rate. (E) Example of a transgenic hybrid aspen (clone 717-1B4), grown in eastern Oregon, showing minor damage in response to the 1× spray rate. (F) Highly tolerant transgenic hybrid cottonwood (clone 189-434) in eastern Oregon after the second 2× treatment in the second year of the study (total of four treatments), with no visible damage. (G) Demonstration plot at the eastern Oregon site, 1× spray rate. The small non-transgenic trees, shown in the right foreground (next to the shade cards and flagging) died, whereas all of the transgenics survived and grew (upper left). (H) Transgenic hybrid cottonwood at a western Oregon site after the first 1× treatment in the second year of the study (total of three treatments), showing typical *GOX*-associated chlorosis. All pictures were taken 4 weeks after treatment. (After Meilan *et al*., 2002a.)

reduced, despite their elevated levels of GSH. However, the growth reduction seen in the cytosol-expressing lines was less pronounced than for the lines in which γ-ECS was expressed in the chloroplasts (Gullner *et al.*, 2001). The authors concluded that a more promising way to obtain more tolerant plants would be to utilize isoenzymes of GST that are more herbicide-specific.

Glufosinate

Glufosinate (phosphinothricin, or PPT), which is a structural analog of glutamate, is the active ingredient in the herbicides known as Basta®, Buster®, and Liberty®. Glutamine synthetase (GS) is the key enzyme involved in ammonium assimilation by plants, catalyzing the formation of glutamine from glutamate and ammonium. Interaction with PPT irreversibly inactivates GS. Blocking GS activity not only causes the accumulation of ammonium but also leads to depletion of amino acid pools, both of which are lethal to plants. Treatment with glufosinate initially causes necrosis at or near the apical meristem but ultimately spreads throughout the plant (Pascual *et al.*, 2008). The *BAR* gene encodes phosphinothricin acetyltransferase, an enzyme that inactivates glufosinate by acetylating its available amino group (Thompson *et al.*, 1987). Poplar calli expressing *BAR* were able to survive and grow *in vitro* on a PPT-containing medium, suggesting that it may be useful as a selectable-marker gene (Chupeau *et al.*, 1994).

Li *et al.* (2008) produced 32 independent transformation events (i.e. lines) in two *Populus* hybrids (*P. tremula* × *P. alba* and *P. tremula* × *P. tremuloides*) using a binary vector containing the *BAR* gene under the control of the promoter from an *A. thaliana* gene encoding the small subunit of ribulose-1,5-biphosphate carboxylase/oxygenase. They evaluated tolerance to glufosinate in these lines at the beginning and end of an 8-year field study, during which time the trees were coppiced repeatedly. The lines exhibited varying degrees of tolerance, ranging from near-complete sensitivity to complete tolerance. For the purposes of analysis, the lines were assigned to three classes (tolerant, intermediate, and sensitive). There were no cases of tolerance breakdown and the levels of tolerance seen initially remained unchanged by the end of the study. The authors concluded that commercially useful levels of stable HT can be introduced into elite genotypes of poplar in 2–3 years, from cellular transformation to field testing.

Concluding remarks

Genetically engineered HT is being utilized extensively by farmers in the USA. By 2014, soybeans engineered for HT were grown on 94% of total U.S. acreage dedicated to this crop. Similarly, nearly 91% of total U.S. cotton (*Gossypium hirsutum*) crop was HT (USDA, 2018). The main benefits of HT have been: (i) reduced weed-control costs; (ii) simplified weed-management strategies; (iii) more flexibility with respect to the timing of treatments; (iv) greater use of no- and low-till cropping systems, with their associated environmental benefits (e.g. nutrient cycling and erosion control); and (v) the use of more environmentally benign herbicides (Fernandez-Cornejo and Caswell, 2006; Li *et al.*, 2008). Many of these same benefits are expected for plantation-grown HT trees, particularly those grown under short-rotation intensive culture (Strauss *et al.*, 1997). Finally, given the recent reports of weeds becoming resistant to Roundup® and other herbicides, due in part to prevalence of HT crops, it is vital for growers to utilize strategies for effective management of weed resistance (Heap, 2014).

Summary

Genetically engineered pest (generally insects and diseases) resistance is not only commercially important but can also be environmentally beneficial because it will likely lead to reduced reliance on synthetic pesticides. Commonly used pesticides kill a broad spectrum of insects, many of which are beneficial, not just the intended pest, whereas a transgenic approach can be more targeted. Competing vegetation is widely viewed as a type of pest, so HT can be considered in this context.

Insect pests are recognized as one of the major factors limiting productivity in tree plantations. They can hinder establishment, reduce growth, and lead to mortality in managed populations. Two approaches have been used to genetically engineer insect resistance in trees: (i) upregulating the expression of a native resistance gene and (ii) introducing a resistance gene from a heterologous source. A disadvantage of the former approach is that some specialist insects that coevolved with their hosts have developed mechanisms whereby they use plant-derived compounds to synthesize defense chemicals of their own. In addition, coevolution of herbivorous insects and their hosts could cause compounds that previously served as deterrents to become feeding stimulants. Because the target insects have typically not been exposed to compounds produced as the result of heterologous gene expression, their ability to overcome their effects may be restricted. Therefore, the products of non-native gene expression tend to result in higher mortality of the target insects.

Genes encoding insecticidal proteins produced by Bt have been widely used to engineer insect resistance in plants. These toxins are selective in two ways: they only affect insects, and each toxin is only effective against a narrow group of closely related species; hence, they have few non-target effects. More than 150 different Bt toxins have been classified; thus, they provide desirable and beneficial alternatives for protecting tree plantations from a wide array of insect pests.

There is concern that, through their widespread use, insect populations could develop resistance to Bt toxins. Before transgenic plants can be deployed commercially in the USA, the developer must submit a petition for non-regulated status to the Animal and Plant Health Inspection Service (Chapter 11, this volume). The U.S. EPA requires that petitions for plants containing a Bt gene (referred to as a plant-incorporated protectant, or PIP) must include a resistance-management plan, which describes steps that will be taken to reduce the risk of insects developing resistance to the PIP.

A variety of plants have also been engineered to express plant-derived transgenes that encode other insecticidal proteins. These include, among others, enzyme inhibitors, lectins, and hydrolytic enzymes. Attempts have also been made to upregulate native genes, leading to the overproduction of compounds known to deter insect feeding, such as CTs. However, upregulating a native tannin regulatory gene, which led to a dramatic increase in endogenous levels of a CT, has caused engineered plants to have greater susceptibility to two common, generalist insect pests. Thus, overexpression of just one regulatory gene can have unintended consequences.

Unexpectedly, overexpression of a gene that led to enhanced trichome development caused a dramatic increase in the rate of photosynthesis as well as resistance to larvae herbivory of a poplar insect pest. This unconventional approach to plant improvement may be more acceptable to the public and governmental regulators than the incorporation of heterologous genes that encode various types of toxins.

Forest trees are susceptible to attack by a wide range of fungal and bacterial pathogens. Several plant genes have been tested for their ability to impart fungal resistance, but results have been mixed. Many fungal and bacterial pathogens produce compounds, which are involved in pathogenicity, that influence plant metabolism. One approach to engineering resistance is to insert a gene that encodes an enzyme capable of metabolizing the toxin. Chitin is an important structural component of fungal cell walls but is absent from higher plants. Expression of a gene encoding a chitin-binding protein has inhibited the growth of several plant pathogens to much lower levels than other known antifungal proteins. Finally, expression of a polypeptide similar to one found in skin secretions of the African clawed frog inhibited pathogenic fungi.

In addition to insect and disease resistance, another major objective of genetic modification is trees that are HT. Genes imparting this trait are not thought to be present in the genome of any tree, so they must be derived from heterologous sources. Having HT greatly reduces the cost of controlling competing vegetation in plantations, which is vital to achieving maximal tree growth rates. Trees have been engineered for tolerance to a variety of herbicides, but the most common is glyphosate, the active ingredient in Roundup®. The main benefits of HT have been: (i) reduced weed-control costs; (ii) simplification of weed-management strategies; (iii) more flexibility with respect to the timing of treatments; (iv) greater use of no- and low-till cropping systems, with their associated ecological benefits, such as nutrient cycling and erosion control; and (v) the use of more environmentally benign herbicides.

References

Ault, K., Viswanath, V., Jayawickrama, J., Ma, C., Eaton, J., *et al.* (2016) Improved growth and weed control of glyphosate-tolerant poplars. *New Forests* 47, 653–667.

Axelsson, E.P., Hjältén, J., and Leroy, C.J. (2012) Performance of insect-resistant *Bacillus thuringiensis* (Bt)-expressing aspens under semi-natural field conditions including natural herbivory in Sweden. *Forest Ecology and Management* 264, 167–171.

Barbehenn, R.V. and Constabel, P.C. (2011) Tannins in plant–herbivore interactions. *Phytochemistry* 72, 1551–1565.

Bateman, D.F. and Beer, S.V. (1965) Simultaneous production and synergistic action of oxalic acid and polygalacturonase during pathogenesis by *Sclerotium rolfsii*. *Phytopathology* 55, 204–211.

Bauce, E., Kumbasli, M., van Frankenhuyzen, K., and Carisey, N. (2006) Interactions among white spruce tannins, *Bacillus thuringiensis* subsp. *Kurstaki*, and spruce budworm (Lepidoptera: Tortricidae), on larval survival, growth, and development. *Journal of Economic Entomology* 99, 2038–2047.

Bell, H.A., Fitches, E.C., Down, R.E., Ford, L., Marris, G.C., *et al.* (2001) Effect of dietary cowpea trypsin inhibitor (CpTI) on the growth and development of the tomato moth *Lacanobia oleracea* (Lepidoptera: Noctuidae) and on the success of the gregarious ectoparasitoid *Eulophus pennicornis* (Hymenoptera: Eulophidae). *Pest Management Science* 57, 57–65.

Bell, H.A., Down, R.E., Edwards, J.P., Gatehouse, J.A., and Gatehouse, A.M.R. (2005) Digestive proteolytic activity in the gut and salivary glands of the predatory bug *Podisus maculiventris* (Heteroptera: Pentatomidae): effect of proteinase inhibitors. *European Journal of Entomology* 102, 139–145.

Boeckler, G.A., Towns, M., Unsicker, S.B., Mellway, R.D., Yip, L., *et al.* (2014) Transgenic up-regulation of the condensed tannin pathway in poplar leads to a dramatic shift in leaf palatability for two tree-feeding Lepidoptera. *Journal of Chemical Ecology* 40, 150–158.

Brasileiro, A., Tourneur, C., Leple, J.-C., Combes, V., and Jouanin, L. (1992) Expression of the mutant *Arabidopsis thaliana* acetolactate synthase gene confers chlorsulfuron resistance to transgenic poplar plants. *Transgenic Research* 1, 133–141.

Bravo, A., Gill, S.S., and Soberón, M. (2007) Mode of action of *Bacillus thuringiensis* Cry and Cyt toxins and their potential for insect control. *Toxicon* 49, 423–435.

Broekaert, W.F., Marien, W., Terras, F.R.G., Bolle, M.F.C., de Proost, P., *et al.* (1992) Antimicrobial peptides from *Amaranthus caudatus* seeds with sequence homology to the cysteine/glycine-rich domain of chitin-binding proteins. *Biochemistry* 31, 4308–4314.

Carozzi, N.B. and Koziel, M.G. (1997) *Advances in Insect Control: the Role of Transgenic Plants*. Taylor & Francis, New York.

Cessna, S.G., Sears, V.E., Dickman, M.B., and Low, P.S. (2000) Oxalic acid, a pathogenicity factor for *Sclerotinia sclerotiorum*, suppresses the oxidative burst of the host plant. *Plant Cell* 12, 2191–2199.

Chupeau, M.-C., Pautot, V., and Chupeau, Y. (1994) Recovery of transgenic trees after electroporation of poplar protoplasts. *Transgenic Research* 3, 13–19.

Comai, L., Sen, L.C., and Stalker, D.M. (1983) An altered aroA gene product confers resistance to the herbicide glyphosate. *Science* 221, 370–371.

Confalonieri, M., Allegro, G., Balestrazzi, A., Fogher, C., and Delledonne, M. (1998) Regeneration of *Populus nigra* transgenic plants expressing a Kunitz proteinase inhibitor (KTi_3) gene. *Molecular Breeding* 4, 137–145.

Constabel, C.P., Bergey, D.R., and Ryan, C.A. (1996) Polyphenol oxidase as a component of the inducible defense response in tomato against herbivores. In: Romeo, J.T., Saunders, J.A., and Barbosa, P. (eds.) *Phytochemical Diversity and Redundancy in Ecological Interactions*. Plenum Press, New York, pp. 231–252.

Coyle, D.R., Nebeker, T.E., Hart, E.R., and Mattson, W.J. (2005) Biology and management of insect pests in North American intensively managed hardwood forest systems. *Annual Review of Entomology* 50, 1–29.

de Kam, M. (1984) *Xanthomonas campestris* pv. *populi*, the causal agent of bark necrosis in poplar. *European Journal of Plant Pathology* 90, 13–22.

Delledonne, M., Allegro, G., Belenghi, B., Balestrazzi, A., Picco, F., *et al.* (2001) Transformation of white poplar (*Populus alba* L.) with a novel *Arabidopsis thaliana* cysteine proteinase inhibitor and analysis of insect pest resistance. *Molecular Breeding* 7, 35–42.

Dickmann, D.I. and Stuart, K.W. (1983) *The Culture of Poplars in Eastern North America*. Department of Forestry, Michigan State University, East Lansing, Michigan.

DiCosty, U. and Whalon, M. (1997) Selection of Colorado potato beetle resistant to CryIIIA on transgenic potato plants. *Resistant Pest Management Newsletter* 9, 33–34.

Donahue, R.A., Davis, T.D., Riemenschneider, D.E., Michler, C.H., Carter, D.R., *et al.* (1994) Growth, photosynthesis, and herbicide tolerance of genetically modified hybrid poplar. *Canadian Journal of Forest Research* 24, 2377–2383.

Dowd, P.F., Lagrimini, L.M., and Herms, D.A. (1998) Differential leaf resistance to insects of transgenic sweetgum (*Liquidambar styraciflua*) expressing tobacco anionic peroxidase. *Cellular and Molecular Life Sciences* 54, 712–720.

Dutton, M.V. and Evans, C.S. (1996) Oxalate production by fungi: its role in pathogenicity and ecology in the soil environment. *Canadian Journal of Microbiology* 42, 881–895.

Edwards, R., Dixon, D.P., and Walbot, V. (2000) Plant glutathione S-transferases: enzymes with multiple functions in sickness and in health. *Trends in Plant Science* 5, 193–198.

Estruch, J.J., Carozzi, N.B., Desai, N., Duck, N.B., Warren, G.W., and Koziel, M.G. (1997) Transgenic plants: an emerging approach to pest control. *Nature Biotechnology* 15, 137–141.

Fernandez-Cornejo, J. and Caswell, M. (2006) The first decade of genetically engineering crops in the United States. Economic Information Bulletin no. 11. Economic Research Service, USDA, Washington, DC.

Fillatti, J.J., Sellmer, J., McCown, B., Haissig, B., and Comai, L. (1987) *Agrobacterium* mediated transformation and regeneration of *Populus*. *Molecular and General Genetics* 206, 192–199.

Gartland, K.M.A., McHugh, A.T., Crow, R.M., Garg, A., and Gartland, J.S. (2005) 2004 SIVB congress symposium proceeding: biotechnological progress in dealing with Dutch elm disease. *In Vitro Cellular & Developmental Biology—Plant* 41, 364–367.

Gatehouse, A.M.R. and Gatehouse, J.A. (1998) Identifying proteins with insecticidal activity: use of encoding genes to produce insect resistant transgenic crops. *Pesticide Science* 52, 165–175.

Génissel, A., Leplé, J.-C., Millet, N., Augustin, S., Jouanin, L., and Pilate, G. (2003) High tolerance against *Chrysomela tremulae* of transgenic poplar plants expressing a synthetic *cry3Aa* gene from *Bacillus thuringiensis* ssp. *tenebrionis*. *Molecular Breeding* 11, 103–110.

Giorcelli, A., Sparvoli, F., Mattivi, F., Tava, A., Balestrazzi, A., *et al.* (2004) Expression of the stilbene synthase (*StSy*) gene from grapevine in transgenic white poplar results in high accumulation of the antioxidant resveratrol glucosides. *Transgenic Research* 13, 203–214.

Gould, R. (1998) Sustainability of transgenic insecticidal cultivars: integrating pest genetics and ecology. *Annual Review of Entomology* 43, 701–726.

Gruppe, A., Fusseder, M. and Schopf, R. (1999) Short rotation plantations of aspen and balsam poplar on former arable land in Germany: defoliating insects and leaf constituents. *Forest Ecology and Management* 121, 113–122.

Gullner, G., Kömives, T., and Rennenberg, H. (2001) Enhanced tolerance of transgenic poplar plants overexpressing γ-glutamylcysteine synthetase towards chloroacetanilide herbicides. *Journal of Experimental Botany* 52, 971–979.

Hjältén, J., Niemi, L., Wennström, A., Ericson, L., Roininen, H., and Julkunen-Tiitto, R. (2007) Variable responses of natural enemies to *Salix triandra* phenotypes with different secondary chemistry. *Oikos* 116, 751–758.

Hammond, R.W., Foard, D.E., and Larkins, B.A. (1984) Molecular cloning and analysis of a gene coding for the Bowman–Birk protease inhibitor in soybean. *Journal of Biological Chemistry* 259, 9883–9890.

Han, K.H., Ma, C., and Strauss, S.H. (1997) Matrix attachment regions (MARs) enhance transformation frequency and transgene expression in poplar. *Transgenic Research* 6, 415–420.

Han, K.H., Meilan, R., Ma, C., and Strauss, S.H. (2000) An *Agrobacterium tumefaciens* transformation protocol effective on a variety of cottonwood hybrids (genus *Populus*). *Plant Cell Reports* 19, 315–320.

Hao, G.X., Zhu, Z., and Zhu, Z.T. (1999) Study on optimization transformation of *Populus tomentosa*. *Acta Botanica Sinica* 41, 936–941.

Harfouche, A., Meilan, R., and Altman, A. (2011) Tree genetic engineering and applications to sustainable forestry and biomass production. *Trends in Biotechnology* 29, 9–17.

Harrell, M.O., Benjamin, D.M., Berbee, J.G., and Burkot, T.R. (1982) Consumption and utilization of leaf tissue of tissue-cultured *Populus* × *euramericana* by the cottonwood leaf beetle, *Chrysomela scripta* (Coleoptera: Chrysomelidae). *Canadian Entomologist* 114, 743–749.

Harrell, M.O., Benjamin, D.M., Berbee, J.G., and Burkot, T.R. (1981) Evaluation of adult cottonwood leaf beetle, *Chrysomela scripta* (Coleoptera: Chrysomelidae), feeding preference for hybrid poplars. *Great Lakes Entomologist* 14, 181–184.

Hart, E.R., James, R.R., Nebeker, T.E., Robison, D.J., Raffa, K.F., and Wagner, M.A. (1996) Entomological research in North American *Populus* and *Salix*: an overview. In: *Proceedings of the International Poplar Commission*, October 1–4, 1996, Budapest.

Haworth, R. and Spiers, A. (1988) Characterisation of bacteria from poplars and willows exhibiting leaf spotting and stem cankering in New Zealand. *Forest Pathology* 18, 426–436.

Heap, I. (2014) Herbicide resistant weeds. In: Pimentel, D. and Peshin, R. (eds.) *Integrated Pest Management*. Springer, Dordrecht, pp. 281–301.

Hibbs, D., Withrow-Robinson, B., Brown, D., and Fletcher, R. (2003) Hybrid poplar in the Willamette Valley. *Western Journal of Applied Forestry* 18, 281–285.

Hilder, V.A., Barker, R.F., Samour, R.A., Gatehouse, A.M.R., Gatehouse, J.A., and Boulter, D. (1989) Protein and cDNA sequences of Bowman–Birk protease inhibitors from the cowpea (*Vigna unguiculata* Walp.). *Plant Molecular Biology* 13, 701–710.

Hilder, V.A., Gatehouse, A.M.R., and Boulter, D. (1990) Genetic engineering of crops for insect resistance using genes of plant origin. In: Lycett, G. and Grierson, D. (eds.) *Genetic Engineering of Crop Plants*. Butterworths, Boston, Massachusetts, pp. 51–66.

Hjältén, J. and Axelsson, E.P. (2015) GM trees with increased resistance to herbivores: trait efficiency and their potential to promote tree growth. *Frontiers in Plant Science* 6, 279.

Hjältén, J., Axelsson, E.P., Whitham, T.G., LeRoy, C.J., Julkunen-Tiitto, R., *et al.* (2012) Increased resistance of Bt aspens to *Phratora vitelinae* (Coleoptera) leads to increased plant growth under experimental conditions. *PLoS One* 7, e30640.

Hu, J.J., Tian, Y.C., Han, Y.F., Li, L., and Zhang, B.E. (2001) Field evaluation of insect-resistant transgenic *Populus nigra* trees. *Euphytica* 121, 123–127.

James, R.R. (1997) Utilizing a social ethics toward the environment in assessing genetically engineered insect-resistance in trees. *Agriculture and Human Values* 14, 237–249.

James, R.R., Croft, B.A., and Strauss, S.H. (1999) Susceptibility of the cottonwood leaf beetle (Coleoptera: Chrysomelidae) to different strains and transgenic toxins of *Bacillus thuringiensis*. *Environmental Entomology* 28, 108–109.

Jia, Z., Sun, Y., Yuan, L., Tian, Q., and Luo, K. (2010) The chitinase gene (*Bbchit1*) from *Beauveria bassiana* enhances resistance to *Cytospora chrysosperma* in *Populus tomentosa* Carr. *Biotechnology Letters* 32, 1325–1332.

Jongsma, M.A. and Bolter, C. (1997) The adaptation of insects to plant protease inhibitors. *Journal of Insect Physiology* 43, 885–895.

Jouanin, L., Bonade-Bottino, M., Girard, C., Morrot, G., and Giband, M. (1998) Transgenic plants for insect resistance. *Plant Science* 131, 1–11.

Karnosky, D.F., Podila, G.K., Shin, D., and Riemenschneider, D.E. (1997) Differential expression of aroA gene in transgenic poplar: influence of promoter and ozone stress. In: Klopfenstein, N.B., Chun, Y.-W., Kim, M.-S., and Ahuja, M.R. (eds.) *Micropropagation, Genetic Engineering, and Molecular Biology of* Populus. General Technical Report RM-GTR-297. USDA-FS, Rocky Mountain Forest and Range Experimental Station, Ft. Collins, Colorado, pp. 70–73.

Klocko, A.L., Meilan, R., James, R.R., Viswanath, V., Huso, M., *et al.* (2014) Bt-Cry3Aa expression reduces insect damage and improves growth in field-grown hybrid poplar. *Canadian Journal of Forest Research* 44, 28–35.

Knowles, B.H. and Dow, J.A.T. (1993) The crystal β-endotoxins of *Bacillus thuringiensis*: models for their mechanism of action on the insect gut. *BioEssays* 15, 469–476.

Lane, B.G. (1994) Oxalate, germin, and the extracellular matrix of higher plants. *FASEB Journal* 8, 294–301.

Lane, B.G., Dunwell, J.M., Ray, J.A., Schmitt, M.R., and Cuming, A.C. (1993) Germin, a protein marker of early plant development, is an oxalate oxidase. *Journal of Biological Chemistry* 268, 12239–12242.

Leple, J.C., Brasileiro, A.C.M., Michel, M.F., Delmotte, F., and Jouanin, L. (1992) Transgenic poplars: expression of chimeric genes using four different constructs. *Plant Cell Reports* 11, 137–141.

Li, J., Meilan, R., Ma, C., Barish, M., and Strauss, S.H. (2008) Stability of herbicide resistance over 8 years of coppice in field-grown, genetically engineered poplars. *Western Journal of Applied Forestry* 23, 89–93.

Li, J.Y. and Zhang, Z.Y. (2000) Genetic variation in photosynthetic traits of triploid clones of *Populus tomentosa. Journal of Beijing Forestry University* 22, 12–15.

Li, Q., Dong, W., Wang, T., Liu, Z., Wang, F., *et al.* (2013) Effect of β-globin MAR characteristic elements on transgene expression. *Molecular Medicine Reports* 7, 1871–1874.

Liang, H., Catranis, C.M., Maynard, C.A., and Powell, W.A. (2002) Enhanced resistance to the poplar fungal pathogen, *Septoria musiva*, in hybrid poplar clones transformed with genes encoding antimicrobial peptides. *Biotechnology Letters* 24, 383–389.

Liang, H., Maynard, C.A., Allen, R.D., and Powell, W.A. (2001) Increased *Septoria musiva* resistance in transgenic hybrid poplar leaves expressing a wheat oxalate oxidase gene. *Plant Molecular Biology* 45, 619–629.

Lin, S.-Z., Zhang, Z.-Y., Zhang, Q., and Lin, Y.-Z. (2006) Progress in the study of molecular genetic improvements of poplar in China. *Journal of Integrative Plant Biology* 48, 1001–1007.

Liu, C.M., Zhu, Z., Zhou, Z.L., Sun, B.L., and Li, X.H. (1993) cDNA clone and expression of cowpea trypsin inhibitor in *Escherichia coli. Chinese Journal of Biotechnology* 9, 152–157.

Lu, M.Z. (2008) Advance in transgenic research and its biosafety of forest trees in China. In: *Proceedings of the 8th Symposium in Agricultural Biochemistry and Molecular Biology.* Chinese Society of Biochemistry and Molecular Biology. Guiyang, China, pp. 1–2.

Lu, S.F., Zhao, H.Y., Wei, J.H., and Song, Y.R. (2001) Establishment of *in vitro* regeneration system of triploid Chinese white poplar. *Acta Botanica Sinica* 43, 435–437.

Luttrell, R.G. and Caprio, M. (1996) Implementing resistance management. In: Dugger, C.P. and Richter, D.A. (eds.) *1996 Proceedings of the Beltwide Cotton Production Conference.* National Cotton Council, Cordova, Tennessee, pp. 161–163.

Maredia, K.M. (1997) Sustaining host plant resistance derived through conventional and biotechnological means. In: Mihm, J.A. (ed.) *Insect Resistant Maize: Recent Advances and Utilization. Proceedings of an International Symposium held at the International Maize and Wheat Improvement Center.* CIMMYT, Mexico City, Mexico, pp. 175–179.

Maxwell, D.P. and Lumsden, R.D. (1970) Oxalic acid production by *Sclerotinia sclerotiorum* in infected bean and in culture. *Phytopathology* 60, 1395–1398.

McCown, B.H., McCabe, D.E., Russell, D.R., Robison, D.J., Barton, K.A., and Raffa, K.F. (1991) Stable transformation of *Populus* and incorporation of pest resistance by electric discharge particle acceleration. *Plant Cell Reports* 9, 590–594.

McGaughey, W., Gould, F., and Gelernter, W. (1998) Bt resistance management. *Nature Biotechnology* 16, 144–146.

McNabb, H.J. Jr., Ostry, M.E., Sonnelitter, R.S., and Gerstenberger, P.E. (1982) The effect and possible integrated management of *Septoria musiva* in intensive, short-rotation culture of *Populus* in the north central states. In: Zavitkovski, W.I.J. and Manhattan, E.H. (eds.) *Proceedings of the North American Poplar Council Meeting.* Division of Extension, Kansas State University, Manhattan, Kansas, pp. 51–58.

Meilan, R. (2006) Challenges to commercial use of transgenic plants. *Journal of Crop Improvement* 18, 433–450.

Meilan, R., Ma, C., Cheng, S., Eaton, J., Miller, L., *et al.* (2000) High levels of Roundup® and leaf-beetle resistance in genetically engineered hybrid cottonwoods. In: Blatner, K., Johnson, J. and Baumgartner, D. (eds.) *Hybrid Poplars in the Pacific Northwest: Culture, Commerce and Capability.* Washington State University Cooperative Extension Bulletin MISC0272, Pullman, Washington, DC, pp. 29–38.

Meilan, R., Han, K., Ma, C., DiFazio, S., Eaton, J., *et al.* (2002a) The *CP4* transgene provides high levels of tolerance to Roundup® herbicide in field-grown hybrid poplars. *Canadian Journal of Forest Research* 32, 967–976.

Meilan, R., Auerbach, D., Ma, C., DiFazio, S., and Strauss, S. (2002b) Stability of herbicide resistance and GUS expression in transgenic hybrid poplars (*Populus* sp.) during four years of field trials and vegetative propagation. *HortScience* 37, 277–280.

Mentag, R., Luckevich, M., Morency, M.J., and Seguin, A. (2003) Bacterial disease resistance of transgenic hybrid poplar expressing the synthetic antimicrobial peptide D4E1. *Tree Physiology* 23, 405–411.

Merkle, S.A. and Nairn, C.J. (2005) Hardwood tree biotechnology. *In Vitro Cellular & Developmental Biology—Plant* 41, 602–619.

Merkle, S.A., Andrade, G.M., Nairn, C.J., Powell, W.A., and Maynard, C.A. (2007) Restoration of threatened species: a noble cause for transgenic trees. *Tree Genetics & Genomes* 3, 111–118.

Miranda, M., Ralph, S.G., Mellway, R., White, R., Heath, M.C., *et al.* (2007) The transcriptional response of hybrid poplar (*Populus trichocarpa* × *P. deltoides*) to infection by *Melampsora medusae* leaf rust involves induction of flavonoid pathway genes leading to the accumulation of proanthocyanidins. *Molecular Plant–Microbe Interactions* 20, 816–831.

Mittler, R., Shulaev, V., and Lam, E. (1995) Coordinated activation of programmed cell death and defense mechanisms in transgenic tobacco plants expressing a bacterial proton pump. *Plant Cell* 7, 29–42.

Mohamed, R., Meilan, R., Ostry, M.E., Michler, C.H., and Strauss, S.H. (2001) Bacterio-opsin gene overexpression fails to elevate fungal disease resistance in transgenic poplar (*Populus*). *Canadian Journal of Forest Research* 31, 1–8.

Moore, L.M. and Wilson, L.P. (1983) Recent advances in research of some pest problems of hybrid *Populus* in Michigan and Wisconsin. *General Technical Report North Central Forest Experimental Station* 91, 94–101.

Newhouse, A.E., Polin-McGuigan, L.D., Baier, K.A., Valletta, K.E., Rottmann, W.H., *et al.* (2014) Transgenic American chestnuts show enhanced blight resistance and transmit the trait to T1 progeny. *Plant Science* 228, 88–97.

Newhouse, A.E., Schrodt, F., Liang, H., Maynard, C.A., and Powell, W.A. (2007) Transgenic American elm shows reduced Dutch elm disease symptoms and normal mycorrhizal colonization. *Plant Cell Reports* 26, 977–987.

Nordman, E.E., Robison, D.J., Abrahamson, L.P., and Volk, T.A. (2005) Relative resistance of willow and poplar biomass production clones across a continuum of herbivorous insect specialization: univariate and multivariate approaches. *Forest Ecology and Management* 217, 307–318.

Nwanze, K.F., Seetharama, N., Sharma, H.C., and Stenhouse, J.W. (1995) Biotechnology in pest management: improving resistance in sorghum to insect pests. *African Crop Science Journal* 3, 209–215.

Ostry, M.E. and McNabb, H.S. Jr. (1983) Diseases of intensively cultured hybrid poplars: a summary of recent research in the north central region. *General Technical Report North Central Forest Experimental Station* 91, 102–109.

Ostry, M.E. and McNabb, H.S. Jr. (1985) Susceptibility of *Populus* species and hybrids to disease in the North Central United States. *Plant Disease* 69, 755–775.

Pascal, S., Gullner, G., Kömives, T., and Scalla, R. (2000) Selective induction of glutathione *S*–transferase subunits in wheat plants exposed to the herbicide acifluorfen. *Zeitschrift für Naturforschung C* 55, 37–39.

Pascual, M.B., Jing, Z.P., Kirby, E.G., Cánovas, F.M., and Gallardo, F. (2008) Response of transgenic poplar overexpressing cytosolic glutamine synthetase to phosphinothricin. *Phytochemistry* 69, 382–389.

Pasteels, J.M., Rowel-Rahier, M., Braekman, J.C., and Dupont, A. (1983) Salicin from host plants as a precursor of salicylaldehyde in defensive secretion of Chrysomeline larvae. *Physiological Entomology* 8, 307–314.

Plett, J.M., Wilkins, O., Campbell, M.M., Ralph, S.G., and Regan, S. (2010) Endogenous overexpression of *Populus MYB186* increases trichome density, improves insect pest resistance, and impacts plant growth. *Plant Journal* 64, 419–432.

Polin, L.D., Liang, H., Rothrock, R.E., Nishii, M., Diehl, D.L., *et al.* (2006) *Agrobacterium*-mediated transformation of American chestnut (*Castanea dentata* (Marsh.) Borkh.) somatic embryos. *Plant Cell, Tissue and Organ Culture* 84, 69–78.

Powell, W.A., Catranis, C.M., and Maynard, C.A. (1995) Synthetic antimicrobial peptide design. *Molecular Plant–Microbe Interactions* 8, 792–794.

Powell, W.A. and Maynard, C.A. (1997) Designing small antimicrobial peptides and their encoding genes. In: Klopfenstein, N., Chun, Y.W., Kim, M.-S., and Ahuja, M.R. (eds.) *Micropropagation, Genetic Engineering and Molecular Biology of Populus*. Rocky Mountain Forest and Range Experiment Station, Fort Collins, Colorado, pp. 165–172.

Powell, W.A., Maynard, C.A., Boyle, B., and Seguin, A. (2006) Fungal and bacterial resistance in transgenic trees. In: Fladung, M. and Ewald, D. (eds.) *Tree Transgenesis*. Springer, New York, pp. 235–252.

Pu, J.W., Song, J.L., Xie, Y.M., and Gu, R.J. (2002) Characteristics of lignin structure of triploid clones of *Populus tomentosa* Carr. *Journal of Beijing Forestry University* 24, 211–215.

Raffa, K.F. (1989) Genetic engineering of trees to enhance resistance to insects: evaluating the risks of biotype evolution and secondary pest outbreak. *Bioscience* 39, 524–534.

Raikhel, N.V. and Lee, H.-L. (1993) Structure and function of chitin binding proteins. *Annual Review of Plant Physiology and Plant Molecular Biology* 44, 591–615.

Ray, T.B. (1984) Site of action of chlorsulfuron: inhibition of valine and isoleucine biosynthesis in plants. *Plant Physiology* 75, 827–831.

Riemenschneider, D., Haissig, B., Sellmer, J., and Fillatti, J. (1988) Expression of an herbicide tolerance gene in young plants of a transgenic hybrid poplar clone. In: Ahuja, M.R. (ed.) *Somatic Cell Genetics of Woody Plants*. Kluwer Academic Publishers, Boston/London, pp. 73–80.

Riemenschneider, D.E. and Haissig, B.E. (1991) Producing herbicide-tolerant *Populus* using genetic transformation mediated by *Agrobacterium tumefaciens* C58: a summary of recent research. In: Ahuja, M.R. (ed.) *Woody Plant Biotechnology*. Plenum Press, New York, pp. 247–263.

Roush, R. (1997) Bt-transgenic crops: just another pretty insecticide or a chance for a new start in resistance management? *Pesticide Science* 51, 328–334.

Roush, R. and Shelton, A. (1997) Assessing the odds: the emergence of resistance to Bt transgenic plants. *Nature Biotechnology* 15, 816–817.

Ryan, C.A. (1989) Proteinase inhibitor gene families: strategies for transformation to improve plant defense against herbivores. *BioEssays* 10, 20–24.

Ryan, C.A. (1990) Protease inhibitors in plants: genes for improving defenses against insects and pathogens. *Annual Review of Phytopathology* 28, 425–449.

Schnepf, E., Crickmore, N., van Rie, J., Lereclus, D., Baum, J., *et al.* (1998) *Bacillus thuringiensis* and its pesticidal crystal proteins. *Microbiology and Molecular Biology Reviews* 62, 775–806.

Strauss, S.H., Knowe, S.A., and Jenkins, J. (1997) Benefits and risk of transgenic, Roundup Ready cottonwoods. *Journal of Forestry* 95, 12–19.

Su, X.-H., Zhang, B.-Y., Huang, Q.-J., Huang, L.-J., and Zhang, X.-H. (2003) Advances in tree genetic engineering in China. In: *XII World Forestry Congress*, Quebec City, Canada. Available at: www.fao.org/docrep/article/wfc/xii/0280-b2.htm (accessed June 27, 2019).

Thompson, C.J., Movva, N.R., Tizard, R., Crameri, R., Davies, J.E., *et al.* (1987) Characterization of the herbicide-resistance gene *bar* from *Streptomyces hygroscopicus*. *EMBO Journal* 6, 2519–2523.

Thompson, M.A., Schnepf, H.E., and Feitelson, J.S. (1995) Structure, function and engineering of *Bacillus thuringiensis* toxins. In: Setlow, J.K. (ed.) *Genetic Engineering*, Vol. 17. Plenum Press, New York, pp. 99–117.

Ting, Z.W., Ming, H.M., Jing, L., Jing, Q.W., and Quan, H.Z. (2011) Rabbit defensin (NP-1) genetic engineering of plant. *African Journal of Biotechnology* 10, 9218–9224.

USDA (2018) Recent trends in GE adoption. Economic Research Service, USDA, Washington, DC. Available at: www.ers.usda.gov/data-products/adoption-of-genetically-engineered-crops-in-the-us/recent-trends-in-ge-adoption.aspx (accessed January 14, 2018).

Wang, J.H. and Constabel, C.P. (2004) Polyphenol oxidase overexpression in transgenic *Populus* enhances resistance to herbivory by forest tent caterpillar (*Malacosoma disstria*). *Planta* 220, 87–96.

Wang, Y. and Pijut, P.M. (2014) Improvement of *Agrobacterium*-mediated transformation and rooting of black cherry. *In Vitro Cellular & Developmental Biology — Plant* 50, 307–316.

Wang, Y.F., Gao, B.J., Zheng, J.B., and Zhang, J.H. (2002) The test on the resistance of transgenic hybrid poplar 741 on *Apriona germari* (Hope).

Journal of the Agricultural University of Hebei 25, 53–56 (in Chinese with English abstract).

Welch, A.J., Maynard, C.A., Stipanovic, A.J., and Powell, W.A. (2007) The effects of oxalic acid on transgenic *Castanea dentata* callus tissue expressing oxalate oxidase. *Plant Science* 172, 488–496.

Wu, N.F., Sun, Q., Yao, B., Fan, Y.L., Rao, H.Y., *et al.* (2000) Insect-resistant transgenic poplar expressing *AaIT* gene. *Chinese Journal of Biotechnology* 16, 129–133.

Xing, X.T. and Zhang, Z.Y. (2002) Genetic control of air-dried wood density, mechanical properties and its application for veneer timber breeding of new triploid clones in *Populus tomentosa* Carr. *Forestry Studies in China* 4, 52–60.

Yang, D., Bernier, L., and Dessureault, M. (1994) Biological control of *Septoria* leaf spot of poplar by *Phaeotheca dimorphospora*. *Plant Disease* 78, 821–825.

Zhang, B., Chen, M., Zhang, X., Luan, H., Diao, S., *et al.* (2011) Laboratory and field evaluation of the transgenic *Populus alba* × *Populus glandulosa* expressing double coleopteran-resistance genes. *Tree Physiology* 31, 567–573.

Zhang, B., Oakes, A.D., Newhouse, A.E., Baier, K.M., Maynard, C.A., and Powell, W.A. (2013) A threshold level of oxalate oxidase transgene expression reduces *Cryphonectria parasitica* -induced necrosis in a transgenic American chestnut (*Castanea dentata*) leaf bioassay. *Transgenic Research* 22, 973–982.

Zhang, Q., Zhang, Z.Y., Lin, S.Z., and Lin, Y.Z. (2005) Resistance of transgenic hybrid triploids in *Populus tomentosa* Carr. against 3 species of lepidopterans following two winter dormancies conferred by high level expression of cowpea trypsin inhibitor gene. *Silvae Genetica* 54, 108–116.

Zhang, Z.Y., Li, F.L., and Zhu, Z.T. (1992) Chromosome doubling and triploid breeding of *Populus tomentosa* Carr. and its hybrid. *Journal of Beijing Forestry University* 14 (Suppl.), 52–58.

Zhang, Z.Y., Li, F.L., and Zhu, Z.T. (1997) Doubling technology of pollen chromosome of *Populus tomentosa* and its hybrids. *Journal of Beijing Forestry University* (English edn.) 6, 9–20.

Zhang, Z.Y., Yu, X.S., and Zhu, Z.T. (2000) Sexual reproduction of hybrid triploids in *Populus tomentosa*. *Journal of Beijing Forestry University* 22, 1–4.

Zhao, J.Z., Fan, X.L., Shi, X.P., Zhao, R.M., and Fan, Y.L. (1997) Gene pyramiding: an effective strategy of resistance management for *Helicoverpa armigera* and *Bacillus thuringiensis*. *Resistance and Pest Management* 9, 19–21.

Zhao, S.M., Zu, G.C., Liu, G.Q., Huang, M.R., Xu, J.X., and Sun, Y.R. (1999) [Introduction of rabbit defensin NP-1 gene into poplar (*P. tomentosa*) by Agrobacterium-mediated transformation.] *Acta Genetica Sinica* 26, 711–714 (article in Chinese, abstract in English).

Zheng, J.B., Liang, H.Y., Gao, B.J., Wang, Y.F., and Tian, Y.C. (2000) Selection and insect resistance of transgenic hybrid poplar 741 carrying two insect-resistant genes. *Scientia Silvae Sinicae* 36, 13–19 (article in Chinese, abstract in English).

Zhu, Z.T. and Zhang, Z.Y. (1997) Status and advances of genetic improvement of *Populus tomentosa* Carr. *Journal of Beijing Forestry University* (English edn.) 6, 1–7.

11   Regulatory Affairs

Introduction

Growing genetically engineered (GE) plants in the field for either research or commercial purposes, or moving them across state/national borders, is carefully regulated by the U.S. government. This is done because there is concern among federal regulators and the general public about the potential effects that transgene expression products or spread of the transgene itself might have on human health and the environment. Substantial fines can be imposed on individuals who are not in compliance. This chapter describes which agencies are involved and the process for getting the required paperwork and approvals.

Governmental Oversight

The U.S. Department of Agriculture (USDA), Environmental Protection Agency (EPA), and Food and Drug Administration (FDA) are the three federal agencies responsible for regulating the testing, movement, and commercial deployment of transgenic plants and their products. Oversight authority for these agencies was outlined by the Office of Science and Technology Policy in 1986 in the Federal Coordinated Framework for the Regulation of Biotechnology. The USDA, via a unit within the Animal and Plant Health Inspection Service (APHIS) known as the Biotechnology Regulatory Services (BRS), has the authority to regulate the importation, interstate movement, and environmental release of transgenic plants by virtue of the Plant Protection Act. The EPA, through the Federal Insecticide, Fungicide, and Rodenticide Act (FIFRA), regulates the growth of transgenic plants into which genetic material has been inserted to impart a pesticidal property (a plant-incorporated protectant, or PIP). In addition to registering the use of a PIP, the EPA also establishes tolerances (i.e. safe levels for pesticides in food) or exemptions from the requirement of a tolerance, through the Federal Food, Drug, and Cosmetic Act (FFDCA). The EPA is also responsible for registering the use of pesticides on crops. Finally, the FDA is granted authority under the FFDCA to regulate transgenic food and feed crops or products from transgenic crops that may come in contact with food or feed. A description of their monitoring protocol is provided in their *Statement of Policy: Foods Derived from New Plant Varieties* (FDA, 1992). Thus, depending on the trait for which a crop has been engineered, more than one federal agency may be involved in its regulation.

What follows is a brief description of each agency's involvement in regulating transgenic plants. Because of its overarching authority, the role played by APHIS will be emphasized.

The role of APHIS

The BRS implements the regulations of APHIS for GE organisms that may pose a risk to plant or human health and the environment, in coordination with other federal agencies, as part of the Federal Coordinated Framework for the Regulation of Biotechnology. Details regarding the biotechnology regulatory process can be found at www.aphis.usda.gov/brs/index.html (accessed July 12, 2019).

A Regulated Article is defined in 7 CFR §340.1 as:

> Any organism which has been altered or produced through genetic engineering, if the donor organism, recipient organism, or vector or vector agent belongs to any

genera or taxa designated in §340.2 and meets the definition of plant pest, or is an unclassified organism and/or an organism whose classification is unknown, or any product which contains such an organism, or any other organism or product altered or produced through genetic engineering which the Administrator, determines is a plant pest or has reason to believe is a plant pest. Excluded are recipient microorganisms which are not plant pests and which have resulted from the addition of genetic material from a donor organism where the material is well characterized and contains only non-coding regulatory regions.

Other useful definitions can be found at www.ecfr.gov/cgi-bin/text-idx?SID=1d-931bf53bd5cf5540556ca78245b6d1&mc=true&node=se7.5.340_11&rgn=div8 (accessed July 12, 2019).

Under certain circumstances, a transgenic plant may not be considered a regulated article by APHIS. For example, a plant that contains a non-plant pest sequence that imparts pest resistance and was introduced via biolistics (Chapter 6, this volume) may not be regulated. The APHIS has developed a process known as "Am I Regulated" (AIR) for developers to determine whether their transgenic plant is considered a regulated article and is thus regulated. More information about the AIR process can be found at www.aphis.usda.gov/aphis/ourfocus/biotechnology/am-i-regulated (accessed July 12, 2019).

Currently, two mechanisms are available for the importation, interstate movement, and release (field testing) of GE plants. Initially (between June 1986 and April 1993), transgenic materials could be moved and/or tested only under a "permit." The permit application review process is lengthy (minimum of 180 days) because an Environmental Assessment (EA) is needed for each application. In spring 1993, APHIS instituted its "notification" system for a wide range of plant species, which streamlined the process (i.e. a 10-day review for interstate movement notifications and a 30-day review for notifications requesting environmental release) for organisms and traits with which APHIS had familiarity in managing risk. Transgenic plants requiring more careful scrutiny are still subject to the permitting process (e.g.

those plant species that are unfamiliar to APHIS or those plants containing coding sequences: (i) derived from a human or animal pathogen; (ii) encoding the production of a pharmaceutical/industrial compound; or (iii) whose function is unknown or unfamiliar to APHIS). During their review, applications for permits and notifications are forwarded to the appropriate regulatory authority within the affected state(s), which then approves the request, denies it (with justification), or imposes requirements more stringent than those specified by APHIS.

It should be noted that before any living organism can be imported into the USA, a permit must be obtained from another division of APHIS, Plant Protection and Quarantine (PPQ). However, PPQ has no authority to regulate transgenic material. Thus, to import a transgenic organism, permits are need from both PPQ and BRS.

Detailed instructions for filing a notification with BRS can be found on the following page of the APHIS website: www.aphis.usda.gov/aphis/ourfocus/biotechnology/permits-notifications-petitions/ct_submissions_home (accessed July 12, 2019). Briefly, a notification must contain the following information:

- Responsible Party (the applicant, who is legally liable for non-compliance);
- Duration of Introduction (notifications are usually valid for only 12 months);
- Recipient Organisms (scientific names of species transformed; limited to a single species);
- List of Phenotypic Designations (including line designation(s); description of gene inserted; expected phenotype; designation of genetic construct(s) used to transform plant(s); and genetic components that went into assembling construct(s), such as promoters, coding sequences, terminators, and selectable markers);
- Mode of Transformation (e.g. *Agrobacterium tumefaciens*);
- Introduction (point of origins, destinations, and/or release sites); and
- Certification (statement in which the responsible party certifies that the regulated

article will be handled in accordance with the relevant Performance Standards (see below) and regulations set forth in the Coordinated Framework).

If permission to move and/or release transgenic materials under notification is authorized, a letter of acknowledgement is sent to the applicant. In the letter, BRS specifies that the responsible party must comply with a list of Performance Standards, which are listed below, and represent a slightly modified version of what is actually used:

1. If the plants or plant materials are shipped, they must be shipped in such a way that the viable plant material is unlikely to be disseminated while in transit and must be maintained at the destination facility in such a way that there is no release into the environment.

2. When the introduction is an environmental release, the regulated article must be planted in such a way that they are not inadvertently mixed with non-regulated plant materials of any species which are not part of the environmental release.

3. The plants and plant parts must be maintained in such a way that the identity of all material is known while it is in use, and the plant parts must be contained or devitalized when no longer in use.

4. There must be no viable vector agent associated with the regulated article. The field trial must be conducted such that:

   a. The regulated article will not persist in the environment; and

   b. No offspring can be produced that could persist in the environment.

5. Upon termination of the field test:

   a. No viable material shall remain which is likely to volunteer in subsequent seasons; or

   b. Volunteers shall be managed to prevent persistence in the environment.

Poplars (*Populus* spp.) were originally included in the notification system for field release, but in 2007, BRS went back to requiring a permit for them. Permits for field testing of transgenic poplars, or any other tree, are only valid for 3 years and then the plants must be "devitalized." A renewal can be obtained if the trees are coppiced (cut at ground level) at the end of 3 years. In general, poplars have a 5–10-year juvenile period (Braatne *et al.*, 1996). Coppicing the trees maintains them in a juvenile condition, mitigating concerns related to transgene flow and persistence in the environment.

In addition to the information that is required for a notification, the following must be included in a permit application:

1. Destination or Release Description: a detailed description of the intended destination (including final and all intermediate destinations), uses and/or distribution of the regulated article (e.g. locations of greenhouses; laboratories; growth chambers; field trials; pilot projects; production, propagation, and manufacture areas; and proposed sales and/or distribution centers).

2. Confinement Protocols: a detailed description of the proposed procedures, processes, and safeguards that will be used to prevent escape and dissemination of the regulated article at each of the intended destinations.

3. Final Disposition Method: destruction/devitalization, storage in contained facility, or other fate of the material.

4. Final Disposition Description: an explanation of how the choice made for item 3 (above) will be met.

ePermits is a web-based system that allows users to apply for a notification or permit for the importation, interstate movement, or release of a regulated article. It also allows users to track their application(s), apply for amendments and renewals, where applicable, and receive copies of their authorizations. In general, ePermits uses a USDA-wide system for login known as USDA eAuthentication. To register for an eAuthentication account, go to: https://identitymanager.eems.usda.gov/Registration/index.aspx (accessed July 22, 2019).

The role of the EPA

The EPA is involved in regulating GE plants when they have been modified to contain a PIP that imparts resistance to a specific pest.

Their evaluation considers the extent to which the introduced PIP is toxic to the target pest, as well as the potential for unintended exposures through the expression of PIPs in GE plants. The EPA also assesses the potential for targeted pest populations to acquire resistance to the PIP as a result of its widespread use. All petitions involving plants containing a gene from *Bacillus thuringiensis* (Bt) must include a resistance-management plan that details steps that will be taken to minimize the risk of insects becoming resistant to the PIP. The EPA's role in regulating GE plants also encompasses: (i) establishing tolerance, or exemptions from the requirement of a tolerance, for the PIP; (ii) registering herbicides to which crops are engineered to be tolerant, if the product was not previously registered for use with the species that was engineered; and (iii) assessing the risks associated with new exposures to herbicide residues.

The role of the FDA

The FDA regulates human food and animal feed derived from all plant varieties, and holds products obtained from GE plants to the same rigorous safety standards required of all other foods. These efforts can include collecting data on allergenicity, digestibility, and toxicology of the protein encoded by the transgene. Information on safety and regulatory issues related to GE plants can be viewed at the FDA website (www.fda.gov/Food/GuidanceRegulation/GuidanceDocuments-RegulatoryInformation/Biotechnology/default.htm) (accessed July 12, 2019).

Regulatory reform

In January 2004, APHIS announced its intention to prepare a programmatic Environmental Impact Statement (EIS) in which the agency would evaluate its biotechnology regulations and consider several possible changes (www.govinfo.gov/content/pkg/FR-2004-01-23/pdf/04-1411.pdf, accessed July 22, 2019). The goal was to develop a multi-tiered system to replace the current permit and notification systems, and to enhance the deregulation process to provide greater flexibility in approving products, as well as long-term monitoring. Changes to the regulations were to be science- and risk-based, although they were also likely to be influenced by consumer issues, trade considerations, and politics. In July 2007, APHIS published a draft EIS, and in October 2008, it proposed a revision of its regulations regarding the importation, interstate movement, and environmental release of certain GE organisms. The public-comment period was initially scheduled to end in November 2008, but was extended to June 2009. A meeting to discuss issues related to the proposed rule changes was convened in April 2009, but the final rule was never published. In March 2015, APHIS withdrew its proposed regulatory changes and published another proposed rule to revise its biotechnology regulations from January 19, 2017. On November 7, 2017, APHIS withdrew this second proposed rule in order to re-engage stakeholders. Information and documents related to that proposed rule can be viewed at www.aphis.usda.gov/aphis/ourfocus/biotechnology/biotech-rule-revision/2016-340-rule/ (accessed July 22, 2019). Following re-engagement with stakeholders, APHIS announced a proposed rule on June 5, 2019 entitled: "Movement of Certain Genetically Engineered Organisms," which is intended to reduce the regulatory burden for developers of organisms that are unlikely to pose plant pest risks (for details, see: https://www.aphis.usda.gov/brs/fedregister/BRS_20190606.pdf, accessed July 22, 2019).

The current Coordinated Framework was established in 1986, prior to many biotechnological advancements that are being used widely. Critics raised questions about the adequacy of current laws to protect human health and the environment, in light of rapidly emerging GE applications, such as plant-based pharmaceutical production. This was in addition to the 2006 discovery of an unapproved variety of GE rice in commercial U.S. rice supplies and the 2008 discovery of an unapproved GE cotton (*Gossypium* spp.) variety comingled with an approved variety. For more information about these cases of non-compliance, readers are encouraged to visit: www.aphis.usda.gov/aphis/ourfocus/biotechnology/sa_compliance_and_inspections/ct_compliance_history (accessed July 12, 2019). Finally, the existing regulatory structure governing federal biotechnology policy is fragmented and there is a perceived lack of transparency.

Therefore, on July 2, 2015, the Obama Administration issued a memorandum calling for an overhaul of the Coordinated Framework. This memorandum initiated a process that was intended to achieve the following objectives: "(1) update the Coordinated Framework to clarify the agencies' roles and responsibilities to regulate biotechnology products; (2) formulate a long-term strategy to ensure that the regulatory system can adequately assess any risks associated with future products of biotechnology while 'increasing transparency and predictability and reducing unnecessary costs and burdens'; and (3) commission an external, independent analysis of the future landscape of biotechnology products" (Congressional Research Service, 2015).

Commercialization

In general, before transgenic plants can be grown for commercial purposes, a petition for "non-regulated status" must be submitted to APHIS. This is done only after extensive evaluation that typically involves several years of field tests. Instructions for submitting a petition can be found at the APHIS website (www.aphis.usda.gov/brs/petguide.html, accessed July 12, 2019). These include requirements for both a complete molecular characterization of the transgenic plants and data on potential environmental impacts. Once a petitioner's package is submitted, APHIS conducts a formal EA and/or EIS, as described under the National Environmental Protection Act, which includes one or more public-comment periods, before deciding whether to approve the developer's request for non-regulated status.

After the petition is approved, the transgenic plant is no longer regulated (i.e. unrestricted distribution and planting is granted).

Between 1987 and 2017, APHIS approved more than 19,200 applications to conduct field trials, most of which were for crop plants harboring genes that impart resistance to specific insect pests or tolerance to certain herbicides. However, other traits being developed through genetic engineering include, tolerance to drought, salt, and temperature extremes; improved rates of photosynthesis; and nitrogen-use efficiency. APHIS has granted non-regulated status to approximately 123 different crop-species combinations (a complete listing can be found at: www.aphis.usda.gov/aphis/ourfocus/biotechnology/permits-notifications-petitions/petitions/petition-status, accessed July 12, 2019). Although 16 tree species have been field-tested, only transgenic papaya (*Carica papaya*), plum (*Prunus domestica*), and apple (*Malus pumila*) have been granted non-regulated status (see below and Chapter 7, this volume).

Helpful hints (for commercialization)

While seeking approval for commercialization of engineered plants, potential developers should be mindful of the following:

- Early consultation with the appropriate regulatory agencies can save a lot of time and needless actions and delays.
- A petition for non-regulated status (APHIS) can include more than one independent line (transformation event).
- Species-gene combinations are evaluated on a case-by-case basis.
- With the appropriate permission, it is possible to use safety data submitted by previous developers.
- Gene flow is not considered a risk *a priori*; it must be evaluated to determine whether a perceived risk is genuine.
- Under rare circumstances, it may be possible to commercialize a product derived from a transgenic plant without deregulation by APHIS. However, the developer must still abide by permit and notification requirements and other regulations

specified under 7 CFR, section 340, if after undergoing the AIR process, the article is considered regulated. In addition, if the inserted gene encodes a PIP or the plant has any potential food or feed use, the EPA and FDA still remain involved. In general, this alternative approach is unwise and leaves the developer open to significant liability and financial risk.

- Pharmaceutical/industrial compound-producing plants might always be regulated in some way (i.e. they may never be fully deregulated by USDA).

Commercializing herbaceous crops

In general, herbaceous annual crops have been heavily domesticated and have few, if any, wild relatives with which they are interfertile. This is not true for perennial grasses, which are being genetically engineered, at both public and private institutions, for a variety of traits, including: tolerances to herbicides, drought, disease, and cold; enhanced quality (e.g. digestibility, nutritional content); reduced stature; and lignin content. The products harvested from many herbaceous crops (e.g. seed, fruit, pollen) are generally derived from the flowers. Thus, there is neither a need nor a desire to prevent flowering in these species. Other than cotton, which is intentionally grown as an annual, few transgenic herbaceous perennial crops have been commercialized. Herbicide tolerance creeping bentgrass (*Agrostis stolonifera*) is an example of an herbaceous perennial that has received non-regulated status. For a detailed history of what led up to this decision and the continuing controversy, readers are encourage to see Meilan (2006), and www.aphis. usda.gov/aphis/ourfocus/biotechnology/sa_ compliance_and_inspections/ct_compliance_history (accessed July 12, 2019).

Commercializing woody perennials

Although numerous transgenic herbaceous crops are currently being grown for commercial purposes (see http://www.aphis.usda.gov/brs/, accessed July 12, 2019), papaya represents the only commercially deployed transgenic woody perennial. Its release resulted from a special effort to save an entire industry from the ravages of the ubiquitous ringspot virus in Hawaii (Gonsalves, 1998). This unique case involved virtually no environmental risk because papaya originally had been introduced to Hawaii (i.e. there are no interfertile wild relatives) and because the Pacific Ocean is an effective physical barrier to transgene escape.

In June 2007, APHIS deregulated a variety of plum known as "HoneySweet" that was engineered, using RNA interference (Chapter 6, this volume) to be resistant to plum pox virus (Scorza *et al.*, 2013). In December 2010, a biotechnology firm in British Columbia, Okanagan Specialty Fruits, petitioned APHIS for approval of its GE variety of apple known as "Arctic® Apple." It has been engineered to resist browning after being sliced. In February 2015, APHIS deregulated this GE apple.

Because poplar has been used in more transgenic research than any other tree species and because of its commercial potential, it will be highlighted in the remainder of this section.

Transgene Confinement

Presently, other than the three described above, all other transgenic tree species in the USA are being grown for research purposes only, for three reasons: (i) existing regulations were written with the perspective of agronomic row crops, which are highly domesticated and have few, if any, wild relatives; (ii) these plant systems present ecological concerns that are different from annual row crops; and (iii) biotechnological techniques have been slower to develop for trees (mainly the last reason). Although APHIS has never stated that transgene escape will not be allowed, it has made clear that efforts must be taken to mitigate the risk of transgene spread from any regulated article, at least during the early stages of development.

With the current state of technology, it is impossible to guarantee that complete sterility can be achieved via transgenesis, for any plant species. However, from a purely scientific or even a risk-reduction perspective, complete sterility may not always be needed before transgenic trees can or should be grown commercially. This is especially true for riparian species, such as poplar, that can be grown under fertigation in xeric environments, such as east of the Cascade Mountains in the Pacific Northwest. Native poplars are largely absent from this landscape. Given their reproductive isolation and the half-life of pollen under conditions that prevail when it is shed, there may be no justification for preventing the establishment of even fully fertile transgenic trees in that region. Triploid genotypes of poplar are available that have greatly reduced innate fertility; transgenics produced in these clones might safely be grown in areas where no interfertile wild relatives are present.

The need for sterility depends on the trait being exploited; the environment within which the transgenics will be grown; the particular species; and various social, political, and ethical considerations. Each case must be analyzed individually. In some situations, the introduced gene imparts a competitive disadvantage; transgenic trees of this sort growing outside an intensively managed plantation would be unlikely to survive. For example, trees engineered to deposit lower levels of lignin in cell walls might be more susceptible to pest attack. Nevertheless, because of the uncertainty, many research groups around the world are exploring techniques to genetically engineer flowering control. The various approaches to engineering bioconfinement are described in Chapter 7 (this volume).

Apart from transgene confinement, engineered reproductive sterility may be advantageous for other reasons. First, rapid (juvenile) rates of growth can be maintained. During the early phase of its life, a plant utilizes most of its photosynthate for vegetative growth; after it undergoes the transition to maturity, significant amounts of carbohydrate are allocated to reproductive effort. This diversion of energy-rich compounds causes a reduction in vegetative growth rates, particularly in trees (Eis *et al.*, 1965; Tappeiner, 1969). Therefore, the earlier the flowering process is interrupted, the greater the potential benefit for production forestry.

One major handicap for researchers working in the area of flowering control in trees is the long juvenile period. Uncovering the genes involved in the regulation of floral development will not only expedite sterility research but also potentially speed the progress to be gained through conventional breeding, by inducing precocious flowering. Conversely, delayed flowering may itself serve as a confinement strategy, especially when trees are grown under short-rotation intensive culture. Thus, a great deal of effort is being made to identify genes that regulate the transition from the vegetative to the reproductive phase in a tree's life cycle (e.g. Brunner *et al.*, 2014).

The National Research Council (NRC) released a report entitled *Biological Confinement of Genetically Engineered Organisms* (NRC, 2004). The committee that drafted the report was asked to address the following questions:

- What is the status of various bioconfinement strategies?
- What methods are available and how feasible, effective, and costly are they?
- How can bioconfinement failures be mitigated?
- What methods are available for detecting failures?
- What are the probable ecological consequences of large-scale bioconfinement?
- What new information is needed for addressing any of the above questions?

The report's major conclusions are listed below.

- The effectiveness of a confinement method will vary depending on the organism, the environment, and the scale over which it is applied.
- To evaluate efficacy, the genetically engineered organism (GEO) should be compared with its progenitor before release. For trees, it may be necessary to begin

field tests before such comparisons are possible.

- Various confinement methods should be tested separately and in combination, in a variety of appropriate environments, and in representative organisms.
- The need for bioconfinement should be considered early in the development of a GEO.
- Non-food crops should be utilized for genes encoding products that need to be kept out of the food supply.
- An integrated confinement system should be based on risk.
- The stringency of confinement should reflect the consequences of GEO escape.
- It is unlikely that any single confinement method will be 100% effective.
- The use of redundant methods will increase the likelihood of accomplishing the desired confinement level.
- Government regulators should consider the potential effects a U.S. confinement failure could have on other nations, and vice versa.
- International cooperation should be sought for managing GEO confinement.
- Social and ethical values should be considered when assessing the stringency needed for confinement.
- Transparency and public participation are needed to develop and implement the most appropriate bioconfinement approach.

Addressing Other Concerns Related to GE Plants

Several non-governmental organizations have voiced their opposition to the commercial deployment of GE plants, including trees. One of their key concerns is that the spread of particular transgenes might increase the recipient's competitiveness and, therefore, its invasiveness. Another issue is that farmers might lose their ability to use existing pest control measures. For example, organic fruit and vegetable growers can now exogenously apply purified Bt toxin (which is, in fact, produced commercially using recombinant DNA technology) and still maintain an "organic" rating for their crop. It is feared that through widespread use of Bt toxins, insect populations will develop resistance to this management tool, rendering it ineffective. Finally, there are people who are philosophically opposed to the idea of genetic engineering (e.g. it is "wrong" or "unnatural").

The use of GE trees must be viewed in the proper context, relative to practices currently being used. For example, companies now plant a single genotype (i.e. clone) of hybrid poplar on thousands of contiguous acres of land. Because these non-transgenic trees have been propagated vegetatively, all of their cells, including their pollen, contain exactly the same DNA. When this synchronous population starts flowering, it produces a monotypic pollen supply. The resulting pollen cloud can be carried considerable distances by the wind and can ultimately affect the genetic diversity of sexually compatible wild trees. Engineering flowering control alone (not coupled with any other introduced trait) would prevent this gene flow from occurring. Under circumstances where it is important to protect sensitive genetic stocks (not just for trees), engineered flowering control on its own may be beneficial.

Another context-driven consideration pertains to herbicide usage. Some growers currently use environmentally detrimental compounds to control competing vegetation. This problem can be overcome using plants that have been engineered to tolerate glyphosate, which is an herbicide that is not volatile, carcinogenic, teratogenic, or mutagenic. Moreover, this more environmentally benign compound is soil inactivated and biodegradable. Using glyphosate-tolerant trees would not only encourage the use of a more "environmentally friendly" product but also reduce the overall reliance on agrochemicals.

There are also concerns regarding the ecological ramifications of complete sterility. For example, insects or mammals may rely on pollen or seed as a food source. However, this argument is not applicable to species within the genus *Populus*, which are wind pollinated and do not produce nectar or support a large number of insect or vertebrate pollinators. Furthermore, poplar seeds are very small, contain few stored reserves, and have a short lifespan. Thus, they are not a significant source of sustenance for wildlife. However, for species whose pollen is a valuable food supply, it may be necessary to engineer them to produce pollen that, although nutritionally unaltered, is infertile. The intention is not to supplant wild trees with GE sterile versions; the goal is to plant trees that have been domesticated through genetic engineering into intensively managed plantations, on previously disturbed sites, to satisfy society's increasing demand for renewable resources. This strategy will reduce the need to harvest our native forests.

On the other hand, trees that never produce any pollen could be beneficial under certain circumstances. Planting pollenless stock in some urban settings would reduce the pollen loads from species that are highly allergenic to a large proportion of the human population (e.g. *Cryptomeria japonica* and birch (*Betula* spp.)). It may also be desirable, from the perspective of a homeowner, to engineer street trees that are incapable of producing nuisance reproductive structures (e.g. sweetgum (*Liquidambar styraciflua*) and *Ginkgo biloba*). This example highlights the need to consider each crop-trait combination on a case-by-case basis.

As mentioned earlier, to minimize the risk of insects becoming resistant to Bt toxins, the EPA requires that petitions involving plants engineered with a Bt gene contain a resistance-management plan, which details steps that will be taken to reduce the risk of the target pest developing resistance to the PIP. These plans often include the establishment of refugia, which are adjacent areas on which only non-transgenic plants of the same species are grown. Insects cannot become resistant unless they are exposed to the toxin, and the longer the exposure, the greater the chance that resistance will develop. Insects feeding on non-transgenic plants will remain susceptible to the toxin and mate with those that may become resistant by feeding on the transgenic plants, thus maintaining a susceptible population.

Another strategy is to engineer plants with more than one Bt toxin gene (i.e. "gene stacking"), each of which is effective against the target insect via a different mode of action. The likelihood that insects will become resistant to more than one toxin gene is the product of the probabilities of them becoming resistant to each toxin individually. Furthermore, non-target insects are much more likely to be exposed to exogenously applied Bt toxin (as is practiced by growers of organic food crops) than to a toxin produced inside plant cells (i.e. the insect actually has to eat some of the transgenic plant in order to become exposed). Again, we must put this issue in the proper context: the alternative to a PIP is the use of broad-spectrum insecticides that kill all insects, both targeted and beneficial.

Finally, there can be substantial risk associated with doing nothing. With the recent increase in international trade agreements but a simultaneous lack of funding for federal agencies charged with enforcing import laws, the potential for environmental catastrophes has risen tremendously. Given this, circumstances can occur under which it would be desirable to allow a transgene to spread into the wild in order to contain a newly introduced environmental threat. For example, the emerald ash borer (*Agrilus planipennis* Fairmaire) is an aggressive exotic pest that arrived from Asia and is attacking and killing all of the North American ash (*Fraxinus* spp.) trees it invades. First identified in Michigan in 2002, *A. planipennis* has since invaded much of the eastern USA and Canada (www.emeraldashborer.info/about-eab.php, accessed July 12, 2019). To stem the spread of *A. planipennis*, affected areas are put under quarantine to restrict the transportation of ash trees, branches, nursery stock, logs, and firewood. This pest is fatal to all trees it attacks, no effective means are available for its control and evidence suggests that, unchecked, it will spread throughout North America. If nothing else is done to eradicate this highly destructive pest, American ash species may go the way of American chestnut (*Castanea dentata*) and elm (*Ulmus americana*). Genetic engineering may be the only effective tool for managing this threat. Releasing fully fertile ash trees that have been engineered to resist *A. planipennis* attack and allowing them to flower will result in wild offspring that contain the resistance gene. This approach would help preserve the genetic diversity of *Fraxinus* spp. across the landscape.

The Need for Field Trials

Transgenic plants need to be field-tested for several reasons. The first is proof of concept: it must be shown that the inserted gene is having the expected effect. *Agrobacterium tumefaciens* is widely used to insert genes into plant cells, but these insertions occur at random locations within the recipient plant's genome. The DNA surrounding the site of insertion influences the efficiency with which the transgene is expressed (i.e. "positional effects"). In addition, transgenes can be differentially expressed under various conditions (e.g. Callahan *et al.*, 2000). Thus, it is essential to screen independent lines (plants derived from cells that have undergone separate gene-insertion events) to verify that the transgene is being expressed at commercially useful levels. It may also be necessary to conduct field trials to determine whether there is sufficient value for end users for them to be willing to pay a biotechnology premium (e.g. a licensing fee) for use of a GE plant. Finally, long-term field studies are needed to detect somaclonal variation and assess the stability of transgene expression (Meilan *et al.*, 2002, 2004).

International Implications

In 2017, GE crops were planted on 457.4 million acres worldwide, by an estimated 18 million farmers in 26 countries. Nearly 91% of that total was located in five countries: the USA (about 40%), Brazil, Argentina, Canada, and India. The most widely planted GE crops were soybean, maize, cotton, and canola (ISAAA, 2017). In the USA, 92% of the acreage on which maize was grown and 94% of the soybean acreage supported GE plants (www.statista.com/statistics/217108/ level-of-genetically-modified-crops-in-the-us/, accessed July 12, 2019).

Regulations in the USA do not require labeling or the segregation of GE crops from those produced by conventional means, as long as they are deemed "substantially equivalent." The approach to regulating biotechnology in the USA differs from that of its major trading partners. For instance, the European Union (EU), Japan, South Korea, New Zealand, and Australia either already have or are in the process of establishing mandatory labeling requirements for products that contain ingredients derived from GE crops (Congressional Research Service, 2015).

In May 2003, the USA, Canada, and Argentina jointly filed a complaint with the World Trade Organization (WTO) for what they considered to be a *de facto* moratorium on the approval of new GE crops by the EU. Agricultural interests in the USA maintained that this moratorium not only blocked importation of its GE products by the EU but also bolstered unfounded concerns regarding their safety. The USA and its partners also argued that the EU moratorium violated WTO rules, which state that a member country's actions to protect public health and the environment must be based on sound science, and that the approval process must be conducted without undue delay (Congressional Research Service, 2015).

Differences between the USA and its trading partners regarding the growth of GE crops, the importation of GE products, and labeling of foods containing GE ingredients have led to contentious discussions during the negotiation of international trade agreements (e.g. the Transatlantic Trade and Investment Partnership, and the Trans-Pacific Partnership). Bridging this chasm will likely necessitate harmonizing the regulatory policies of the USA and its trading partners. However, to date, there has been little movement toward a common position.

Summary

Growing GE plants in the field for either research or commercial purposes, or moving them across state/national borders, is carefully regulated by the U.S. government. This is done out of concern, among federal regulators and the general public, about the potential effects that transgene expression products or the spread of the transgene itself might have on human health and the environment.

The USDA, EPA, and FDA are the three federal agencies responsible for regulating the testing, movement, and commercial deployment of transgenic plants and their products. The USDA, via a unit within APHIS, known as BRS, has the authority to regulate the importation, interstate movement, and environmental release of transgenic plants. The EPA regulates the growth of transgenic plants in the field and into which genetic material has been inserted to impart a pesticidal property (a plant-incorporated protectant, or PIP). In addition to registering the use of a PIP, the EPA also establishes tolerances (i.e. safe levels for pesticides in food). The EPA is also responsible for registering the use of pesticides on crops. Finally, the FDA has the authority to regulate transgenic food and feed crops or products from transgenic plants that may come in contact with food or feed.

The BRS implements the regulations of APHIS for GE organisms with other federal agencies, as described in the Federal Coordinated Framework for the Regulation of Biotechnology. Under certain circumstances, a transgenic plant may not be considered a regulated article. For example, a plant that contains a non-plant pest sequence that imparts pest resistance and was introduced via biolistics (Chapter 6, this volume) may not be regulated. APHIS has developed a process known as "Am I Regulated" (AIR) for developers to determine whether their transgenic plant is considered a regulated article.

Currently, two mechanisms are available for the importation, interstate movement, and release (field testing) of GE plants. Initially (between June 1986 and April 1993), transgenic materials could be moved and/or tested only under a "permit." The permit application review process is lengthy (minimum of 180 days) because an EA is needed for each application. In 1993, APHIS instituted its "notification" system for a wide range of plant species, which streamlined the process (i.e. a 10-day review for interstate

movement notifications and a 30-day review for notifications requesting environmental release) for organisms and traits with which APHIS had familiarity in managing risk. Transgenic plants requiring more careful scrutiny are still subject to the permitting process.

Poplars (*Populus* spp.) were originally included in the notification system for field release, but in 2007, BRS went back to requiring a permit for them. Permits for field-testing transgenic poplars, or any other tree, are only valid for 3 years and then the plants must be "devitalized" (i.e. killed). A renewal can be obtained if the trees are coppiced (cut at ground level) at the end of 3 years. Coppicing the trees maintains them in a juvenile condition, mitigating concerns related to transgene flow and persistence in the environment.

Before any living organism can be imported into the USA, a permit must be obtained from another division of APHIS, PPQ. However, PPQ has no authority to regulate transgenic material. Thus, to import a transgenic organism, permits are need from both PPQ and BRS.

The current Coordinated Framework was established in 1986, prior to many biotechnological advancements that are being used widely today. Critics raised questions about the adequacy of current laws to protect human health and the environment, in light of rapidly emerging GE applications, such as plant-based pharmaceutical production. On 2 July 2015, the Obama Administration issued a memorandum calling for an overhaul of the Coordinated Framework. This memorandum initiated a process that was intended to achieve the following objectives: "(1) update the Coordinated Framework to clarify the agencies' roles and responsibilities to regulate biotechnology products; (2) formulate a long-term strategy to ensure that the regulatory system can adequately assess any risks associated with future products of biotechnology, while 'increasing transparency and predictability and reducing unnecessary costs and burdens'; and (3) commission an external, independent analysis of the future landscape of biotechnology products."

In general, before transgenic plants can be grown for commercial purposes, a petition for "non-regulated status" must be submitted to APHIS. This is done only after extensive evaluation that typically involves several years of field tests. Once a petitioner's package is submitted, APHIS conducts a formal EA and/or EIS, as described under the National Environmental Protection Act, which includes one or more public-comment periods, before deciding whether to approve the developer's request for non-regulated status.

Presently, transgenic forest-tree species are being grown only for research purposes in the USA, for three reasons: (i) existing regulations were written with the perspective of agronomic row crops, which are highly domesticated and have few, if any, wild relatives; (ii) these plant systems present ecological concerns that are different from annual row crops; and (iii) biotechnological techniques have been slower to develop for trees (mainly the latter). Although APHIS has never stated that transgene escape will not be allowed, it has made clear that efforts must be taken to mitigate the risk of transgene spread from any regulated article, at least during the early stages of development.

Regulations in the USA do not require labeling or the segregation of GE crops from those produced by conventional means, as long as they are deemed "substantially equivalent." The approach to regulating biotechnology in the USA differs from that of its major trading partners. For instance, the EU, Japan, South Korea, New Zealand, and Australia either already have or are in the process of establishing mandatory labeling requirements for products that contain ingredients derived from GE crops. Differences between the USA and its trading partners regarding the growth of GE crops, the importation of GE products, and labeling of foods containing GE ingredients, have led to contentious discussions during the negotiation of international trade agreements (e.g. the Transatlantic Trade and Investment Partnership, and the Trans-Pacific Partnership). Bridging this chasm will likely necessitate harmonizing the regulatory policies of the USA and its trading partners. However, to date, there has been little movement toward a common position.

References

Braatne, J.H., Rood, S.B., and Heilman, P.E. (1996) Life history, ecology, and conservation of riparian cottonwoods in North America. In: Stettler, R.F., Bradshaw, H.D. Jr., Heilman, P.E. and Hickley, T.M. (eds.) *Biology of* Populus *and Its Implications for Management and Conservation.* NRC Research Press, Ottawa, Canada, pp. 57–85.

Brunner, A.M., Evans, L.M., Hsu, C.Y., and Sheng, X. (2014) Vernalization and the chilling requirement to exit bud dormancy: shared or separate regulation? *Frontiers in Plant Science* 5, 732.

Callahan, A.M., Scorza, R., Levy, L., Damsteegt, V.D., and Ravelonandro, M. (2000) Cold induced dormancy affects methylation and post-transcriptional gene silencing in transgenic plums containing plum pox potyvirus coat protein gene. In: *Sixth International Conference of Plant Molecular Biology*, 18–24 June 2000, Quebec, Canada (abstract).

Congressional Research Service (2015) Agricultural biotechnology: background, regulation, and policy issues. Available at: www.everycrsreport.com/reports/RL32809.html#_Toc425858971 (accessed June 27, 2019).

Eis, S., Garman, E.H., and Ebell, L.F. (1965) Relation between cone production and diameter increment of Douglas-fir [*Pseudotsuga menziesii* (Mirb.) Franco], Grand fir [*Abies grandis* (Dougl.) Lindl.], and western white pine [*Pinus monticola* (Dougl.)]. *Canadian Journal of Botany* 43, 1553–1559.

FDA (1992) Statement of Policy: Foods Derived from New Plant Varieties. Available at: https://www.fda.gov/regulatory-information/search-fda-guidance-documents/statement-policy-foods-derived-new-plant-varieties (accessed July 12, 2019).

Gonsalves, D. 1998. Control of papaya ringspot virus in papaya: a case study. *Annual Review of Phytopathology* 36, 415–437.

ISAAA (2017) Biotech Crop Highlights in 2017. Pocket K No. 16. ISAAA Brief No. 53. ISAAA, Ithaca, New York. Available at: http://www.isaaa.org/resources/publications/pocketk/foldable/Pocket%20K16%20(English)%202018.pdf (accessed July 12, 2019).

Meilan, R. (2006) Challenges to commercial use of transgenic plants. *Journal of Crop Improvement* 18, 433–450.

Meilan, R., Auerbach, D.J., Ma, C., DiFazio, S.P., and Strauss, S.H. (2002) Stability of herbicide resistance and *GUS* expression in transgenic hybrid poplars (*Populus* sp.) during several years of field trials and vegetative propagation. *HortScience* 37, 277–280.

Meilan, R., Ellis, D., Pilate, G., Brunner, A., and Skinner, J. (2004) Accomplishments and challenges in genetic engineering of forest trees: reflections on 15 years of steady progress. In: Strauss, S.H. and Bradshaw, H.D. (eds.) *Forest Biotechnology: Scientific Opportunities and Social Challenges*. Resources for the Future Press, Washington, DC, pp. 36–51.

NRC (2004) *Biological Confinement of Genetically Engineered Organisms*. National Academies Press, Washington, DC.

Scorza, R., Callahan, A., Dardick, C., Ravelonandro, M., Polak, J., *et al.* (2013) Genetic engineering of Plum pox virus resistance: 'HoneySweet' plum—from concept to product. *Plant Cell, Tissue and Organ Culture* 115, 1–12.

Tappeiner, J.C. (1969) Effect of cone production on branch, needle and xylem ring growth of Sierra Nevada Douglas-fir. *Forest Science* 15, 171–174.

12 Protecting Intellectual Property

Richard Meilan,[1] Kannan Grant,[2] and Vincent Shier[3]

[1]*Department of Forestry and Natural Resources, Purdue University, Indiana, USA;* [2]*Office of Technology Commercialization, University of Alabama in Huntsville, Alabama, USA;* [3]*Patent Attorney and Partner, Oblon, McClelland Maier and Neustadt, L.L.P., Virginia, USA*

Introduction

Food is a fundamental human need. Plants and their by-products have always comprised a large part of the human diet, and we have been altering them to improve their nutritional value for tens of thousands of years. It started with selection of naturally occurring mutants and traditional breeding to make wild types more palatable and safe for human consumption. This was followed by more advanced breeding efforts to develop plants needed to feed an ever-growing human population. Within the last few decades, innovations, including genetic engineering, have allowed us to improve plants in a much shorter time frame and to possess specific traits that benefit humans. Collectively, we call the processes involved genetic modification.

Plants have also been genetically engineered to synthesize non-native molecules, such as pharmaceutical and industrial compounds, or their precursors (Chapter 11, this volume), reducing our dependence on the hydrocarbons present in crude oil as a source of carbon skeletons. Scientists are also engineering plants to improve their utility as a feedstock for the sustainable production of renewable fuels (Chapter 8, this volume). These biofuels will reduce our reliance on crude oil as a source of transportation fuel and reduce atmospheric levels of CO_2, which is contributing to climate change. Moreover, researchers are engineering plants to be tolerant of more environmentally benign pesticides. Utilizing these plants is leading to a reduction in the overall use of pesticides (Chapter 10, this volume).

These technological innovations are important for driving economic growth and maintaining industrial competitiveness. They also can yield societal benefits, be favorable for the environment, and improve the human condition. In this chapter, we will describe the need for and the ways in which innovations such as these can be protected, with an emphasis on plant biotechnology.

The Need to Protect Intellectual Property

Intellectual property (IP) protection laws were established to stimulate invention and fuel innovation. Many technological advances are achieved through the research and development (R&D) conducted in both the public and private sectors. These efforts often require significant investment of resources, including time, know-how, and financial investment. It is necessary to protect the resulting IP to prevent competitors from taking (unfair) advantage of investments made by the developers of technology. These protections can also provide (financial) incentive for researchers and inventors (risk takers) to produce new and/or improved technologies and products (see below), to make them available for public consumption at the earliest possible time.

Protecting plant biotechnology is becoming increasingly important because of the biological and digital tools and techniques that are rapidly emerging. For example, using RNA sequencing, along with tools in the rapidly evolving field of bioinformatics, it is now easier than ever to clone genes and incorporate them into plants directly, to impart desired phenotypes. In addition, with advances in genome sequencing, it is now possible to identify molecular markers that can be used for marker-assisted breeding, obviating the need for large, expensive, and time-consuming field trials that are typically associated with conventional breeding.

Laws concerning IP protection vary among countries, and negotiations to integrate them can be contentious. Countries with extensive innovation capabilities (technological sources) generally prefer strong IP protection, which stimulates novelty by granting innovators temporary market exclusivity for their inventions. However, countries with less-developed innovation systems (technological sinks) often favor weaker protections, because it allows them to access IP developed elsewhere without having to pay licensing fees that may be beyond their ability (or desire).

Types of IP Protection for Plant Biotechnology

Plant variety protection (PVP)

When the Plant Variety Protection Act (PVPA) was passed in 1970, the general view of the U.S. courts was that life was not patentable (Goeringer, 2013). (This view was no longer the case a decade later, following the landmark Supreme Court decision described in the section on utility patents). The PVPA allowed breeders of sexually reproduced or tuber-propagated plants to obtain IP protection for their innovations. The PVPA provided patent-like rights (in the form of a PVP certificate, not a patent) to plant breeders. As with patents, the primary goal of the PVPA was to encourage the development of new varieties of plants and their introduction into the marketplace.

In the USA, PVP is offered to plant breeders through the Plant Variety Protection Office, which is within the U.S. Department of Agriculture. Outside the USA, PVP is more commonly known as plant breeders' rights (PBRs). Because PVP was so variable among countries, the Union Internationale pour la Protection des Obtentions Végétales (International Union for the Protection of New Varieties of Plants, or UPOV) was developed as a minimum global standard, and many current PVP systems are derived from one of the UPOV Conventions (see below). Plant breeders' rights grant developers the exclusive right to market a new plant variety or its reproductive structures for 20 years (25 years for trees or vines), and precludes others from obtaining PVP or utility patent rights (see below). To qualify for this kind of protection, a plant variety must be novel, genetically uniform, and stable through successive generations.

US plant patents

Section 161 of Title 35 of the U.S. code (35 USC § 161) originated as an amendment to the pre-existing patent statute, the Plant Patent Act (PPA) of 1930. As enacted, the "invents or discovers" requirement limited patent protection to plants "that were created as a result of plant breeding or other agricultural and horticultural efforts and that were created by the inventor" (www.uspto.gov/, accessed July 14, 2019). It was an innovative means of protecting germplasm, and enacted primarily to encourage plant breeding for horticultural crops. The PPA provides protection for asexually reproduced, novel plant varieties but excludes tuber-propagated and uncultivated plants. One condition the U.S. negotiators imposed for becoming a party to the UPOV agreement was that there would continue to be protection for asexually propagated plants under pre-existing U.S. laws (i.e. the PPA). Plant patents remain in effect for 20 years from the earliest U.S. filing date (35 USC § 154). They recognize the inventor's right to prevent others from asexually propagating the protected plant, and from using, offering for sale, or selling it or any of its parts in the USA, or from importing the reproduced plant, or any of its parts, into the USA (35 USC § 163).

The U.S. Patent and Trademark Office (USPTO) denies patent applications for plants with a UPOV-based certificate if the certificate request was filed more than 1 year before submitting the U.S. plant-patent application. Thus, consistent with the long-standing U.S. view with respect to personal grace periods for inventors, public disclosure of a new plant variety, or an approved patent for the new variety in another country does not prevent issuance of a patent in the USA, provided that the foreign disclosure occurred less than 12 months prior to filing for a plant patent in the USA.

Unlike utility patents (see below), for which the USPTO may require deposition of the claimed plant into a public germplasm repository, no such deposit is necessary to secure a plant patent. Also, unlike utility patents, plant patents can only present a single claim in the otherwise proscribed omnibus format (e.g. "A new and distinct variety of hybrid tea rose plant, substantially as illustrated and described herein"). Accordingly, the specification(s) and the illustrations of a plant patent are intended to provide, in as complete a manner as possible, a full and accurate description, expressed in botanical terms, of the new plant variety, including: the origin or parentage of the plant variety sought to be patented, the Latin name (generic and specific epithets) of the plant claimed, the location or place of business of where the plant has been asexually reproduced, and the manner in which the variety of plant has been asexually reproduced.

Utility patents (as applied to plant biotechnology)

As mentioned earlier, for most of the 1970s, it was a generally held belief that man-made living organisms (e.g. recombinant microorganisms) could not be patented. That all changed in 1980, when the U.S. Supreme Court approved, by a 5:4 majority, in *Diamond v. Chakrabarty*, a patent on a genetically altered bacterium (i.e. a man-made living organism). Since then, utility plant patents have been issued to the breeder or inventor under prevailing patent laws by the USPTO (under 35 USC § 101.)

The city state of Venice was the first to introduce the concept of patents (Nard and Morriss, 2006). The first U.S. Patent Act was passed on April 10, 1790, which was based on patent law originating in England in 1623, called The Statutes of Monopolies. It specified that a utility patent shall provide protection for any new and useful manufacture, or composition of matter, or any new and useful improvement thereof (35 USC § 101). In addition to meeting the eligibility requirements under 35 USC § 101, any plant invention that meets the statutory requirements of novelty, utility, and non-obviousness may obtain utility patent protection. Given this, the PPA is not an exclusive form of protection, and does not conflict with the granting of a plant utility patent[1]. Thus, inventions claimed under 35 USC § 101 may include the same asexually propagated plant that is claimed under 35 USC § 161, along with the plant materials and processes involved in producing or using the protected plant material. Although the application process is expensive and complex, utility patents provide the strongest protection for plant materials improved through biotechnology.

Utility patents have been issued in the USA for inbred and hybrid plants, as well as those engineered to possess certain attributes (e.g. cold tolerance, salt tolerance). This approach is particularly beneficial when the invention is not restricted to a specific variety. However, the more common approach to using utility patents in a plant context is for protection of parts derived from the plants (e.g. seeds, pollen, fruits, and flowers); biotechnology methods; genes; and many other novel plant-derived products. They may also protect DNA sequences, expressed proteins, isolated substances, processes used to produce transgenic plants, uses of the plants, extracts, parts or materials contained therein, in addition to the plants themselves, and uses thereof. In short, utility patents allow plant breeders the right to control the use, sale, import, and reproduction of protected plants. This includes preventing others from using a protected plant for subsequent breeding.

Any utility- or plant-patent application filed on or after June 8, 1995, has a term that begins on the date the patent is issued and ends on the date that is 20 years from the date on which the application for the patent was effectively filed in the USA. Finally, it should be noted that all IP rights afforded to innovators are negative rights (i.e. innovators only have rights to prevent others from making, using, and selling their inventions); the innovators themselves may not be able to utilize their invention based on the IP protection afforded to them by the U.S. government.

Patenting of genetic sequences and other nature-based products

A contentious question that needed to be answered is whether a genetic sequence can be considered an invention. In 1999, the rules of the European Patent Convention (EPC) were amended (Official Journal of the EPC, 1999) to indicate that biological material isolated from its natural environment or produced through a technical process, even if it already exists in nature, can be patented (Rule 23(c); Official Journal of the EPC, 1999). However, in order to patent the sequence, it must be novel and non-obvious, and the process of isolating and characterizing it must also be patentable (Harfouche *et al.*, 2011).

A similar principle was applied in the USA following its Supreme Court's acknowledgement that Congress, through legislative history, intended statutory subject matter to "include anything under the sun that is made by man." (see *Diamond v. Chakrabarty*, 447 US 303, 309; 206 USPQ 193, 197 (1980)). From this decision in 1980 until *Association for Molecular Pathology v. Myriad Genetics, Inc.*, 569 US 576; 106 USPQ2d 1972 (2013), the USPTO liberally applied the eligibility requirements of 35 USC § 101, largely abiding by the premise that if man's hand was involved, then it was distinct from a product of nature. This was, perhaps, no more evident than in the context of gene patents. Even if a particular sequence was obtained from a natural source (e.g. plant, fungus, bacterium, animal), so long as it was prefaced with certain key words such as: "recombinant," "isolated," or "purified," the claim was considered to pass muster under 35 USC § 101.

Following the *Myriad* ruling, however, the concept of patent eligibility of natural products has become very murky and, at times, frustratingly indecipherable. Presently, the general operating premise is that, during examination before the USPTO, a natural product, as it exists in nature, cannot be patented. It no longer applies that if a naturally occurring product has been isolated, purified, or otherwise altered, it can be patented. This question now turns on a complicated series of tests set forth in the currently developed Utility Guidelines and Examples set forth by the USPTO (www.uspto.gov/patent/laws-and-regulations/examination-policy/subject-matter-eligibility, accessed July 14, 2019), which are revised on an evolving basis. These tests now bring into patent eligibility determination issues previously reserved for novelty and non-obviousness analysis, including the state of the art, and even novelty itself. The larger guiding premise directing patent eligibility of nature-based products is whether they exhibit markedly different characteristics from the naturally occurring counterpart. The areas in which to look for markedly different characteristics are in the product's structure, function, and/or other properties.

Because the USPTO considers genes to be naturally occurring chemical compounds, they are potentially patentable but increasingly difficult to get past the gatekeeper of 35 USC § 101. The *Myriad* ruling made it clear that cDNA (Chapter 6, this volume) is patent, eligible because its structure differs from what would be found in nature, but when looking at fragments (of any size) of genomic DNA, distinctions in chemical structure of the isolated or excised fragment are not considered for purposes of patent eligibility. Therefore, patent eligibility rests entirely on the sequence information itself. Generally, if the sequence is the same as that in nature, it is no longer eligible for patent protection, even if previously never discovered and even if prefaced by terms such as "recombinant," "isolated," or "purified." To be patent-eligible, generally the sequence must be different from the naturally occurring sequence; there must be some further structural modification other than at the free 5′ and 3′ ends that would naturally occur, or the applicant must make some non-natural modification (e.g. labels or additional sequence at the 5′ and/or 3′ ends), or some other change relative to the naturally occurring sequence.

A similar analysis to the foregoing is also being applied to other nature-based products, including proteins and parts of plants. Genetic sequences tend to be the most difficult to pass muster under 35 USC § 101 following *Myriad*, but all nature-based products are being examined for whether they possess markedly different characteristics from their counterpart in nature. For proteins, this is often easy to show by changes in the structure or that a fragment/domain as claimed has a new and unique property or utility, compared with the same fragment/domain in the full protein in nature.

Even after genetic sequences or other nature-based products have been found patent-eligible under 35 USC § 101, they still face further scrutiny for written description, enablement, novelty, and non-obviousness.

Material transfer agreements (MTAs)

Historically, researchers working in an academic setting felt they had unfettered access to fundamental knowledge and products, and few, if any, formal agreements existed. However, because patent laws contain no unambiguous exemptions for academia, administrators at publicly funded institutions began utilizing MTAs to permit the use of protected property in their R&D programs. An MTA is a legally binding agreement between two parties, and typically stipulates that: (i) the specified items are to be used for research purposes only; (ii) a separate agreement is needed for commercial purposes; and (iii) the provided item(s) may not be given to a third party without prior written permission from the developer. A standard version, known as the Uniform Biological MTA (UBMTA), is now commonly used, but additional conditions may be stipulated. For example, there may be restrictions on what modifications or improvements can be made. Ownership of the altered material is determined in accordance with the appropriate inventorship laws and may be specified in an amended UBMTA. This may include the first right of refusal to negotiate a non-exclusive license to market products that constitute improved versions of the original material.

Trademarks, trade secrets, know-how, and geographical designations

There are sundry other methods for protecting IP that may be more appropriate under certain circumstances. The first is a trademark, which is a way to distinguish a developer's product(s) from those produced or marketed by other individuals or firms. Registered trademarks can only be used by the entity to which it was awarded. They are granted by national governments, and this protection extends only to the country within which they are issued. They are sought because much of the value of a product may reside not in its intrinsic properties but in the identity and/or reputation of the manufacturer.

Trade secrets are a form of IP protection that help maintain confidentiality, and penalties can be imposed if secret information is shared inappropriately. For example, when an individual goes to work for a competing firm, the previous employer may require that the former employee not reveal certain information for a specified interval. No application is needed to protect trade secrets, but two conditions must be met: (i) the secret must have commercial value; and (ii) an effort must already have been made to keep the information secret. Common exceptions are that the confidential information: (i) was known to the receiving party prior to being obtained from the former employer; (ii) becomes publicly known, through no fault of the receiving party; (iii) was obtained from a third party without an obligation of confidence; (iv) was developed by the receiving party independent of any disclosure by the former employer; and (v) has to be disclosed to comply with a court or administrative subpoena.

Trade secrets may be useless without the technical know-how that is required to implement them. Guarding its know-how may be the only form of IP protection an entity has, absent a patent. Technical know-how may be a combination of processes, methodologies, and formulations. Know-how is often revealed during the patent-licensing process or within an MTA.

Geographical designation is another form of product protection. An item may be associated with a specific locale, region, or territory, where a certain feature, reputation, or other attribute is linked to its origin. Geographical designations may also be used to protect plant varieties that were bred to include alleles imparting traits that may be advantageous in a given environment. These designations can prevent unauthorized users from misrepresenting or exploiting the reputation of a product or its producer.

Patent Reform Act of 2011

On September 16, 2011, President Barack Obama signed into law the Patent Reform Act of 2011, also called the Leahy-Smith America Invents Act (AIA). It was the first major patent reform since the Patent Act of 1952. Parts of the law were enacted immediately, while the rest went into effect on March 13, 2013. Some goals of the law were to: (i) harmonize U.S. patent law with the rest of the world; (ii) spur innovation; (iii) hire more patent examiners to speed up the patent process and reduce the backlog of applications; and (iv) curtail the activities of patent trolls, limiting the number of defendants that can be included in patent-infringement litigation.

One of the major changes introduced by the AIA was to replace "first to file" with "first inventor to file." Previously, the first inventor would not necessarily get the patent if s/he was beaten to the USPTO by another inventor who happened to file first. This had resulted in a rush to file the patent first at the USPTO. The AIA also gives the Patent Trial and Appeal Board (PTAB) authority to adjudicate any disagreements. Key changes made by the AIA are summarized in Table 1.

The impact of AIA is still being studied, but initial and early indications are that it has not really had the impact that the lawmakers had wanted. However, such impact can only be determined after an extended period of time.

International IP Protection Agreements

Plant research is widely dispersed. In addition, plants can be propagated vegetatively and seeds from adjoining fields can be saved. Thus, global IP protection can be challenging. Over the past three decades, many multi-lateral agreements have been negotiated and implemented in an effort to coordinate and strengthen enforcement efforts. The three most significant agreements are: (i) the International Union for the Protection of New Varieties of Plants (UPOV); (ii) the Trade-related IP Rights (TRIPS) agreement; and (iii) the International Treaty on Plant Genetic Resources (ITPGR). Each of these will be described briefly below.

UPOV

Because measures for protecting plant-related IP varied so widely among countries, there was a need for a set of global, minimal safeguards for protecting plant products. Thus, the UPOV agreement was negotiated and implemented. Member states that are signatories to this convention agreed to guarantee that the standards set forth in this document would be met through their own legislation. The UPOV was first adopted in 1961 but revised in 1972, 1978 and 1991. Currently, both the 1978, and 1991 versions are in effect. Although these two acts are similar, some important changes were included in the 1991 version. The most noteworthy modifications were: (i) the definition of propagating material was tightened; and (ii) farmers were allowed to collect and sow seed for personal use (Sechley and Schroeder, 2002). Although many countries have accepted standard PVP under UPOV, its provisions have not been uniformly adopted; in particular, countries with long histories of protecting farmers' privilege have yet to fully comply with this convention (Cullet, 1999).

Table 12.1. Relevant changes to the patent law brought about by the AIA.

Part of the law	Change(s)
Assignee filing (patent filing by the entity that has the property rights for the patent)	The company to which the invention is assigned may now file a patent application. The company may also file statements in lieu of the inventor's oath or declaration, although an inventor's oath or declaration still must be filed as a condition for granting a patent
Best mode (a written description of the invention, and of the manner and process of making and using it, in such full, clear, concise, and exact terms as to enable any person skilled in the art to which it pertains, or with which it is most nearly connected, to make and use the same, and shall set forth the best mode contemplated by the inventor of carrying out his invention (35 USC § 112))	The "best mode" to practice the invention is still required to be presented in the specification; however, if the patent does not disclose "best mode," it would not be cause for the patent to be declared invalid or unenforceable. "Best mode" is still arguably a condition for entitlement of foreign priority
Oath requirement (declaration from inventor(s) as to the facts of the patent)	No requirement to indicate that the inventor is the first inventor. Requirement to list the inventor's citizenship country has been relaxed
Prior use (any evidence that subject matter of the patent application was already publicly known or in use, in whole or in part, before the patent filing date)	AIA has broadened the scope of "prior use" to include public sale or use anywhere in the world. However, a limited "university exception" creates a prior-user rights exception for universities
Patent marking (a way to label your patented product so the public is informed about your IP rights)	Permits virtual marking
Derivation proceedings (a trial proceeding conducted by the PTAB to determine whether: (i) an inventor named in an earlier application derived the claimed invention from an inventor named in the petitioner's application; and (ii) the earlier application claiming such invention was filed without authorization)	Replaces patent interference proceedings. There is a limited window to file the derivation proceedings and there is a higher burden of proof
Post-grant review (a trial proceeding conducted at the PTAB to review the patentability of one or more claims in a patent on any ground that could be raised under § 282(b)(2) or (3) (defenses in any action involving the validity or infringement of a patent))	Available for patents with priority date later than March 15, 2013. Filed by a third party and handled by the PTAB. Has to be filed within 9 months of patent issuance or reissuance. All grounds of rejection available, except "best mode." Broad-sweeping estoppel if petitioner loses
Inter partes review (a trial proceeding conducted at the PTAB to review the patentability of one or more claims in a patent only on a ground that could be raised under § 102 (novelty) or § 103 (non-obvious), and only on the basis of prior art consisting of patents or printed publications)	Conducted by the PTAB. Filed by a third party. Scope of challenge is limited to patents and printed prior art. It has to be filed 9 months after the issuance of the patent, or if a post-grant review is initiated, it cannot be filed until conclusion of the post-grant review. Estoppel limited to any prior art raised or that reasonably could have been raised as to the challenged claims
Third-party pre-issuance submission	Provides for third parties to submit prior art that is of relevance to the patent application but only during a narrow window during examination. Limited opportunity to explain relevance of prior art

Continued

Table 12.1. Continued.

Part of the law	Change(s)
Supplemental examination	Mechanism provided to the patent owner to request the USPTO to consider or reconsider or correct information believed to be pertinent to the patent. If the information raises a new issue, re-examination is ordered. To take advantage of the cleansing effect of inequitable conduct, no action can be initiated until after conclusion of the supplemental examination and any ordered re-examination
Advice of counsel	Failure of infringer to obtain advice from counsel cannot be used to prove willful infringement
Human organism	Any patent claims that are directed to or encompassing human organisms have been specifically excluded

TRIPS

The TRIPS agreement, which went into effect on January 1, 1995, is presently the most detailed global pact on IP (Sechley and Schroeder, 2002). Enforcement of TRIPS is expected to promote and disseminate technological advancements for the benefit of both its producers and users, in a way that enhances social and economic welfare, while balancing rights and obligations (www.wto.org/english/docs_e/legal_e/27-trips.pdf, Articles 7 and 8, 1996, accessed July 22, 2019).

The tools regulated by TRIPS include: copyrights, trademarks, geographical designations, trade secrets, and patents. The minimum level of protection that should be provided in each of these areas by each member is defined in the agreement. For each of these devices, the key aspects of protection are defined, including: (i) the item to be safeguarded; (ii) the rights to be conferred, as well as possible exceptions to those rights; and (iii) the duration of protection. Member countries that have ratified this agreement are allowed to decide on the most appropriate way to implement its provisions through their own legislation. Under TRIPS, any invention, whether a product or a process, in all fields of technology, are patentable, provided that they are new, involve an inventive step, and are suitable for industrial application (Article 27(1), 1996). Although these requirements are consistent with those that need to be met to secure a patent, obtaining the latter for biotechnological inventions is frequently too controversial in many developed countries.

ITPGR

This treaty was derived from and eventually replaced the International Undertaking on Plant Genetic Resources (IU), which was adopted in 1983 by the Food and Agriculture Organization of the United Nations. Participation in the latter was voluntary and was not a legally binding agreement. The IU was the first attempt to ensure that plant genetic resources of economic and social interest would be evaluated, preserved, and made available globally for breeding and research purposes. The ITPGR provides a general framework for the conservation and sustainable use of plant genetic resources.

IP-related Technology Transfer

Academic institutions are well equipped for conducting research but often lack the infrastructure and resources needed to conduct the extensive trials needed to commercialize a product. In addition, public-sector scientists are increasingly being pressured by funding agencies to cooperate with the private sector in order to be eligible to apply for funds. In fact, some grant programs require an industrial match. This is intended to leverage public money and demonstrate to elected officials that resources used to support these programs have been well spent. It also helps to ensure that discoveries make it to the marketplace, and that society benefits from these investments.

In the USA, the establishment of Technology Transfer Offices and the promotion of the biotechnology industry were facilitated by the passage of the Bayh-Dole Act in 1980 (University and Small Business Patent Procedures Act, 1980). This legislation allows universities to assume ownership of IP generated through the use of federal funding. Prior to the enactment of this law, the federal government retained ownership of the IP generated through its funding and, as a result, few of these inventions were ever licensed. Now, royalties generated through licensing fees are used to recover patent-filing costs and generate revenue for the institution, to subsidize other initiatives. A portion of the proceeds is often used to reward the investigator(s), as a way of providing financial incentive for them to produce commercially useful products, and the departments within which the work was performed.

Summary

Plants and their by-products comprise a major part of the human diet, and we have been altering them to improve their nutritional value for millennia. It began with the selection of naturally occurring mutants and conventional breeding to make plants more palatable and safe for human consumption, and was followed by more advanced breeding to develop varieties needed to feed a rapidly expanding human population. More recently, genetic engineering has allowed us to improve plants much more quickly and to impart specific traits that benefit humans. Developments such as these are important for driving economic growth and maintaining

industrial competitiveness, but if their investments are not protected, inventors will have a disincentive to innovate.

Intellectual property protection laws were established to stimulate invention and fuel innovation. Many technological advances are achieved through the R&D conducted in both the public and private sectors. These R&D efforts often require significant investment of resources, including time, know-how, and money. It is necessary to protect the resulting IP to prevent competitors from taking (unfair) advantage of investments made by the developers of technology. Protecting plant biotechnology is becoming increasingly important because of the biological and digital tools and techniques that are rapidly emerging.

There are many types of IP protection for plant biotechnology. The PVPA, which was passed in the USA in 1970, provides patent-like rights to plant breeders to encourage the development of new varieties of sexually reproduced or tuber-propagated plants. Because PVP was so variable among countries, UPOV was developed as a minimum global standard. The PPA provides protection for asexually reproduced, novel plant varieties but excludes tuber-propagated and uncultivated plants. One condition U.S. negotiators imposed for becoming a party to the UPOV agreement was that there would continue to be protection for asexually propagated plants under pre-existing U.S. laws.

When the PVPA was passed in 1970, the general view of the U.S. courts was that life was not patentable. In 1980, the U.S. Supreme Court approved a patent for a man-made living organism. Since then, utility plant patents have been issued in the USA for plants engineered to possess certain attributes. They also provide protection for plant parts (e.g. seeds, pollen, fruits, and flowers); biotechnology methods; genes; and many other novel plant-derived products.

Historically, academic researchers felt they had unrestricted access to various types of IP and no formal agreements existed. However, because patent laws contain no unambiguous exemptions for academia, administrators at publicly funded institutions began utilizing MTAs. An MTA typically mandates that: (i) the specified items are to be used for research purposes only; (ii) a separate agreement is needed for commercial purposes; and (iii) the provided items may not be given to a third party without prior written permission from the developer. A standard version, known as the UBMTA is now commonly used, but additional conditions may be stipulated.

Other methods for protecting IP may be more appropriate under certain circumstances. Registered trademarks are a way to distinguish a developer's product from those manufactured or marketed by others. Trademarks can only be used by the entity to which it was awarded. They are granted by national governments, and this protection extends only to the country within which they are issued. Trade secrets are a form of IP protection that help maintain confidentiality, and penalties can be imposed if secret information is shared inappropriately. Trade-secret agreements are commonly used when an individual goes to work for a competing firm, to prevent an employee from revealing certain information obtained from a former employer, for a specified interval. Geographical designations are another form of product protection. An item may be associated with a specific locale, region, or territory, where a certain feature, reputation, or other attribute is linked to its origin. These designations can prevent unauthorized users from misrepresenting or exploiting the reputation of a product or its producer.

Plant research is widely dispersed. In addition, plants can be propagated vegetatively and seeds from adjacent fields can be saved. Over the past few decades, many multi-lateral agreements have been negotiated and implemented in an effort to coordinate and strengthen global enforcement efforts.

The three most significant agreements are: (i) UPOV, (ii) TRIPS, and (iii) ITPGR.

Note

[1] *Ex parte Hibberd*, 227 USPQ 443 (Bd. Pat. App. & Int. 1985) (affirmed by the Supreme Court in *J.E.M. Ag Supply, Inc. v. Pioneer Hi-Bred International, Inc.*, 534 US 124, 122 S. Ct. 593, 60 USPQ2d 1865 (2001)).

References

Cullet, P. (1999) Revision of the TRIPS agreement concerning the protection of plant varieties—lessons from India concerning the development of a *sui generis* system. *Journal of World Intellectual Property* 2, 617–656.

Goeringer, P. (2013) Understanding the Plant Variety Protection Act. In: *2013 Agricultural Policy and Outlook Conference*, Annapolis, Maryland. Available at: https://agresearch.umd.edu/sites/agresearch.umd.edu/files/_docs/programs/canrp/PaulPVPA.pdf (accessed July 14, 2019).

Harfouche, A., Meilan, R., Grant, K., and Shier, V.K. (2011) Intellectual property rights of biotechnologically improved plants. In: Altman, A. and Hasegawa, P.M. (eds.) *Plant Biotechnology and Agriculture: Prospects for the 21st Century*. Elsevier, Amsterdam, pp. 525–540.

Nard, C.A. and Morriss, A.P. (2006) Constitutionalizing patents: from Venice to Philadelphia. *Review of Law and Economics* 2, 223–231.

Official Journal of the EPC (1999) Decision of the Administrative Council of 16 June 1999 amending the Implementing Regulations to the European Patent Convention. Available at: https://archive.epo.org/oj/issues/1999/07/p437/1999-p437.pdf (accessed July 22, 2019).

Sechley, K.A. and Schroeder, H. (2002) Intellectual property protection of plant biotechnology inventions. *Trends in Biotechnology* 20, 456–461.

Index

This book is published by **CABI**, an international not-for-profit organisation that improves people's lives worldwide by providing information and applying scientific expertise to solve problems in agriculture and the environment.

CABI is also a global publisher producing key scientific publications, including world renowned databases, as well as compendia, books, ebooks and full text electronic resources. We publish content in a wide range of subject areas including: agriculture and crop science / animal and veterinary sciences / ecology and conservation / environmental science / horticulture and plant sciences / human health, food science and nutrition / international development / leisure and tourism.

The profits from CABI's publishing activities enable us to work with farming communities around the world, supporting them as they battle with poor soil, invasive species and pests and diseases, to improve their livelihoods and help provide food for an ever growing population.

CABI is an international intergovernmental organisation, and we gratefully acknowledge the core financial support from our member countries (and lead agencies) including:

Discover more

To read more about CABI's work, please visit: **www.cabi.org**

Browse our books at: **www.cabi.org/bookshop**,
or explore our online products at: **www.cabi.org/publishing-products**

Interested in writing for CABI? Find our author guidelines here:
www.cabi.org/publishing-products/information-for-authors/